D0850444

BRIDGING DEEP SOUTH RIVERS

The Life and Legend of Horace King

John S. Lupold and Thomas L. French Jr.

BRIDGING

Published in cooperation with
the Historic Chattahoochee Commission
and the Troup County Historical Society

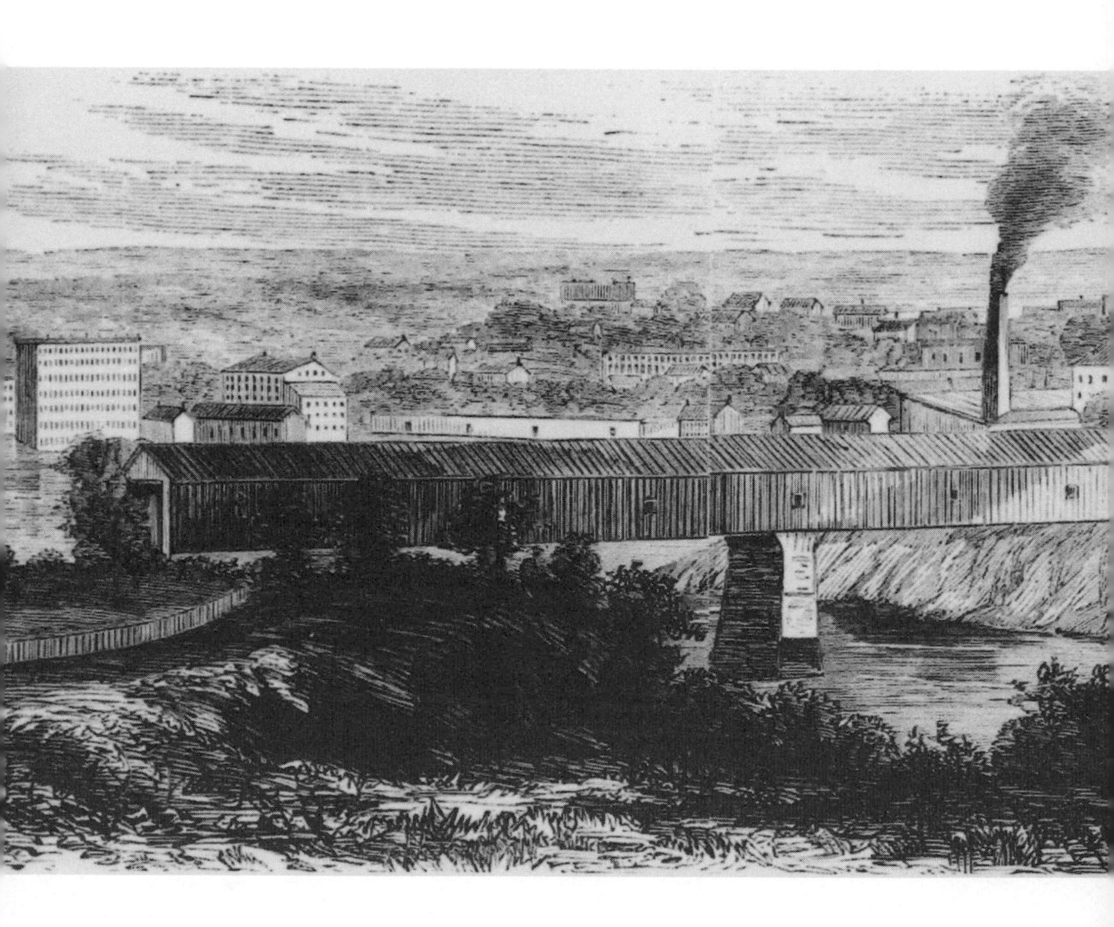

DEEP SOUTH RIVERS

The Life and Legend of HORACE KING

The University of Georgia Press ATHENS AND LONDON

Athens, Georgia 30602

Designed by Sandra Strother Hudson
Set in Bell MT with Franklin Gothic Display
by Graphic Composition, Inc.
Printed and bound by Maple-Vail
The paper in this book meets the guidelines for permanence
and durability of the Committee on Production Guidelines
for Book Longevity of the Council on Library Resources.
Printed in the United States of America
08 07 06 05 04 c 5 4 3 2 1

Library of Congress Cataloging-in-Publication Data
Lupold, John S., 1942–
Bridging deep south rivers : the life and legend of Horace
King / John S. Lupold and Thomas L. French Jr.
p. cm.
"Published in cooperation with the Historic Chattahoochee
Commission and the Troup County Historical Society."
Includes bibliographical references and index.
ISBN 0-8203-2626-7 (alk. paper)
1. King, Horace, 1807–1885. 2. King, Horace, 1807–1885—
Influence. 3. African Americans—Biography. 4. African Ameri-
can engineers—Biography. 5. Civil engineers—United States—
Biography. 6. Bridges—Southern States—Design and construc-
tion—History—19th century. 7. African American
legislators—Alabama—Biography. 8. Slaves—South Carolina—
Biography. I. French, Thomas L. II. Historic Chattahoochee
Commission. III. Troup County Historical Society. IV. Title.
E185.97.K49L87 2004
328.761′092—dc22
2004001271

British Library Cataloging-in-Publication Data available

CONTENTS

ILLUSTRATIONS

MAPS

FIGURE

PHOTOGRAPHS

PREFACE

The trail of information leading back to Horace King has been a long and winding one. The Chattahoochee Valley's fascination with Horace King dates back more than a century to the biography of the Rev. Francis L. Cherry, but the effort to transcend the myths and to document King's life and work began about thirty years ago. Since that time, many people have contributed to the real story of Horace King by uncovering valuable sources relating to him.

Thomas L. French Jr. has played the pivotal role in collecting King materials. As a researcher, he has followed the trail of "the prince of bridge builders" in six states, investigating numerous courthouses and libraries and following leads offered by local folks about this man of color who built bridges all over the Deep South. Equally important, he has become the archivist for Horace King, assiduously collecting the documents he found, as well as those discovered by others. His work with William H. Green, Tom Lenard, and Karl-Heinz Reilmann, which is described in chapter 16, certainly added to the storehouse of King knowledge.

French also collected oral histories from the King family. Between 1979 and 1996 he interviewed the great- and great-great-grandchildren of Horace King. Great-granddaughter Theodora Thomas of LaGrange supplied the most valuable information. She was interviewed by French and William H. Green together several times and separately by Forrest Clark Johnson of LaGrange. Her family sketch is available at the Troup County Archives and Columbus Museum. Tom French also consulted several times with Horace's great-grandson Horace H. King of Philadelphia. His most significant memories related to his grandfather John T. King of LaGrange. Other descendants who were farther removed from Horace—David H. King, Rebecca King Rosenberg, Lois King Carroll—could not provide specific information about the historic family, but they did demonstrate the pride in Horace King shared by his descendants.

The family's oral traditions became one of many pieces of the evidence collected for this book. For every county where the Kings lived from 1810 until 1880 or wherever John and Ann Godwin resided, the census records were analyzed, as well as the deed and estate records. The search area stretched from Anson County, North Carolina, to Lowndes County, Mississippi. The Web site of genealogist Kenneth Vance Smith also helped establish the family relationships.

The search to determine which bridges Horace actually built began with local history sources. For every county where he may have built a span, a search was made of the published histories of the county, city, or both, as well as vertical file materials in local libraries and contemporary newspapers. In the relevant counties and cities we also investigated the extant city hall or courthouse records—bridge books, inferior court (early county commissioners) and city council minutes, deed books, and other relevant court proceeding. The lack of records for Chesterfield District, South Carolina, and Baldwin County (Milledgeville), Georgia, as well as gaps in the late nineteenth- and early twentieth-century records in Russell County, Alabama, hampered research.

Beyond those efforts, research in archival facilities uncovered a variety of relevant documents. The South Carolina Department of Archives and History provided information about the nascent town of Cheraw; it also held petitions signed by John Godwin as a young builder. The deed records of the Marlboro County Courthouse yielded an unusual contract between the property owners and Godwin detailing the house he was to build for them. The William R. Godfrey Collection in the South Caroliniana Library contained about one hundred items—receipts, balance sheets, and lists of supplies and materials—pertaining to the building of the 1824 Cheraw bridge, probably the most significant cache of material about the early construction of wooden bridges in the South.

Most nineteenth-century bridges were chartered by the state legislature. The Georgia Legislative Documents in the Digital Library of Georgia site in University System of Georgia's Galileo online library provided easy access to the legislation chartering all the bridges in the state from 1790 until 1900.

In Savannah the records of the Central of Georgia Railroad in the repository of the Georgia Historical Society contained the minutes of the Mobile and Girard Railroad and the contract for King to build its postwar bridge in Columbus. The connections of John Thomas King and his son, Horace H. King, with Clark University were confirmed by materials in the archives department of the Atlanta University Center in the Robert W. Woodruff Library. Also the *Bulletin of Atlanta University* served to document John King's role in building the Negro Building at the Atlanta Exposition in 1895.

In the Hargrett Rare Book and Manuscript Library at the University of Georgia, the John Fontaine Collection contained important sources about two of King's building projects with his occasional partner James Meeler.

The Columbus State University archives probably contains the greatest collection of articles containing the Horace King legend, especially the newspaper and magazine articles of W. C. Woodall. Drafts by the writers of the Works Progress Administration are also available in the Alva Smith Collection. This CSU archives also holds several valuable county records. The archives personnel, Craig Lloyd, Reagan Grimsley, Adabelio Garcia, and Lewis Powell IV, provided valuable assistance. In a similar fashion, Dalton Royer expedited finding sources in the Genealogy Room of the W. C. Bradley Memorial Library.

However, the most important source for King material was the Troup County Archives because of the detailed research performed by its personnel. Kaye Lanning Minchew and Forrest Clark Johnson should probably be considered coauthors of the two chapters dealing with LaGrange. In addition to collecting materials, archive staff, especially Clark Johnson, have indexed the local newspapers, thereby providing excellent documentation for the work and community activities of the Kings.

A search of the R. G. Dun Collection, Baker Library, Harvard Business School, yielded nothing about Horace but allowed some insight into the career of his son Washington W. King. The Dun Collection also documented the business climate of Whitesburg, near Moore's Bridge, in which the Kings owned a third share. The family history and their Civil War experiences are detailed by David Evans in *Sherman's Horsemen.* King's 1878 interrogatories or testimony to the federal commissioners from the National Archives was provided by Ken Thomas Jr., historian for the state of Georgia's Historic Preservation Division.

Lee Formwalt, a history professor at Albany State University, documented King's role in building Nelson Tift's bridge house in Albany. He also discovered the correspondence in the Freedmen's Bureau Papers about King's attempt to establish a freedmen's colony in Coweta County, Georgia, in 1867.

The Alabama Department of Archives and History provided two significant pieces of information that were discovered by other researchers. Richard Bailey found the bill introduced by Jemison to exempt King from being reenslaved if the Alabama legislature passed such a general law. Tom Lenard found the entry in the Reverend Cherry's diary that listed his appointment with Horace, and the telegrams concerning Horace's whereabouts during the Civil War. King's legislature was researched using materials at the state archives and the Special Collections at Auburn University.

The most significant manuscripts used in preparing this book were the Robert Jemison Jr. papers in the W. S. Hoole Special Collections at the University of Alabama. More than any other source its letters documented King's life. Clark Center and Ellen Garrison greatly facilitated access to Jemison's letterbooks. Elise Stephens Hopkins provided copies of Horace's letters to Jemison.

The architect Samuel H. Kaye and his research associate Carolyn Burns-Neault provided a history of King's bridge in Columbus, Mississippi. More information on bridges in the area came from Gary Lancaster and the Historic American Engineering Record (HAER) study by Dan Clement on the bridges in the Upper Tombigbee River Valley. This publication, along with numerous photographs of historic structures, compiled by HAER and the Historic American Building Survey (HABS), was accessed through the Art and Architecture Collection Finder of the Library of Congress American Memory Home Page.

The initiative for Thomas L. French Jr. and John S. Lupold to collaborate on a full-length biography came from Douglas C. Purcell of the Historic Chattahoochee Commission and Kaye Lanning Minchew of the Troup County Archives. They provided the financial support to complete the research and write this work. As early as the mid-1970s Lupold, as an assistant professor at Columbus College researching the city's early industry, began finding references dealing with King and shared them with French. Beginning in 1998 Lupold completed the necessary research to begin writing this work. Then, in close collaboration with French, Lupold wrote the text. Both of them take responsibility for any errors of commission or omission, as well as the assertions sprinkled throughout this volume.

This biography presents the life of Horace King as revealed by the evidence found to this date. At any time, the discovery of a letter in a seemingly unrelated manuscript collection or a contract or a court case in a courthouse basement may answer one of the many remaining mysteries about his life. Was he related to Ann Wright Godwin's family? Exactly why was he freed in 1846? When and why did he experience his severe financial reversals? Why did he and his family move to LaGrange? This book represents only a temporary bridge along the journey to document and reveal the career of Horace King. We encourage others to continue the search for sources that will reveal his extraordinary life.

Acknowledgments

Without the foundation built by countless researchers investigating Horace King, covered bridges, the King family, the Godwins, and related topics, we could never have assembled our biography of this extraordinary bridge builder. Many of the major contributors are noted in the introduction, the last chapter, and the notes; an even larger number of researchers, although not cited by name, have our sincere appreciation.

Our special thanks go to Doug Purcell and Kaye Minchew. Without their scholarly support, as well as the financial underwriting of their respective organizations—the Historic Chattahoochee Commission and the Troup County Historical Society—this biography would never have appeared. We are also indebted to our copyeditor, Anne R. Gibbons, and to the staff of the University of Georgia Press, especially Jennifer Reichlin, Sandra Hudson, and Nancy Grayson, for the pleasant and supportive way they facilitated the publication of the book.

Most of all, however, we need to recognize the support and sacrifices made by our wives as we labored to raise this edifice. For decades, Rose French tolerated Tom's research trips, his ever-growing boxes and file cabinets of Horace documents, and his endless talk about Horace, who became an honorary member of their family. For years, Lynn Willoughby tolerated John's discussion of bridges and free blacks, all the while encouraging, proofing, and guiding the preparation of the manuscript. *Bridging Deep South Rivers* would never have made it to press without the support of Rose and Lynn. Thanks!

INTRODUCTION

Meet Horace King at Ulysses Smiths' Drug Store, 10 A.M., 10-3-1883
Rev. Francis L. Cherry Diary

It seemed as if summer would never end. It was still hot, too hot to be October 3, even along the banks of the Chattahoochee River, whose muddy waters carved the boundary between Georgia and Alabama. Rain had been so scarce that no water tumbled over the gleaming new granite block dam of the Eagle and Phenix Mill. The factories diverted what little water did come downstream to turbines that turned the spinning frames and drove the looms. Because of the limited power, only a skeleton crew reported to the mills at 6 A.M. that day. Some of them crossing the factory bridge at Franklin (Fourteenth) Street noticed a distinguished gentleman of color, dressed in a velvet lapelled coat, standing at the portal of the covered bridge. Many of them called him by his first name, and the older ones probably wondered why he was there, knowing he had lived in LaGrange for the past ten years.[1]

The bridge builder, a gray-haired gentleman with a small tuft of hair under his lower lip, had probably spent the night at his sister's place less then a mile away, just west of Browneville, the mill village in Alabama. He had an appointment at 10:00 A.M. to talk about the early history of the region. This interview became one of the foundation piers of the Horace King legend. Knowing that later in the day he would reminiscence about his earlier life, Horace might have killed time by inspecting his bridges and buildings. He could have strolled through the city along this route.

After sixty years of arriving at construction sites at dawn, he still rose early even if his schedule did not demand it. He had several hours to kill before his appointment. As morning light broke, the seventy-six-year-old bridge contractor standing on the factory bridge discerned the web of one of his railroad bridges, four blocks to the north. A little farther upstream he could barely trace the silhouette of the grist mill he built, which clung to the Georgia side of the river. As he strolled into his covered bridge he examined his work; his skilled, worn hands instinctively testing the diagonal pine boards and his fist tapping the oak treenails to check the soundness of the web. The bridge had already stood for twenty years and would stand for another twenty. Content with the condition of his bridge, he walked on into Columbus.

A block or so east of the bridge, he turned south on Broad Street, the city's main commercial artery where merchants, bankers, grocers, and workers—both black and white—greeted him as a friend. After stopping to visit at several stores, he continued his perambulations, moving in the general direction of the lower bridge, called Dillingham Street Bridge, stopping on Front Avenue at the Fontaine Warehouse, another of his achievements. There, he and W. C. Bradley, an ambitious twenty-three-year-old clerk in the firm, talked about adding some decorative parapets to modernize the building's appearance.

After visiting old friends at the Columbus Iron Works, where he had worked during the Civil War, and checking the condition of his Mobile and Girard Railroad bridge, the town's southernmost span, he crossed back into Alabama on his Dillingham Street Bridge. The construction of that bridge had brought him to Columbus in 1832. In addition to the first crossing there, he had reconstructed it twice and had maintained its piers, sills, siding, and flooring for four decades. As he contemplated his appointment on that October day, his mind filled with fifty-one years of memories of building, of contracting, of speculating, of family, and of politics as he passed through the Dillingham Bridge and on into Girard, Alabama, a town he helped to pioneer.

As he left the bridge he turned right to complete his circuit of Columbus and its Alabama suburbs. Just north of the Dillingham Bridge, at Holland Creek, he stopped to talk with Hiram Williams, the foreman of the crew replacing his short wooden span with an iron and steel one. As the era of wooden bridges, in which he excelled, seemed to be passing, he must have wondered if his sons could still find work building covered bridges.

At 10:00 A.M. he arrived at his destination. The proprietor, Ulysses H. Smith, greeted him warmly and showed him to a table where a white man waited. Because of the heat, a crowd might have already gathered to seek relief at the soda fountain. As they sipped their cool drinks, they would not have been surprised to see a distinguished, light-skinned mulatto sitting with a white man whose long, somewhat scraggly, beard accentuated his bald head. Smith's regular customers were not shocked in October 1883 to see a black man and a white man sitting together because the rigid lines of segregation had yet to be fixed firmly in the postbellum South. Too, they recognized this gentleman of color and knew his life had been an exception to the usual racial mores. In most southern communities, a former black member of the general assembly conferring with someone might bring images of a radical Republican cabal. But they knew Horace King was no radical. They were more likely to wonder who the other man was, assuming he was there to talk about a new

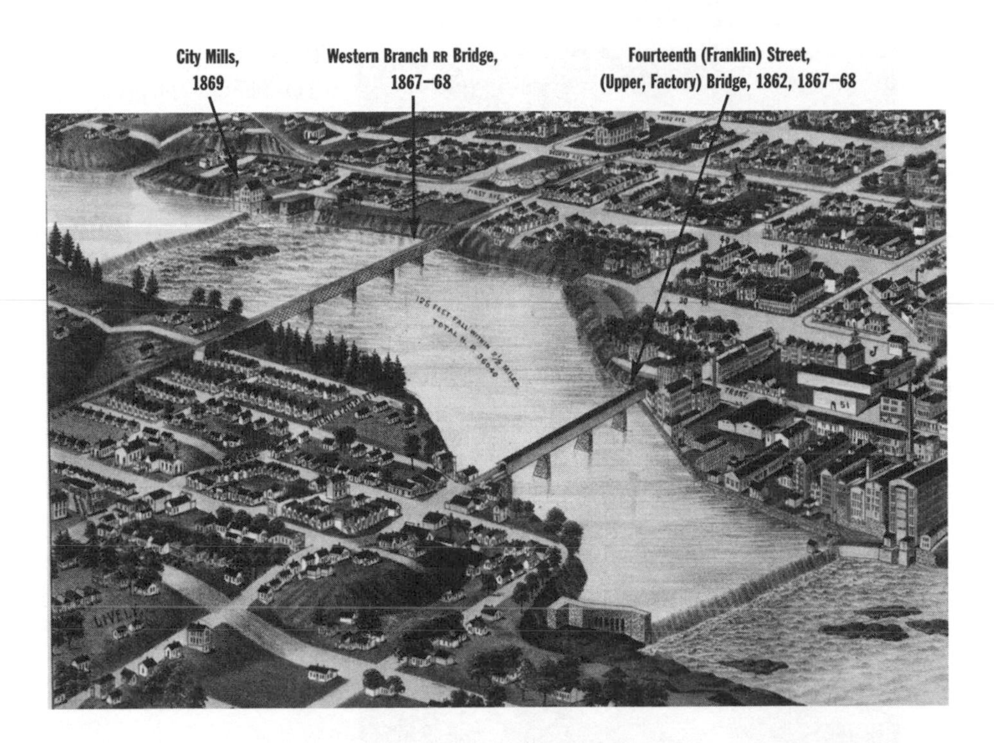

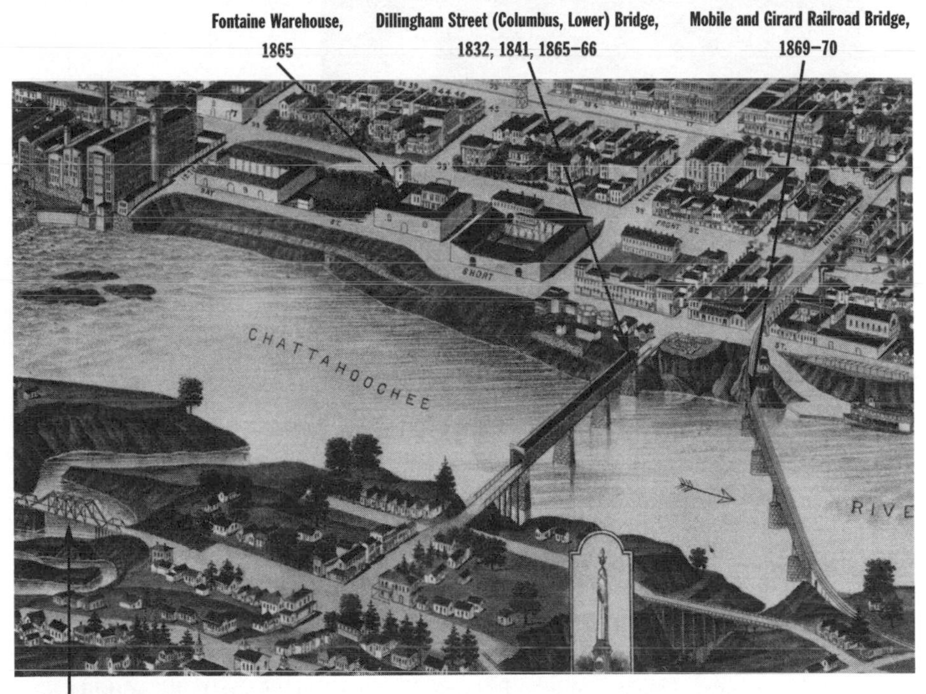

Perspective map of Columbus, Georgia, 1886. Drawn only a year after King's death, these riverfront views indicate the extent of his construction there. *H. Wellge, Beck, and Pauli Lith. Co. Library of Congress, Prints and Photographs Division, Map Collection, 1500-2003, 75693191.*

Horace King, ca. 1870. This photograph probably dates from King's service in the Alabama legislature, 1869–72. He was approximately ten years older when he met the Reverend Cherry at Smith's drugstore in Browneville. *Collection of Thomas L. French Jr.*

bridge. He scribbled notes as Horace talked, and the way the white man's eyes squinted through his narrow wire-rim glasses made the interviewer look as if he were carefully weighing every word Horace said. Undoubtedly he was.

In actuality, the Reverend Francis L. Cherry, who wrote under the pen name "Okossee," was interviewing King for his local history column in the Opelika newspaper that began appearing two days later. Since King was the last surviving initial resident of Girard, Cherry wanted to hear his reminiscences and include them in his articles. Cherry had worked as a journalist, publisher, and preacher. At that time he was serving as a minister at the First Methodist Church in Opelika, the next major town west of Browneville, about thirty miles up the rail line.

The preacher questioned King about his early life and probably reviewed what he already knew about the former slave—his birth in South Carolina, his moving to Girard, his bridge-building credits, his being freed, his serving in the legislature, his moving to LaGrange, Georgia, about 1872, and his relationship with the Godwin family. Horace probably talked more about his bridges than he did about his life as a free black. The minister also coaxed from him a story Cherry relished: King's erecting a grave marker for his former master, John Godwin, only about a mile west of Smith's drugstore before the Civil War. That personal act of love and the moderate actions of the former slave allowed Cherry to assert in his columns that the "character" of King "is worthy of imitation by the colored people."[2]

After leaving the Reverend Cherry, King may have reflected back. King should have been proud of his achievements. The bridges and buildings along the Chattahoochee in Columbus testified to his skills as an architect and engineer of heavy timber structures. Similar structures in Troup County and around the square in LaGrange, as well as bridges across Georgia, Alabama, and northeast Mississippi, enhanced his reputation. His extraordinary accomplishments as a builder were obvious to his contemporaries and would be to future generations as well.

What King could never have imagined on that hot October day was that 120 years later he would rank as the best-known nineteenth-century builder and the most-recognized African American in the Chattahoochee Valley. The legend of Horace King that began during his lifetime continues to grow in the twenty-first century. King would also be surprised to see his name being proliferated throughout the Chattahoochee Valley today. Surely he would shake his head at the recently dedicated Horace King Friendship Bridge, about four blocks north of the first bridge he built in Columbus. Today, a new bridge in LaGrange, a picnic shelter on Lake West Point, and a park in Valley, Alabama, all bear his name. In February 2003 a ceremony at the Alabama capitol honored King and placed his portrait on display for Black History Month.

Horace, the Reverend Cherry, and the King family created the legend that local journalists and historians have repeated and amplified over the years. Horace himself laid the foundation for the legend in several ways. He was an extraordinary craftsman, and his race made him unique as a bridge builder. Then, his 1859 monument to his master captured the interest of contemporary editors. One account of this memorial, entitled "All Honor to Horace," was reprinted throughout the Deep South.

Cherry's 1883 account stressed Horace's accomplishments but also noted the Godwin marker and painted King's character as one that other blacks should emulate. Cherry's telling of the story was repeated by local authors who perpetuated the legend.

After Horace's death in 1885 the King family kept his legend alive. His sons continued to receive recognition for their building accomplishments, and Horace's descendants continued to recount his life story to any local historian or journalist who would listen. Unfortunately, the King family burned what few papers and documents had survived in a general cleaning in the late 1920s following the death of John T. King. At that time, the family possessed a brief,

The Rev. Francis Lafayette Cherry, ca. 1870. A Methodist minister who had also worked as a newspaperman, Cherry interviewed King in 1883 and published the first biographical sketch of the bridge builder. In the process he laid the foundation of the Horace King legend. *Courtesy of Alabama Department of Archives and History, Montgomery, Alabama.*

written family history much of which apparently became incorporated into a sketch about the Kings collected by 1930s writers for the Works Progress Administration. By that date, the Kings were already the poster children for black history. The LaGrange family members, especially Theodora Thomas, continued to preserve and share their oral tradition into the 1980s.[3]

The legend as amplified after Horace's death involved several inaccuracies and embellishments. The most exaggerated themes involve the state of Ohio—that Horace gained his freedom, joined the Masons, and attended Oberlin College there. A related claim had him participating in a white Masonic lodge in Columbus before the Civil War. Many later sources overstate how he cared for the Godwin family after the death of his master. Local writers and chambers of commerce proudly claim their Horace King bridge or building even when little historical evidence exists to verify their claim. The extent and durability of his work has also been overemphasized. Authors have assumed his bridges lasted forever, even though several of his major spans failed within a decade.

King's legendary status stemmed from several factors. It rested on his exceptional skills and his acceptance among the white community. For example, when the Columbus City Council requested bids to rebuild the burned City Bridge at the end of the Civil War, they told all the potential contractors "that whoever undertakes the contract, . . . will of necessity be compelled to have the services of Horace King."[4] Few African Americans enjoyed such stature among conservative white southern leaders in 1865. Beyond his skills, his experience appeared to embody slavery at its best, a kinder and gentler form of servitude. In the 1850s and later King's life served to lessen the collective guilt about the South's peculiar institution. Slavery must not have been that bad, local historians implied, if this slave erected a monument to his master. King unwittingly became an apologist for slavery because of this marker.

In recent years, King has been co-opted as a black Confederate, an African American who supported the southern cause. If his Unionist testimony reflected his true belief toward the Lost Cause, Horace King would have abhorred any association with the Confederacy and detested the fact that a C.S.A. emblem now marks his grave.

Finally, Horace's legend was enhanced by the romance of covered bridges. It became immaterial to the legend that *all* major wooden bridges were covered, not just Horace's.

In seeking to move beyond the Horace King legend, we have two major objectives: to document King's professional career and to separate the facts from the fiction in the Horace King lore. A third important objective is to chronicle his life as a slave and freedman. But the nature of the King sources limited our explorations of his personal life. The bulk of the surviving sources deal with King's life as a builder. A wealth of newspaper articles, city council minutes, county records, contracts, deed records, as well as informative correspondence in the Robert Jemison Jr. Collection at the University of Alabama, document his construction activities. What has not survived is his personal correspondence or diaries.

After the Civil War, King claimed to be a Unionist and petitioned the federal government for damages inflicted on his property by U.S. troops. The federal commissioners denied his claim, but the associated interrogatories are an important source of information about King's life. Given the lack of personal papers, Cherry's account and King's 1878 testimony, despite their brevity, are crucial in reconstructing his life.[5]

King was not a good oral history subject; he had trouble remembering dates. To both Cherry and the federal commission he dated his emancipation as 1848. The Alabama legislature actually freed him in 1846. This error either speaks volumes about how the Godwins treated him as a slave—that legal freedom did not really matter—or it says a great deal about a mind that could quickly figure the quantity and dimensions of the lumber needed for a bridge and yet had trouble remembering dates. Actually, the entire King family had trouble with chronology. In the family Bible someone recorded Horace's death as occurring in 1887 when it actually happened in 1885. Fortunately for the traveling public, the Kings never made such numerical errors in constructing their bridges.

The level of documentation for King contrasts sharply with two free blacks whose lives have been remembered because of their published biographies—William Johnson, "the barber of Natchez," and William Ellison, a free black cotton gin maker in Stateburg, South Carolina. While King only received passing references from academic historians and then only as "Horace," Johnson and Ellison received full-length works because they left a legacy of written materials.[6]

Johnson and Ellison were lost to history until their biographies appeared; now their voices speak to later generations. King, on the other hand, remained a historical figure—even a historical celebrity—in the Chattahoochee Valley because of a succession of brief sketches in newspapers and local histories, many of them based on Cherry's account. So King the builder became legendary, but King the person, unilluminated by letters or diaries, remains mute.

Beyond the difference in the level of documentation, King's life was dissimilar from that of Johnson's and Ellison's in other ways. They were all entrepreneurs, but Johnson and Ellison probably accumulated more wealth than King. The other two used their money to purchase slaves, a frequent act for affluent free blacks. King never purchased slaves nor did he buy freedom for his family members who remained enslaved. One reason why free blacks acquired slaves was to establish a reputation as being safe and trustworthy among their white neighbors. Being a slave owner made free blacks less threatening because they supported the status quo and would be unlikely to start a revolt. Becoming a fellow slaveholder made a free black respectable.[7] King may have erected the marker to Godwin as a means of emphasizing his support of slavery at the time when the Alabama legislature was debating whether to reenslave all the free blacks within the state.

All free blacks who operated within the general economy needed good reputations. The geographical area where King operated was much larger than Johnson's and Ellison's, maybe larger than any other free black in the South. Johnson needed to be respected within the city of Natchez, Mississippi, whereas Ellison had to maintain his good name as a craftsman in several districts in the middle of South Carolina. King's fine reputation as a man and contractor extended at least from Milledgeville, Georgia, through west Georgia, across all of Alabama, and on to Columbus, Mississippi. Given the range of his operations and his skills as an engineer and architect, a full-length biography of Horace King is certainly warranted despite the paucity of primary sources.

In the process of documenting King, we explore several other topics: the significance of bridges to the development of the cotton South; the problems involved in maintaining a bridge as both a physical and political entity; bridge building as it relates to the local history of the Chattahoochee Valley; and the life of John Godwin, Horace's master.

Bridging major rivers developed along with the new American republic. Very few bridges spanned major rivers during the colonial period. The profession of bridge building emerged as result of the development of wooden trusses and was relatively new when Godwin and King entered it. Although they built many small county bridges, they established their reputations by "throwing" massive spans over wide rivers for private investors and municipalities.

In the Deep South the development of the cotton economy and the emergence of cotton trading towns, especially at the fall line, depended on these new wooden bridges. These crossings were not minor affairs. They were extremely expensive, with most being financed by private investors. Bridges generated large revenues for some investors and cost others their fortunes. They nearly always engendered debates about their location, their tolls, and their maintenance. Godwin and King were more than just ordinary builders; they were contractors and engineers for the central element in a town's economic health. The importance of bridges helps to explain the notoriety of Horace King.

We illustrate the significance of these spans by focusing in detail on the 1824 Cheraw (South Carolina) bridge, the Columbus (Georgia) City Bridge, and other crossings within the Chattahoochee Valley. The construction records of the Cheraw bridge have survived. The Columbus City Council minutes document the problems of maintaining a covered bridge as well as years of political controversy over the city's most important source of revenue. The histories of the bridges in Eufaula (Irwinton), Alabama, and Florence, Georgia, illustrate the problems faced by private investors, as well as the vulnerability of these structures.

Without John and Ann Godwin there never would have been a Horace King. His purchase by the Godwins saved Horace from a life of anonymity as a skilled slave on a Carolina plantation or as a skilled carpenter in Charleston. The Godwins allowed King to develop and use his talents, to serve as a contractor while still a slave, and, ultimately, to become a free man. The Godwins' early actions are significant in documenting King's life because only King's brief testimony about his family illuminates his life before 1840. The contours of King's life during the previous decade can only be speculated by interpreting the Godwins' actions.

Finally, a return to the arena of speculation explores one last detail about the 1883 conference between Cherry and King. If King were not on an extended visit to Girard or Columbus, then the meeting place was inconvenient for a seventy-six-year-old man; it seems to illustrate southern social convention relating to race. Why did Cherry pick a drugstore in a mill village, even if it was adjacent to King's old home? If King came to Browneville for that meeting alone, then he probably rode the train from LaGrange down the Atlanta and West Point Railroad. He had been riding this line regularly since the late 1850s, when he established a second home in Carroll County, Georgia. The train always stopped in Opelika, where Cherry lived. There, in 1883, King boarded another train for the thirty-mile trip to Browneville. Why did he not simply visit the Reverend Cherry at his home in Opelika? Perhaps Cherry did not want to upset his neighbors by bringing Horace in the front door nor insult Horace by bringing him in the back door. While the actual voice of Horace is somewhat mute, his second-class citizenship within the South is often apparent. Given his unsurpassed skills as a craftsman and the nature of his personality, King, more than most contemporary African Americans, overcame many of the racist restrictions and prejudices of nineteenth-century society, but he never completely overcame the ingrained racism of American society. The amazing degree to which he succeeded represents the triumph of his life; the fact that prejudice limited his accomplishments embodies the tragedy.

CHAPTER ONE

PEE DEE ORIGINS

I was born in South Carolina a Slave, the property of Edward King. He died and his heirs sold me to Jennings Dunlop of Cheraw. He sold me to John Godwin.

Horace King's Testimony before the U.S. Commissioners on Claims, February 1878

The ancestry of Horace King reflected the racial dynamics of the southern frontier where three ethnic groups melded to forge a distinctive culture. King's deepest American roots came from his Indian forebearers, his name stemmed from his European ancestors, but his African heritage defined his legal and social status within this multiracial society.

King's complexion showed more "Indian blood than any other," according to the Reverend Francis L. Cherry. But the exact origin of his "Indian blood" and even its tribal identity remains a mystery. King's mother, Susan or Lucky—who apparently answered to either name—was born about 1790. As a slave belonging to the Kings, she married Edmund King, another King family slave. Cherry, presumably because of his interview with Horace, noted his mother's people as being Catawbas and cited her birthplace as the Cheraw district, South Carolina. In the 1870 census, however, when Susan lived with her daughter in Alabama, someone in the household gave her nativity as being Virginia, which would eliminate the Catawbas as the source of her Indian roots. Even so, Horace identified with the Catawbas. Perhaps his Indian ancestors gave him more than his light coloration and his high cheekbones.[1]

Early in the eighteenth century the English began using the tribal name of Catawba to refer to all the Indian groups living near the Catawba River in the Carolinas.[2] After the colonial government granted them several hundred acres of land, these Indians accommodated themselves to living in proximity with white settlers. Even so, on several occasions the Catawba national council discussed the possibility of leaving their river and joining the Creeks. Such a migration would have brought King's great-grandfather to the banks of the Chattahoochee River, but the Catawbas decided to stay in the Carolinas, where they avoided removal.

Rather than becoming Christians and farmers as the eastern Cherokees did, the Catawbas rented their land to white farmers, three hundred of them by 1791. The Catawbas also became peddlers, selling a variety of wares—moccasins, baskets, table mats, and pottery—to the settlers. Pottery—made by Catawba women—became their most profitable trade good. Their new role as landlords, potters, and peddlers forced the once warlike Catawbas to develop different attitudes in their relations with whites. They learned to be deferential while at the same time forceful enough to collect the rent or conclude a profitable sale. Patience and deference, as well as similar traits inherited or learned from his Indian forebearers, would be necessary for Horace King to make his way in a business world dominated by white elites. Cherry noted that King "is choice in the selection of words without appearing to be so." This trait is more consistent with Indian rather than European or even African personalities.

While Cherry praised King's capabilities—his intelligence, honesty, and the respect he garnered from whites—he also noted King "never lost sight of his social position by intruding himself beyond his legitimate sphere, which exalted rather than diminished him in public esteem."[3] Such condescension was a necessary skill for Catawbas in the Carolinas or a free mulatto in the South.

The exact source of King's English or European blood is just as hazy as that of his Indian forefathers. His father was Edmund King, presumably a slave and a mulatto, but he must have had Indian roots as well. Horace provided no other information about him. Might he have been a carpenter who transferred his skills to his son? The oral traditions have ignored the possible role of Edmund in shaping the bridge-building life of his son.[4]

Edmund's last name obviously came from his master's family. In the case of most mulattos in the antebellum South, the white parentage was usually assumed to be the master's family. Exactly who fathered a particular slave child often remained a mystery, as in the case of Sally Hemings's children.

William Johnson, the free black barber of Natchez took the name of the man who freed him as did his mother and sister, but in thousands of diary pages he never conclusively identified his father. In a similar fashion, the exact identity of the father of William Ellison, the skilled South Carolina gin maker, has never been established but is presumed to be a member of the Ellison family.[5]

Frederick Douglass, the most prominent free black in antebellum America who escaped and then bought his freedom from his Maryland family, wrestled throughout his life with the identity of his white father. When he first wrote his autobiography, Douglass combined his father and his master. Ten

The Peedee Region, 1830

years later he wrote, "I say nothing of *father* for he is shrouded in a mystery I have never been able to penetrate."[6] Horace King might have had similar thoughts about his paternal grandfather or his other white progenitors.

Horace's recorded statements about his origins were cryptic. In discussing his 1807 birth, Horace stated, "I was born in South Carolina, a slave, the property of Edward King. He died and his heirs sold me [in about 1830]." The exact identity of Edward remains a mystery—because William T. Sherman's troops burned the Chesterfield District courthouse. Edward's name does not appear on any census during the period from 1800 to 1830 for the Chesterfield District or any surrounding region. Researchers have linked Edward with Miles King (1758–1823), who had both a brother and a son named Edward. Miles, the patriarch of this prominent family, was born in Virginia, studied medicine at the University of Pennsylvania, became a Revolutionary War soldier, and lived in North Carolina, where he served in its House of Representatives before arriving at Society Hill, South Carolina, about 1790. Ten years later he moved into the somnolent village of Cheraw where he resided for his last twenty-three years.[7] These Kings are related to the Wrights of the Marlboro District, the family of John Godwin's wife. This link represents the only feasible means of establishing kinship between Horace and his masters.[8]

Edward, Miles's brother, might have migrated from Virginia and lived in his brother's household and thus never appeared on the census. Horace's mother having been born in Virginia provides a very slender thread of evidence to support that theory.[9] Several Horace King researchers have misidentified Edward, the son of Miles King, as Horace's master. But he did not die about 1830 in the Chesterfield District, as was the case for Horace's master. Genealogists concentrating on this family have proved that Edward, Miles's son, left Chesterfield County and moved to Alabama during the 1820s or 1830s.[10] Thus the identity of Horace's first master remains a mystery.

Given Horace's extraordinary skills as a builder, it could reasonably be suggested that his first master may have been a contractor who moved from place to place. However, the chronology Horace traced for the Reverend Cherry does not indicate a peripatetic childhood. Horace apparently spent his first twenty-three summers fanning away gnats and walking in the hot, sandy soil of Chesterfield District. He was born there in 1807 and sold from Edward King's estate in about 1830.

Becoming part of an estate could spell hardship for a slave; owners, concerned with maximizing profits, had no incentive to care about preserving families.[11] Furthermore, a slave adept at craftsmanship might have been sold to someone in Charleston and become one of hundreds of skilled black builders whose names were never recorded for history. Horace, his mother, Susan (or Lucky), his sister, Clarissa, and his brother, Washington, fared better than many slaves. John Jennings Dunlap of Cheraw purchased all of them.[12]

Because of the paucity of Chesterfield records, little is known about John Jennings Dunlap. He appeared in the 1830 census for Chesterfield District at about the time he purchased Horace and his family. Dunlap's household consisted of his wife, another male about the same age as Jennings (between twenty and thirty), four children, and only one female slave between the ages of thirty-six and fifty-five. Dunlap was not a slave trader. His family may have been in the building trades. John Jennings, who could have been a maternal relative of Dunlap, advertised in June 1823 that he could supply brick in such quantity as to "explode the practice of building with wood."[13] If this brickmaker were Dunlap's grandfather or uncle, then it strengthens the idea that Horace was already identified as a craftsman in 1830.

In the mid-1820s Dunlap joined responsible citizens of Cheraw in signing a successful petition to the legislature requesting that the seat of the equity court be shifted to Cheraw. On the same document appeared the signature of

John Godwin, the man to whom Dunlap eventually sold Horace. Dunlap's name never appeared in any Chesterfield censuses after 1830, so he may have joined the westward migration, perhaps to Alabama.[14] Maybe the sale of Horace financed Dunlap's move.

Other than King's intuitive skills as a craftsman or designer, no other factor played a more significant role in the life of Horace King than his being purchased by John and Ann Godwin. Few records document the Godwins in Chesterfield District, but the Godwins and their in-laws did leave footprints in adjacent counties. John Godwin, the father of Horace's master, became involved in Sneedsboro, North Carolina, an ill-fated town twelve miles upstream from Cheraw. This experience, which involved rival interstate speculation over river transportation at the fall line, would be an accurate harbinger for the lives of the younger John Godwin and his slaves.[15]

Throughout the South, from Virginia to Alabama, the fall line (running midway between the coast and the mountains) marks the dividing line between the wide, flat Coastal Plain and the hills of the Piedmont. Large rivers were generally navigable to the fall line. In the case of the Great Pee Dee River, the head of navigation was at Cheraw, just south of the boundary between North and South Carolina. William Johnson, who owned a gristmill and cotton gin in Sneedsboro, and Archibald D. Murphey, a Hillsborough lawyer who promoted internal improvements in North Carolina, planned a canal to extend navigation on the Pee Dee just across the North Carolina line to Sneedsboro, which they envisioned as a great inland port.[16]

Chartered in 1795, this would-be boom town, floundered at birth. In the nine miles south of the state line, the river fell thirty feet and in dry weather only drew fifteen inches of water.[17] Johnson and Murphey conceived of a long canal that skirted those rapids and continued into Sneedsboro. Rather than being adjacent to the river, the town was sited approximately two miles west of the river, close to a mineral spring. As construction started on this improbable waterway, Johnson laid out a grid pattern of streets and sold sixty-four half-acre lots.

The older John Godwin, the father of Horace's master, paid William Johnson sixty dollars for four lots—an entire town block—apparently in the heart of Sneedsboro judging from the sound of the street names that circumscribed Godwin's land: Broad, Meeting, Water, and South.[18] Was the elder John a builder who purchased these lots for speculative houses or were these future home sites for his four sons: James, Thomas, Wells, and John? In 1790 John Godwin apparently lived in Anson County, the site of this speculative port town, so they may have been some of the first residents, but the town never prospered. The elaborate canal, the heart of the scheme, was never finished and a typhoid epidemic eliminated any chance of the community's mineral spring attracting invalids. By 1830 Sneedsboro had become a ghost town, but by that time the Godwins had moved south.[19]

John Godwin the elder died before 1800. In that year his widow, Mary, still lived in Anson County with five children (her daughter and four sons) and no slaves. Judging by the lack of slaves, the Godwins were not particularly wealthy. John Godwin, the son, first appeared in public records in 1823, as a resident of Cheraw, at which time he and his older brother, Wells, lived together.[20]

By 1823 Cheraw was growing beyond a sleepy little village. In the 1770s it had consisted of a few houses, a couple of stores, a courthouse, and a jail.[21] The region did not share in the early agricultural prosperity stemming from rice and indigo along the South Carolina coast. Starting in the 1790s the boom associated with green-seed upland cotton cultivation increased business activity in the Carolina interior. By the late 1810s the state began improving navigation on the Pee Dee from Cheraw to the coast. By 1819 obstructions had been cleared from Georgetown inland for about one hundred miles to Cheraw. One steamboat and two team boats pulled by eight mules operated on the Pee Dee by 1823.[22]

In that year, the superintendent of public works for the state praised the economic progress of Cheraw. He boasted of the leap in Cheraw's commerce in only three years. A total of 3,264 bales of cotton shipped in 1819–20 had grown to 15,192 by 1822–23. "In the beginning of 1819, three or four houses formed the whole village; there are now [1823] 250 houses in it, and the population has increased from thirty-five persons, to more than eight hundred."[23]

Even allowing for exaggerations of its meteoric growth, Cheraw was a house builder's dream. Constructing two hundred houses using 1820s technology was no easy feat. The mass production method of balloon framing with two-by-fours and machine-made nails had yet to evolve. Thus the major frame pieces of every house consisted of hand-hewn timbers with mortise and tenon joints pegged together. Labor was cheaper than nails in the 1820s. John Godwin and his brother Wells certainly made money as house contractors. Some of John's profits were used to purchase Horace King and his family. Maybe King learned the skills of a carpenter and a heavy-timber frame craftsman during this period. He and Godwin may have become friends because of their shared profession.[24]

The year 1823 would not have been a good one for Horace: the South was still reeling from the terror of Denmark Vesey's slave conspiracy in Charleston. After this free black's plot was discovered in the summer of 1822, Vesey and his coconspirators were hanged. Then, authorities severely limited the movement of free blacks and slaves in Charleston and the state. If Horace had experienced any special liberties as a slave, the local community after the Vesey scare would have restricted these freedoms. However, the year 1823 brought a significant change for John Godwin. On September 2, 1823, he married Ann H. Wright from the adjacent Marlborough District. John increased his wealth because of this marriage. By the time the 1830 census taker reached the house of John and Ann Godwin, they owned seven slaves.[25]

The inventory of Ann's grandmother's estate in 1817 totaled $3,264, including fifty acres of land and nine slaves. Sarah Wright willed $100 to her granddaughter Ann. Perhaps from 1823 and certainly by 1837, Ann's uncle William C. Wright acted as a trustee for Ann's property, a typical arrangement in this period.[26] When a woman married, her husband gained control over her property, and she lost all rights to her land, slaves, and other possessions. Thus wealthy families transferred ownership of women's properties to a male trustee, usually a relative, so wives could maintain some degree of economic independence and so property did not escape the family. The existence of such an arrangement between William and Ann Wright indicates the wealth of the Wrights.

Even if John Godwin's wife, who was ten years younger than he, did bring more money into the marriage, he had established himself as a respectable citizen of Cheraw by 1823. During that period he signed petitions to the legislature requesting an equity court for Cheraw and, more relevant to his occupation, permission to build a dam on Thompson's Creek. John and six other men, including William Wells, proposed a four-foot dam on the property of John Pervis. They built a simple, partial wing dam; the impounded water powered both grist- and sawmills. Godwin's involvement may have been limited to constructing the dam and mills, or the sawmill may have produced his lumber.[27]

On September 5, 1826, John Godwin probably joined a local Baptist church. Mason Risley Lyon, a newspaper publisher and a Baptist minister, marked the occasion by presenting Godwin with a Bible.[28] John must have enjoyed an excellent reputation among local builders by 1828 because a very legalistic couple, James and Maria B. Moffett, hired him to build their house at Level Green, about twenty miles east of Cheraw. James Moffett graduated with a medical degree from the University of Edinburgh in 1822, then settled in Fayetteville, North Carolina, and in 1825 served as an escort for the Marquis de Lafayette as he traveled from that town to Cheraw. The Moffetts then moved to the Cheraw area. The physician and his wife prepared a contract for their house and took the unusual step of filing it in the deed records.[29]

John Godwin agreed to build them a Carolina or Marlboro cottage for the sum of $1,125. The story-and-a-half, side-gabled house measured forty-eight

"Carolina Cottage," Cheraw, South Carolina, ca. 1900. The details of this house resemble the home John Godwin built for the Moffetts at Level Green in 1828. *Photograph by unidentified traveler, Cheraw (S.C.), #9 in Photograph Album, 1899–1904. Courtesy of the Rare Book, Manuscript, and Special Collections Library, Duke University, Durham, North Carolina.*

feet deep and forty-two feet wide with four rooms and a center hall on the first floor and two rooms in the loft space. The windows on the first floor, four on every side, contained eighteen panes of glass; the three dormers on the front roof, the single rear dormer, and the gable windows all had twelve lights. The porches, an essential feature for catching any breeze in this flat landscape, consisted of a full-width integral "piazza" and a smaller portico in the rear. Two interior brick chimneys served every room on the main floor, and brick piers, exactly three-and-a-half-feet high, supported the house. Godwin furnished all the lumber, bricks, nails, glass, hinges, locks, and fastenings.[30]

Executed on August 1, 1828, the contract called for a completion date of October 17, 1828. Given that time for construction, a little over two months, Godwin could have built several homes a year of this design and scale—not a grand house, but a substantial one and typical of many surviving dwellings in the region.[31]

In order to reach the Moffetts' house site in Level Green from Cheraw, Godwin crossed the Pee Dee River. At that location entrepreneurs erected a bridge in 1824; it washed away in 1826, and another was thrown across the river in 1828. Godwin's and King's roles in building the second bridge remain one of the mysteries of their lives. They probably learned their trade by working on this span, but the construction history of the 1828 bridge remains unknown. The erection of the 1824 crossing, however, is well documented. Ithiel Town designed it according to his 1820 patent, and he traveled to Cheraw to consult with the builders, William Warren, George T. Hearsey, and Seth King.[32]

CHAPTER TWO

ITHIEL TOWN'S BRIDGE

This grand and extensive work across the Pee Dee River opposite this place is now completed . . . constructed on the self supporting principle invented by Mr. Ithiel Town. . . . In point of elegance and extent, it by far surpasses any in this State, and is believed to be little inferior to any in the Union.

Cheraw Intelligencer and Southern Register, June 18, 1824

According to the Rev. Francis L. Cherry's biography, "Horace early developed an extraordinary capacity for bridge-building, the first he ever built being across the Pee Dee River in South Carolina before he came to Girard [Alabama]." The records detailing the construction of the first Cheraw bridge (1823–24) do not substantiate that King, the slave, or his future master Godwin, the house builder, played a role in this project.[1]

A flood washed away the 1824 crossing in 1826, and another span replaced it in 1828. The 1824 version is more important to history in a general sense: Ithiel Town, the inventor of the most-used bridge building system, came to Cheraw to consult with its builders. Detailed records exist for this bridge; they enumerate every expense. No manuscripts and no Cheraw newspapers have survived to illuminate the 1828 span, on which Godwin and King probably worked.

Even if they never operated an auger or supplied a piece of lumber in 1823 or 1824, Godwin and King certainly watched intently as Ithiel Town supervised the building of this prototypical bridge. At a minimum, King and Godwin observed and learned as coffer dams diverted the muddy Pee Dee and piers rose from the damp clay bottom, as the long uncovered approach (or land) bridge appeared on the low-lying west side of the river, as slaves hewed beams, and as other hands bored holes in the multitude of sawn pine or cypress boards that became the timber tunnel held together by hardwood treenails. Perhaps Godwin and King first met while inspecting the pristine white span. They almost certainly would have joined the first crowd to test the bridge, walking from Cheraw over to Marlboro County.

This bridge was the model for King's lifework. For the next fifty years he replicated Town's lattice trusses over other southern rivers. The construction of the 1824 Cheraw bridge—the quintessential American bridge—is so well documented that it warrants a detailed examination.

Ithiel Town invented the perfect structure to span the rivers of the expanding new republic. Its characteristics shouted American pragmatism. This bridge, designed for utilitarian rather than aesthetic purposes, could be erected with fewer skilled workers and fabricated anywhere on the frontier. It served the middle-class needs of average Americans in a hurry to make money in a new place.

Town's prototype had been created not by chance but by invention. The latter part of the eighteenth and early nineteenth centuries represented a period of innovation in bridge construction throughout the Western world and especially in the United States. Inventors used new materials—cast- and wrought-iron arches and chains. They also shaped a very old material into new forms—wood fashioned into trusses. They were searching for a universal bridge, one capable of spanning any stream and bringing the patent holder a steady stream of wealth. The moneymaking potential of bridges attracted designers and wealthy financiers.

Wood had always been used for short spans. These structures, known as beam bridges, consisted of vertical posts supporting horizontal beams. The length of the span determined the number of posts and beams. However, such bridges were unsuited for a large bridge over navigable waters, in which boats would require an open passage. As a result, designers sought cheaper materials to span greater distances. Chronologically, the media used by bridge innovators began with iron, then moved to wooden trusses, then iron supplementing wooden trusses, then iron and steel trusses. These utilitarian, universal versions always competed with more elegant, more expensive one-of-a-kind bridges.

The arched masonry bridge—the most long-lived design—with its roots in the Roman world creates an impressive, singular structure. In the decade preceding Town's Cheraw bridge, John Rennie and his son spanned the River Thames with nine graceful granite arches (each measuring 120 feet) and piers decorated with a pair of Doric columns. Visitors came to London from around the world just to see Waterloo Bridge.[2] The Cheraw Bridge Company had neither the money nor the masons to build such a monument. Their objective was to attract wagons loaded with cotton, not visitors gawking at their bridge. Besides, a decorative stone arch could not reach as great a length as a wooden truss, and longer spans were necessary to clear the southern waterways plied by steamboats.

In the late eighteenth century iron began competing with stone and wood as a bridge-building material. In the iron-producing region of Shropshire,

England, Abraham Darby III erected the Ironbridge (1777–79) at Coalbrookdale. This gracefully arched one-hundred-foot span is credited as the world's first major cast-iron bridge.[3] The lure of making money from this novel material attracted inventors from a range of professions on both sides of the Atlantic. In the 1790s Thomas Paine, the Englishman who served as America's revolutionary pamphleteer, designed a four-hundred-foot cast-iron arched bridge for Philadelphia. Perhaps his earlier career as a corset maker taught him about fashioning supports. His objective was a universal span long enough to reach across ice-clogged American rivers without building a middle pier. Paine traveled to England to supervise the fabrication of his arches, but his financial backer withdrew. Then, attracted by revolution, Paine migrated to France and forgot about bridges.[4]

Builders also experimented with iron chains for large-scale suspension bridges. Small suspension spans carrying foot traffic had existed for centuries, especially in Asia. By 1800 the Industrial Revolution made larger ones possible. James Finley—a Pennsylvania farmer, judge, and state legislator—patented a design in 1808 that incorporated two wooden towers from which chains supported a rigid roadway. Finley built thirteen small bridges, several of which failed. Heavy snow collapsed one and a herd of cattle destroyed another. At the same time, English builders were creating monumental suspension bridges. From his American context, Finley compared these "majestic" structures with his "simple contrivances" and found he much preferred his. "Happy for me," Finley wrote, "utility[,] economy[,] and dispatch, are the ruling passions of the day, and will always take preference over expense, idle elegance and show." Finley's design was not the epitome of utility, economy, and dispatch, but those characteristics were the ruling passions of men who built American bridges. Iron, however, remained an unsuitable medium in the early nineteenth century. Paine's long arches could have spanned the Pee Dee or the Chattahoochee, but the cost of making and shipping them from Shropshire, England, or even Pittsburgh made them impractical.[5]

Theoretically, the grand English suspension bridges would have worked for any southern river, but the expense would not warrant the construction. Finley's simple plan would have been feasible for smaller streams if reliable iron chains had been available. The village smithy, however, could never supply uniform wrought-iron links for the necessary chains. The use of iron and steel bridges emerged in the United States after the Civil War with the creation of a national market based on the railroads. This expanding network of rail lines allowed access to the entire nation and created companies specializing in iron and steel bridges.

In the 1820s and 1830s "utility, economy, and dispatch" in bridge building meant wooden trusses. Even though wooden bridges dated from the dawn of time, the use of wooden trusses was still being perfected in the 1820s, specifically by Ithiel Town at Cheraw. Several men sought the same objective with wood as Finley had tried to achieve with iron chains—a universally adopted bridge that brought royalties to its designer.

The American economy was fueled by and built with wood. From its colonial roots, American timber represented an important export commodity. Americans used wood where Europeans utilized stone for canal elements, factories, mansions, and churches. This practice baffled foreign visitors. With a constantly moving frontier, nothing in America was permanent. Wood was cheap and plentiful, and this surfeit led to waste. Inexpensive wood, consumed in gargantuan amounts, powered steamboats, railroads, and the hungry machines that created mass-produced consumer goods by the midcentury.[6]

Given the nation's preference for wood as a building material, American inventors in the early nineteenth century naturally explored using timber to fabricate inexpensive bridges. A simple wooden span with no arch or truss could span sixty feet—too short a distance for most American rivers. At the beginning of the nineteenth century, two new types of wooden bridges were developed. One employed an arch stiffened by a truss, and the other consisted of a truss without an arch. The wooden arch mimicked the idea of a masonry arch, but the wooden version had to be reinforced by triangles in the form of a truss.

In the simplest terms, a truss is merely a series of triangles attached to a horizontal beam. The triangles serve to stiffen the beam so it can span a longer distance without bending. The most basic truss, the king post, consists of one upright support braced by two diagonal beams that form triangles on either side of the horizontal beam, which forms the base of the two triangles. The diagonals transfer the load to the extreme ends of the truss where it rests on a pier or an abutment. Different designers arranged multiple king posts and diagonals in a variety of configurations. The first American bridge inventors experimented with combined arch-truss structures; they tended to be master carpenters who understood the strength of arches.[7]

Timothy Palmer became the first American builder to cover his bridge. In 1806 his "Permanent Bridge" spanned the Schuylkill River where Paine had proposed building his iron bridge. Palmer's three spans of about five hundred feet took five years to build at a cost of three hundred thousand dollars. Following the suggestion of a friend, Palmer covered his bridge as an afterthought. Given that large an investment, the additional cost of covering the bridge to protect the main timbers from the weather certainly was warranted.[8]

Despite the romantic lore as to why bridges were covered, they were sided and shingled for a pragmatic reason—to extend their life by seven or eight

Small king post bridge, Decatur County, Georgia, ca. 1900. Swayed by the power of the Horace King myth, one journalist in praising King deemed him the inventor of the king post truss. This simple use of triangles to extend the reach of a structure beyond the span of a single beam dates from humankind's earliest bridges. Many builders and inventors employed multiple king posts in their trusses. *Courtesy of Georgia Division of Archives and History, Office of Secretary of State.*

times. The purpose was not simply to shed snow since they were just as popular in the South as in the North. Horses and cattle would cross a river on an open bridge. Even though enclosed bridges might have aided courtships, such was not the objective of builders. Making money from a practical design was their passion. A bridge—or more specifically the large beams—that lasted and could be warranted by the builder earned contracts and saved money in replacement cost.

In 1812, shortly upstream on the same Schuylkill River, Lewis Wernwag, a German immigrant, built another massive covered bridge, "the Colossus"—a single arched span extending 340 feet. But Palmer and Wernwag's bridges were too expensive. Only large cities could afford such structures. Towns like Cheraw and Columbus, Georgia, could never finance such extravagant works. Wernwag, recognizing the exorbitant expense of erecting his arch, developed

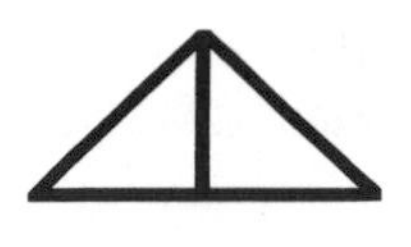

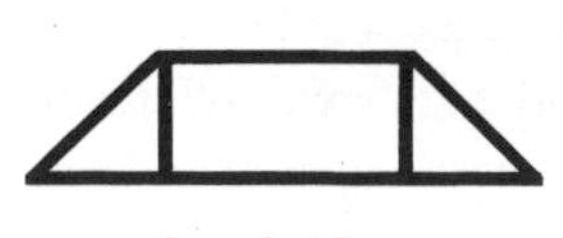

Queen Post Truss

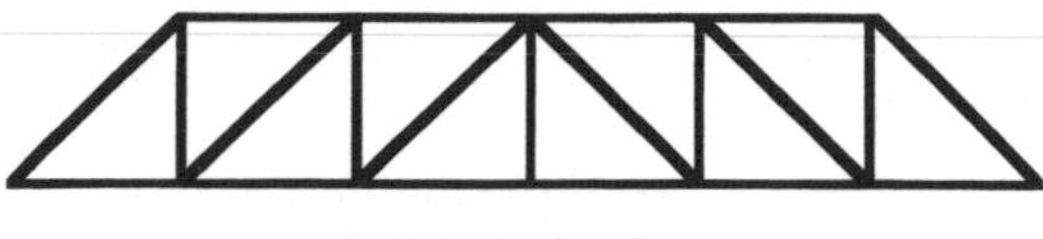

Multiple King Post Truss

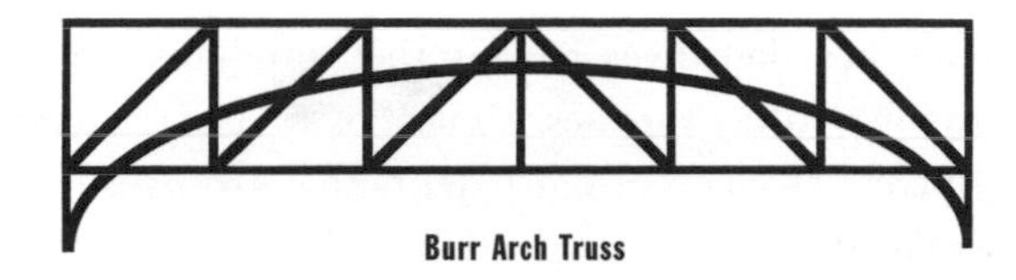

Illustrations of some bridge trusses.

smaller bridges, the simplest of which he called "Economy." But its design employed cast-iron parts that were difficult to produce in the hinterlands. Wernwag secured a patent by 1829, but by that time he had been bested by Theodore Burr and Ithiel Town.[9]

Theodore Burr of New York became the first inventor to derive significant profits from a bridge patent. In 1805 he built a long bridge across the Hudson River at Waterford, New York, by adding an arch to a simple truss. In 1806 Burr patented his arched truss and charged other bridge builders who employed his scheme. A Burr bridge being built over the Congaree River at Columbia, South Carolina, was sketched by Basil Hall, the British naval captain, who later the same year (1828) described the founding of Columbus, Georgia. Hall found it strange that Americans rode and drove on the right side of the road, and he criticized the new bridge. "It is not possible to conceive anything more ungraceful than these huge snail-like housings of bridges." Perhaps he was comparing this wooden tunnel with the elegantly arched Waterloo Bridge in London.[10]

Despite the negative observations of English visitors, Burr collected royalties from his patent. Perhaps that motivated Ithiel Town to produce a simpler, less expensive bridge. The first bridges Town built employed Burr's ideas, but rather than continuing to pay Burr, Town designed his own truss, which quickly surpassed Burr's in popularity. Town's architectural career encompassed both high style architecture and practical bridge building.

Born in Connecticut in 1784, Town worked as a carpenter and then studied architecture with Asher Benjamin, a prominent Boston designer. By 1812 Town was practicing his craft in New Haven. His commissions, some of which he shared with his partner A. J. Davis after 1829, produced public buildings in New York and Boston, state capitols in Connecticut, Indiana, and North Carolina, and a multitude of elegant mansions in the various revival styles—Greek, Gothic, and Egyptian.

Town's bridge designs appeared to be the antithesis of his high style architecture. Some of his own bridges had elegant portals and tollhouses, but the work of Town's bridge disciples was practical and simple in appearance compared to the finely detailed entablatures on Town's churches. He created his architectural work for the elite. His bridges served the needs of average Americans going about their daily lives, getting their products to market, or making their way to town to buy supplies. Typical of an American inventor, Town received more income from his bridge patents than from his high style buildings. Money was made in America by satisfying the needs of the middle class, not by catering to the tastes of the elite, a formula later copied by Henry Ford.

By 1816 Town had built two Burr-style bridges in Massachusetts. From 1819 to 1823, as he worked on the North Carolina capitol in Raleigh, he designed and constructed the Clarendon Bridge on the Cape Fear River at Fayetteville, North Carolina, on the fall line northeast of Cheraw. Another early Town bridge spanned the Yadkin River, the upper reaches of the Great Pee Dee. These two bridges probably lacked arches and represented prototypes of his unique truss structures.[11]

Town patented the "Town Lattice Mode" truss in 1820. He eliminated the arch, a major component of the Burr truss. This laminated arch, formed by joining and bending several pieces of wood, could not be assembled by average carpenters; it required skilled woodworkers plus time. Also, under a heavy load the arch and the truss components tended to shake apart. Town removed the necessity of the arch. His truss consisted of numerous intersecting diagonal boards stiffed by horizontal members called chords. The numerous diagonal members and chords spread the load over the entire lattice and did not strain any one point of the truss.

Town's design also made the construction of the piers and buttresses less critical because they had to withstand less stress. In the case of arched bridges, the load on the end of the arch was exerted outward. Town's trusses, acting as a beam, applied a load in a vertical direction; thus, the piers and abutments could be less substantial and, therefore, less costly. A Town bridge could be built for one-quarter the cost of Burr's arch-truss bridge.[12]

The Burr arch truss, Congaree River bridge, Columbia, South Carolina, 1828. Note the combination of the arch and the multiple king posts. British naval captain Basil Hall sketched this span while it was under construction. Later the same year, he described the founding of Columbus, Georgia. *From Basil Hall,* Forty Etchings: From Sketches Made with the Camera Lucida, in North America, in 1827 and 1828. *Courtesy of John Sheftall.*

Ithiel Town. The career of Horace King rested on the inventiveness of Town, a Connecticut architect who designed high style buildings and state capitols as well as utilitarian bridges. Town's practical wooden truss perfectly filled the need for building inexpensive bridges anywhere in the expanding nation. *Angelo Franco, artist, commissioned by Thomas L. French Jr.*

Side view of Town's truss.

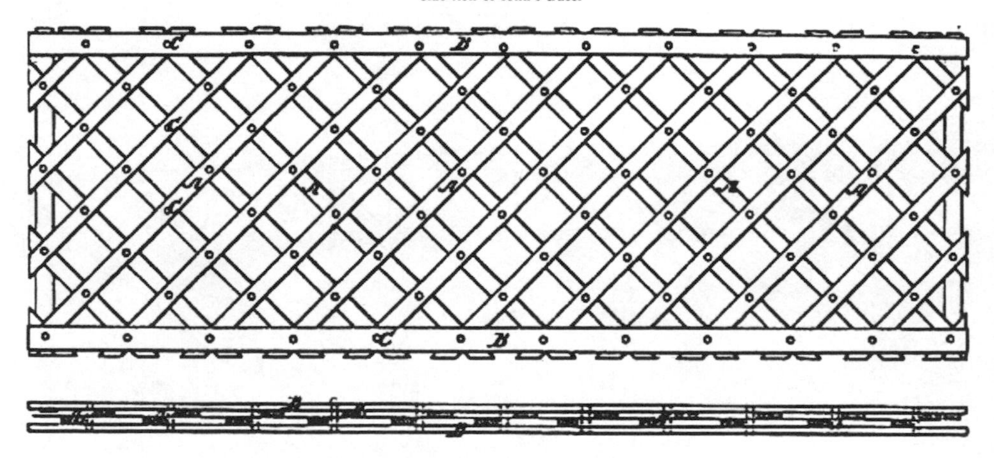

Top view of truss showing the diagonals sandwiched between the stringers.

Plan of Ithiel Town's truss from his original patent papers, 1820. His truss consisted of crossed braces or diagonals sandwiched between horizontal stringers or chords. Town quickly changed his design by placing two treenails in every junction of the diagonals and the stringers; this modification appeared in the Cheraw bridge by 1824. *Collection of Thomas L. French Jr.*

Town sent a model of his 1820 bridge to his New Haven neighbor Eli Whitney. America's best-known inventor gave Town a testimonial in which he praised the bridge's "simplicity, lightness, strength, cheapness, and durability." For the right to erect one of Town's lattice truss bridges, the builder had to pay Town one dollar per foot. If one of Town's agents found a bridge company that had not paid the royalty, they charged that firm two dollars per foot.[13]

The beauty of Town's truss was its simplicity: A skilled carpenter could build one with lumber from any sawmill. Complex mortise-and-tenon joints were not required. The boards were joined with treenails (pronounced trunnels) driven into drilled holes. According to covered bridge lore, Town's trusses could be manufactured by the mile and cut off by the yard. Even if his bridges were easier to build, their cost remained high. For example, in 1823 the Cheraw Bridge Company spent seventeen thousand dollars constructing three covered spans totaling 425 feet, along with 922 feet of uncovered land bridges.[14] Town's spans were durable, but many of them, on an almost regular basis, would succumb to floods and fires. An ordinary carpenter could build small spans over creeks, but large reaches across major rivers required an experienced builder, even to erect the seemingly simple Town trusses—hence the success of Horace King's career.

Building a bridge in the early American republic required more than a plan, a consultant, some workmen, and a little capital for lumber. Even if the builder owned both sides of the river, he had to obtain a charter from the state. At some point before 1821, a group of Cheraw citizens, not including John Godwin or the 1823 builders, petitioned the legislature for the right to build a bridge at the site of the existing ferry.[15]

A major argument for the Cheraw bridge—bringing cotton to market—would later be the same rationale for spanning the Chattahoochee at Columbus, Georgia; the Coosa at Wetumpka, Alabama; and other rivers in other fall line towns. These communities existed because of the adjacent navigable river. Their business leaders usually sought to improve the river and then to expand their commercial hinterlands, the area from which they attracted cotton to their warehouses. Cotton piled high in a town's warehouses meant farmers would spend their profit in that community. Bridges attracted wagons filled with cotton.

Even though towns benefited from bridges, in most cases large bridges were built by investors, either individuals or stockholders, rather than the municipalities themselves. Communities lacked the means to raise the necessary capital. Most contractors did not have the financial wherewithal to fund bridges. Throughout the antebellum period, entrepreneurs provided the money to finance most ferries and bridges and the political clout to gain approval from the legislators. Wade Hampton Sr., the patriarch of one of South Carolina's largest fortunes, built or financed the initial bridges over the Congaree River at Columbia and the Savannah River at Augusta in the 1790s. He joined with James Gunn to span the Ogeechee, near Savannah, in 1790.[16] A few years later, Gunn as a U.S. senator engineered the infamous Yazoo land sale, ruining his political career and the future of the Federalist Party in Georgia. However, bridge building continued to attract prominent speculators. Later Georgia examples include Seaton Grantland and William Towns, both of whom received the right to build toll bridges over the Flint River in 1834, the same year they were elected from contiguous districts to the U.S. Congress.[17]

In the case of Cheraw in 1823, three men provided the capital for the bridge's construction: George T. Hearsey, Ward Coming, and Seth King (who was not related to Edward King).[18] The detailed records of this bridge project provide insight into the finances of later bridges built by Horace. George T. Hearsey, who apparently arrived in Cheraw in 1823 and later established the city's first bank, appeared to be the lead investor.[19] He contributed three-fifths or $10,302 to the construction effort. Ward Coming and Seth King each provided $3,434 toward building the $17,170 bridge. In return, these gentlemen sold goods and services to the company at a profit. Hearsey provided at least $2,700 worth of sawn lumber; King sold various supplies totaling about $1,500; and Coming handled real estate transactions.

Neither King, Hearsey, nor Coming acted as the contractor for the project. The newspaper identified William Warren as the builder, but the records show a prominent role for Russell Warren, presumably a relative. They supervised the laborers and authorized payment for them, but George Hearsey dispersed the wages. Ithiel Town, the inventor of their structure, assisted them. He received at least $1,500 from the Cheraw Bridge Company, more than $1,000 over the dollar-a-foot fee for a 415-foot bridge. The company reimbursed his expenses when he visited Cheraw.[20] Town insured that the construction met his standards since this bridge might serve to attract future builders who used his patent.

The skilled workers needed were unavailable in Chesterfield or Marlboro Districts. The company imported men from Georgetown and Charleston, perhaps master builders as well as men experienced in erecting piers in riverbeds. The Cheraw Bridge Company absorbed expenses for feeding the men, an indication that they were living away from home. The investors purchased blankets, plates, utensils, candles, salted beef, codfish, bacon, pork, coffee, potatoes, butter, meal, sugar, salt, wine, brandy, and gallons of rum to ward off the dampness rising from the Great Pee Dee River.

Labor costs for skilled workers alone amounted to at least $3,236, equaling about one-fifth of the total cost. These wages were paid to men who were

Western end of Great Pee Dee Bridge at Cheraw, ca. 1900. This photograph of the post–Civil War span, which replaced the one burned by William Sherman's troops, evokes the romantic, nostalgic image of covered bridges. *Courtesy of South Caroliniana Library, University of South Carolina, Columbia.*

recorded by their last names. Other lists only gave the first names, such as the fifteen laborers, probably slaves, provided by Hearsey. Their wages ran about fifty cents a day, but of course that pay went to Hearsey. Other owners received compensation for work done by their slaves. Neither Horace nor Godwin appeared on these lists.[21]

The other major expense was wood—hewed timbers and sawn lumber. The trees were cheap, twenty-five cents each; 569 trees and about 150 poles came from a nearby estate. Another subcontractor, William Trantham, or more precisely his slaves, cut the trees and hewed many of them into square timbers, the largest measured fourteen-by-fourteen inches by forty feet in length. These beams formed the piers and the various plates beneath the bridge. The sawn lumber most likely came from different trees than those supplied by Trantham. Hearsey's lumber mill cut 270,735 board feet of lattice members, chords, braces, and other supports, all of which cost $2,700. The company's brickyard fired 3,400 bricks.[22]

Judging by when the company started purchasing supplies, construction probably began in November 1823, a month of reduced stream flow on the Pee Dee and other southern rivers. By May 1824 the floor was laid, and pedestrians and wagons began using the bridge. At the same time, workers nailed split wooden shingles to the roof and vertical boards to the sides.[23] The Cheraw newspaper declared the bridge completed on June 18, 1824 and bragged that "this grand and extensive work. . . . in point of elegance and extent. . . . far surpasses any in this State, and is believed to be little inferior to any in the Union." After describing the construction of the span, the editor noted: "The bridge is handsomely enclosed, and painted white. Its arch does not make more than 3° or 4° of a circle, and is given to it more on account of appearance than necessity."[24] That slight arc imparted camber to the trusses and was an essential part of the design of a wooden bridge. That small bow prevented a bridge from failing under a heavy load. If a bridge sagged below the horizontal, it would never recover its original shape and would eventually fail.

Despite generalizations about how an ordinary carpenter could build a Town lattice truss, large spans demanded a skilled bridge architect because of two essential ingredients—ensuring camber in the trusses and building the midchannel piers. Horace King mastered those skills early in his career and, thereby, became a sought-after bridge builder.

While the camber did not fail in Hearsey, Coming, and King's bridge, the bridge failed to yield a large return on their investment. The accounting records of Ward Coming indicate that through July 27, 1825, the company had only received $102.67 in tolls, a meager half a percent return on their investment the first year. And even that small income was wiped out when a flood washed the bridge away, probably in August 1826. Seth King remained a stockholder in the enterprise. After the disaster, he signed a petition asking the legislature for permission to assume the rights of the ferry, which the bridge company had purchased. Neither Hearsey nor Coming signed that document. King appeared to control the bridge throughout the entire antebellum period. Later, probably 1865, Jane O. King—Seth's niece—was the largest stockholder in the company, controlling 120 of the company's 350 shares.[25]

If Hearsey and Coming deserted the company after the flood, then Seth King financed and controlled the 1828 rebuilding. Given his later relationship with Godwin and Horace King, they presumably worked on this structure. Horace probably identified this span to the Reverend Cherry as the first he built. Since Godwin was an ambitious house builder, his involvement was logical. If the calendar for the 1828 construction mimicked that of the 1823 one, the work would have begun after Godwin had completed the Moffetts' house in Level Green. Godwin had definitely become a bridge builder four years later.

Great Pee Dee Bridge at Cheraw, ca. 1900. One of Horace King's obituaries credited him with inventing the "lattice truss." Though King did not devise this mode, Horace and his sons made a comfortable living erecting them. Along with other builders they used Town's technology for major bridges until 1900 and on smaller ones into the 1920s. *Courtesy of the Rare Book, Manuscript, and Special Collections Library, Duke University, Durham, North Carolina.*

The fraternity of southern bridge building appears to have been small. Seth King continued to finance bridges. By 1861 he owned major shares of spans in Raleigh; Cheraw; Columbus, Mississippi; Wetumpka, and Tuscaloosa, Alabama. He remained a mysterious figure, a wealthy bachelor who was identified in 1861 as a bridge architect in Tuscaloosa, where he spent considerable time, though he was not considered a resident. The Black Warrior River bridge there was the only one of his bridges he designed and built. Horace King designed and built two of Seth's bridges, not counting the one in Cheraw. But Horace's relationship with Seth was never close.[26]

Seth probably introduced Horace to Robert Jemison Jr., who partnered with Seth to finance bridges in Mississippi and Alabama, which Horace built. Seth and Jemison became bitter enemies, and their projects ended in litigation between the two. Jemison and Horace, on the other hand, developed a warm friendship. Jemison, a Tuscaloosa, Alabama, entrepreneur, state legislator, and Confederate senator, ranked second only to Godwin as a promoter of Horace's professional career. In the 1850s and 1860s Jemison wrote Horace seeking Seth's whereabouts; his correspondence clearly indicates that Horace and Seth never enjoyed a close or "family" relationship.[27]

Seth King's 1824 and 1828 Cheraw bridges illustrate the complexities and vicissitudes of bridge building. Fire destroyed the 1828 span in 1836. A third Town lattice truss at the same site survived until 1865 when retreating Confederate forces burned it. The same Cheraw Bridge Company financed a fourth Town lattice truss bridge in about 1866, at a cost of twenty thousand dollars. The city finally bought the bridge in 1899, but a 1908 flood carried away all but one section of the lattice bridge. The city left the timber portion on the Cheraw side of the river and added two iron and steel through-trusses to complete the crossing in 1909. That jury-rigged structure with its wooden section carried a thousand cars a day on U.S. Highway 1 until 1939, when the state replaced it with a concrete structure.[28]

Wooden, covered bridges were essential for southern towns to attract the cotton trade. Securing the right to build a span meant approval from the state legislature. Building them required large amounts of capital, timber, sawn lumber, supplies, and sizable gangs of labor with varying skills. A well-built bridge could bring tremendous income. Its failure, through collapse, fire, or flood, could bring ruin. Such was the world that Godwin and King entered when they left Cheraw to accept a contract to build the Columbus bridge in 1832.

THE MISSING YEARS AND OTHER MYSTERIES

He sold me to John Godwin. I learned the trade of bridge building. Worked at leisure time, made money, and bought my freedom in 1848 and have since been engaged in bridge building.

Horace King's Testimony before the Federal Commissioners on Claims, February 1878

For John Godwin, the years from 1828 to 1832 were formative ones in his professional career: he established himself both as a house builder and a bridge contractor. Between 1830 and 1832 Horace King became Godwin's slave, learned or perfected the craft of heavy timber frame construction, and, more importantly, began developing his unique relationship with his new master. Unfortunately, few records illuminate their lives or work during these seminal years. Instead of facts, persistent myths about Horace have been spun about an apocryphal trip to Ohio. The most fanciful version has him joining the Masons, attending Oberlin College, and receiving his freedom in Ohio before 1832. The documentary evidence, however, and Horace's testimony fail to support these colorful assertions.

In 1828 Godwin was building houses in the Cheraw area. Four years later, while still living in Cheraw, he won the contract to construct a 560-foot bridge across the Chattahoochee River into Alabama Indian territory. He undoubtedly built similar bridges before bidding on this job. Though just four years old, Columbus would not have hired an untried craftsman to span the river at the bottom of the rapids that climaxed the Chattahoochee's thunderous passage over the fall line. Godwin must have worked on the 1828 bridge at Cheraw and other spans in that area.

Perhaps both King and Godwin worked on the 1828 bridge in Cheraw, enabling Godwin to appreciate King's skills, or perhaps Godwin's capital to purchase the King family came from Godwin's involvement with that span. In 1830 Godwin bought Horace, his mother, Susan (Lucky), and perhaps his siblings, Washington and Clarissa, from J. Jennings Dunlap, who had purchased them from Edward King's estate in 1830. The exact date of Godwin's purchase of King's siblings remains uncertain. Cherry's account, presumably based on what Horace told him, only mentioned Horace and his mother being bought by Godwin. According to the 1830 Chesterfield County census, Godwin owned seven slaves. One of the three males between the ages of ten and twenty-four could have been Horace, but Godwin's one female slave does not fit the age of Horace's mother. Perhaps he acquired the Kings after being enumerated by the census taker. The money for Godwin's slaves might have come from Godwin's wife's family, because in 1837 Godwin transferred the ownership of this chattel to his wife, Ann, and her uncle William Carey Wright, who served as her trustee.[1] Horace could have always been Ann's slave.

The two years between Horace's becoming a Godwin possession and their moving to Girard, Alabama, was the most formative period in Horace's life, but no records or correspondence illuminate it. One central question remains: How did Horace gain his skills as a bridge designer and builder?

The primary influence on determining the occupations of nonslave blacks depended on their age at emancipation. African Americans born into freedom or freed in their childhood pursued careers shaped by their free black relatives. The barber of Natchez represents this scenario. Being freed as an eleven-year-old, William Johnson learned his trade from his mulatto brother-in-law.[2] The occupations of nonslave blacks who were emancipated as adults were usually determined by their masters. William Ellison, the South Carolina cotton gin manufacturer, illustrates that point. Before allowing Ellison to buy his freedom, his master provided Ellison with an extended apprenticeship under a master gin maker.

The education of King paralleled that of Ellison. Obviously, learning a craft separated King from 90 percent of the nation's African American males who only performed menial work. But King's training also exceeded that of even the most skilled blacks, who tended to learn a specific job—carpentry, bricklaying, plastering—but not acquire the expertise of a contractor who orchestrated all the various construction tasks. The description of William Ellison's capacity as a gin maker fits King's work as a builder. Ellison "developed the mechanic's ability to conceptualize the whole, to envision each part meshing with all the others."[3]

The foundation of King's skills rested on his innate, intuitive skills as a builder—the abilities to shape wood with precision and to join the pieces into a stable framework. He knew how to mark, cut, and drill pieces for a bridge, and then treenail them together into a single truss that worked. The process of honing Horace's God-given talents probably came from several sources. His father, Edmund King, could have been a skilled carpenter, and his first mas-

Milledgeville, January. 23d, 1829.

I John R. Woolaw Clerk of the Inferior Courts of Baldwin County do hereby certify that Rachael Gould is a free person of colour, and that she has resided in this county for the last Ten years That she has conformed to all the laws & regulations requiring of Free persons of colour in this State and that she has been regularly enrooled, as a free person of colour and recognized as such enjoying all the right & priviledges granted such persons by the Laws of this State.

Given under my hand and private seal (there being no seal of Office) at Office.

John R. Woolaw Clk
I.C.B.C.

Certificate of freedom for Rachael Gould, Baldwin County, Georgia, 1829. The King family WPA history, written in the 1930s, misidentified this document as freedom papers for Horace and Frances. It actually certifies that Rachael, Frances's mother, is free. Rachael's freedom guaranteed freedom for Frances and her children. *Collection of Thomas L. French Jr.*

ter, Edward King, could have been a Cheraw house builder. Horace must have been a master carpenter by the time Godwin purchased him. Godwin could not have elevated King's skill level from being untrained to that of a bridge builder in the span of two years. Godwin just honed King's native ability and facilitated his transformation into a bridge builder. Cherry's narrative, based on King's words, placed his learning about bridge building after Godwin acquired him, between 1830 and 1832. This chronology indicates that Godwin had a major impact on Horace's training.

One version of the King lore developed an elaborate story of King's being educated at Oberlin College in Ohio before coming to Columbus. That assertion leads to other conundrums about when he became free and when he joined the Masons. All these questions revolve around whether or when he went to Ohio.

This rabbit trail began as early as the 1920s, when family members stated that King had gone to Ohio gain his freedom. In 1926 the obituary of his son John T. King said Horace was manumitted in Ohio. A brief 1930s Works Progress Administration (WPA) sketch of the lives of Horace and John asserted that the King family in LaGrange "now have in their possession a treasured heir-loom, the certificate of freedom issued to Horace King and wife—John T's father and mother—in 1829. It is written, signed and sealed by John Godwin, properly witnessed and executed in the State of Ohio—truly an interesting document."[4] The WPA writers never saw this certificate, and the interviewee probably had not read it in years, if ever. John Godwin could not have freed Horace's wife, Frances Gould Thomas King, in 1829. She was born free, and she did not marry Horace until 1839. This document, signed by the clerk of the Superior Court of Baldwin County (Milledgeville), Georgia, is dated 1829 and certifies the freedom of Frances's mother, "Rachael Gould, a free person of color." This handwritten document, magnified by family lore, became freedom papers for Horace and Frances.[5]

According to the WPA sketch, "Godwin desired to free both Horace King and his wife, and in order to expedite their manumission, took them to the free state of Ohio, where the legal formalities in this connection were much simpler than in Georgia." This idea of a passage to Ohio easing the transition to freedom continued to be repeated by a chain of historians and journalists. Cherry's narrative perhaps inadvertently planted the seed of this myth, but it and Horace's 1878 testimony to the federal commissioners include no reference to his freedom coming in Ohio. Over time, what perhaps began as a family legend was repeated and embellished by others until it had the ring of truth.

For slaves yearning for freedom, the land north of the Ohio River occupied the status of a mythical kingdom. The underground railroad from the South terminated at the Ohio River, and there slaves could figuratively throw off their chains. While many slave owners, especially Virginians, did take slaves to Ohio in order to free them, most of those slaves never returned to the South where their status would be ambiguous and even imperiled. They remained in Ohio, where they faced much of the same discrimination that other nonslave blacks tolerated in the South: stiff taxation—including a five hundred dollar bond—without any representation and without most of the civil rights enjoyed by whites.[6]

Even though Godwin could have taken King to Ohio and freed him with ease, what would that have gained for either Godwin or King? Neither Alabama nor Georgia would have recognized a freedom gained in Ohio. Earlier researchers, attempting to reconcile the family's discussion of Ohio with King's emancipation by the Alabama legislature, have asserted that gaining his freedom in Ohio prevented Horace from being subject to the Fugitive Slave Law.[7] Both the original Fugitive Slave Law of 1793 and the more famous one of 1850 were only invoked when a master sought to have a slave returned. Freeing him in Ohio would only save him from his own master, who in this case would be freeing him.

Perhaps the focus on Ohio came from a passage in the Reverend Cherry's brief biography: "Horace was initiated into the sublime mysteries of the ancient order of Free and Accepted Masons, in the State of Ohio, the laws of the craft not permitting it in Alabama at that time." This simple statement might be the origin of all the ideas about Horace and his Ohio adventures. However, this assertion probably means he was inducted into the Masons under the jurisdiction of the Ohio Prince Hall Grand Lodge. An examination of the history of American black masonry illuminates the relationship between Ohio and Alabama.

Horace was not initiated into a white Masonic lodge in the state of Ohio. Nor was he a member of a white lodge in Alabama or Georgia before 1865. If he had joined such a lodge before the Civil War, he would not have joined the Prince Hall Lodge in Montgomery, Alabama, in 1869.

African American Masons throughout the nation belonged to separate Prince Hall Lodges. Prince Hall, a free mulatto from Barbados who migrated to Boston, had been inducted by British soldiers into the Masons during the Revolutionary War. In 1784 Hall petitioned the Grand Lodge of England to organize a Grand Lodge in Philadelphia. The Prince Hall or African American Masonic organization is descended from his original Grand Lodge. The Prince Hall Masons did not arrive in Ohio until after the Alabama legislature granted King his freedom in 1846. The Prince Hall Lodge in Philadelphia chartered the first African American lodge in Ohio in 1847. By 1849 a Grand Lodge existed for the state of Ohio, which then colonized the Midwest.

After 1865 the Ohio leaders of the Prince Hall lodges aggressively spread the Masonic order among African Americans in the South. In 1869 they organized eight Prince Hall lodges in Alabama, the first Masonic organizations for blacks in the state. These lodges "remained under the Ohio jurisdiction until September 24, 1870." King must have been inducted into the ancient mysteries of the Masons before that date. The records of one of those lodges, the King Solomon's Lodge, No. 4, Montgomery, show King as a member in 1870.[8] That initiation, according to the Masonic laws of Ohio or its Grand Lodge, is Horace King's only documented connection to Ohio.

Family tradition also maintains that Horace, his son Marshal Ney, or both were educated at Oberlin College. While that institution led the nation in ed-

Horace King's Masonic medal. Note the initials H. K. on one side and Masonic symbols on the reverse. King was proud of his Masonic membership. As a Mason he laid at least one cornerstone in LaGrange during the 1870s. Unfortunately, this medal was stolen during a burglary of the family home; only these two photographs survive. Horace King became a Mason when he joined a Prince Hall (or African American) Lodge in Montgomery, Alabama, sometime between 1868 and 1873. Contrary to the often-repeated Horace King legend, he was not initiated into the Masons in Ohio nor did he participate in a Columbus, Georgia, Masonic lodge before the Civil War. *Photograph by Thomas L. French Jr.*

ucating African Americans before the Civil War, and some southern planters did send their mulatto offspring to school there, Oberlin's extant records show neither Horace nor Marshal as a student. Starting with the premise that Horace was associated with Ohio, the King family in the twentieth century must have exaggerated his Ohio involvement into his having attended Oberlin. No African American would have made such a claim in Georgia or Alabama in the 1830s. In 1835 the Columbus Town Commissioners added two thousand dollars to the five thousand dollar reward already established by the state of Georgia for the arrest and conviction of Arthur Tappan, the president of the American Anti-Slavery Society. One of Tappan's most notable acts in that decade was a large donation to Oberlin University because of its antislavery orientation. Most white southerners viewed that institution as threatening the foundations of their society.[9]

Some sources have identified King's engineering training with Oberlin. The first African American to attend Oberlin matriculated in 1838; six years later, the first black graduated. So, if Horace received his education there, it could not have occurred before he arrived in Columbus in 1832. Furthermore, Oberlin did not teach engineering; it had a liberal arts curriculum. Its professors never provided technical instruction or the utilitarian skills needed to build a bridge. Only the U.S. Military Academy at West Point was teaching bridge building at that time.[10]

But much more important than whether he journeyed to Ohio, was his unique relationship with John Godwin, which apparently began from the moment they became master and slave. Godwin obviously allowed his slaves to learn to read and write or to continue to practice those skills in spite of the laws forbidding it. King was obviously an apt student, capable of mastering the financing as well as the joinery involved in erecting a bridge. Perhaps King even solved some of the construction dilemmas he and Godwin encountered in their early career. Whatever the motive, Godwin became a unique slave owner when he made King a complete apprentice, as the master taught the slave every aspect of the enterprise. The slave so mastered those abilities that he impressed and inspired confidence in everyone he met.

The Columbus Town Commissioners placed enough faith in Godwin as a builder to award him the contract for the first covered bridge over the Chattahoochee River. Judging from the newspaper account when Godwin won the contract in 1832, he still lived in Cheraw. The migration of his in-laws and many other families from the area of Chesterfield County to Alabama explains why Godwin knew of the opportunity to build a bridge on the eastern border of the state. Godwin's wife's uncle William Carey Wright, who later served as Ann's legal trustee and one of Horace's masters, sold his mother's property in Marlboro County, South Carolina, in 1824.[11] Wright and Ann's father, Joseph, moved to Montgomery, Alabama.

By May 1832 Godwin and King—technically as master and slave but more realistically as fellow builders—arrived on the banks of the Chattahoochee and began erecting the first of many bridges in the area and the first Town lattice truss to join Georgia and Alabama.

BOOM TOWN ON THE CHATTAHOOCHEE

The project took like wildfire; and the advantages of the new city [Columbus, Georgia,] being loudly proclaimed over the land, people flocked from all quarters to see and judge of it for themselves. We arrived, fortunately, just in the nick of time to see the curious phenomenon of an embryo town—a city as yet without a name, and any existence in law or fact, but crowded with inhabitants, ready to commence their municipal duties at the tap of an auctioneer's hammer.
Basil Hall, *Travels in North America in the Years 1827 and 1828*

The state's policy of creating "embryo" towns at the head of navigation on major rivers provided ample work for contractors and carpenters. In the first three decades of the nineteenth century, Georgia created Milledgeville on the Oconee River as its capital in 1804, Macon on the Ocmulgee River in 1822, and Columbus on the Chattahoochee River in 1828. After pressuring the federal government to remove the Indians, the state then distributed this "newly opened" land through a lottery whereby every male head of household had a chance of gaining 202½ acres of property.[1]

Realizing the commercial significance of the land at the navigational head of major rivers, the state exempted or reserved these areas from the lottery system and organized the survey and auction of town lots. In the case of Columbus, the legislature authorized its creation in December 1827 and charged its state-appointed commissioners to take "special regard to the future commercial prosperity of said town." The Chattahoochee provided more easily navigated waters than either the Oconee or the Ocmulgee, and the possibility of attracting cotton from Alabama to Georgia helped to shape the legislature's view of Columbus as a major "trading town." Its potential waterpower, which exceeded any other site in the Deep South, also engendered visions of a booming manufacturing center.

The commissioners selected a Methodist minister and surveyor, Edward Lloyd Thomas, to delineate the streets, lots, and other features. The falls of the Chattahoochee determined the town's location. A level plain on the east bank of the river allowed Thomas to create a rectangular grid of streets with the town's western border contiguous to both falls for manufacturing and slack water for steamboat wharfs. Thomas also planned access to Alabama, envisioning a span at what he labeled Bridge (now Fifteenth) Street. The town never used this site for a crossing. Thomas's bridge location might have been where Creek Indians crossed what they had always called Coweta Falls. As whites migrated into Georgia and Alabama, Indian trails became roads, with one major exception—where they crossed waterways. Indians who moved by foot or horseback along the fall line crossed major rivers at or slightly above the falls, whereas Europeans who moved on wheels could not traverse the rocks in their carriages, wagons, and gigs. These vehicles needed a ferry, which usually operated below the falls where calmer waters allowed a barge to be roped or oared across the waterway.

South of the future site of Columbus at least three ferries operated before 1833, the first two in conjunction with the Federal Road. In 1805 the Creeks agreed to allow the U.S. government to build a post road through their territory. Its construction started in 1811, and it simply followed existing Indian paths from Milledgeville to Montgomery and became the major interior route through the Gulf South. An important stop on the road was Fort Mitchell, just a couple of miles west of the Chattahoochee and about twelve miles south of the falls. There, Joseph Marshall, a mestizo, ran a ferry that served the major artery of the Federal Road. Kinnard's Ferry, also owned by a mixed-race Indian, functioned south of the present location of the Oglethorpe Bridge inside the future limits of Columbus. It carried a small number of travelers who took the northern spur of the Federal Road. By 1827 Kinnard's Tavern, along with a few stores and houses, stood at the eastern edge of this ferry. The creation of Columbus probably closed this ferry, because in 1828 the state assumed control of all waterfront property within the new town.[2]

In 1831 Seaborn Jones and Samuel Ingersoll opened another ferry, aimed at serving local traffic just a mile south of the new town. Jones of Columbus and Ingersoll, eventually of Alabama, represented two of the most prominent entrepreneurs in the area. Their ownership of this ferry illustrates the role of wealthy men in financing major ferries and bridges. Jones might have used this conveyance to reconnoiter the Indian lands he and his colleagues began buying in 1832. But the first land boom came in Columbus in 1828.

British naval captain Basil Hall arrived there four months before the auction but in time to record the birth of the "embryo town." He heard anvils "ringing away at every corner; while saws, axes and hammers were seen flashing amongst the woods all round." Future property owners had small houses built on wheels ready to move to their lots, after they purchased it. Hall also noted an unusual building practice: "At least sixty frames of houses . . . lying in piles on the ground, and got up by carpenters on speculation, ready to answer the call of the future purchaser." An 1833 Swedish visitor, C. D. Arfwedson, reflecting back to 1828 noted, "Carpenters, masons and workmen of every kind were never without employment and could not erect houses fast enough. . . . Most of the houses were of wood, and some of brick: a few in the English style, others again in the Grecian taste."[3]

Columbus was a builder's paradise in those early years, but history has not preserved the identities of these artisans, especially those who worked on the simple wooden houses. John Godwin's skills and those of Horace King would have already ranked them as master builders. While at times both of them constructed small houses, they obviously made their reputations and fortunes as large-scale contractors. By 1832, when he first traveled across two states to launch the infant town's most important building project, Godwin's reputation was well established.

Most of the builders in Columbus lacked the skills to submit a bid for the new bridge. Perhaps only Asa Bates, a heavy contractor, could have competed for it. Bates remained a constant in the lives of Godwin and King as a rival, a partner, and a friend to them both. Bates arrived in Columbus in 1828 from Massachusetts and quickly became involved in civic activities and in building. He erected houses and cotton warehouses.[4]

One of Bates's most impressive buildings was a large theater erected in less than a week in 1832. The impresario Sol Smith and his traveling troupe needed a venue and Bates obliged in record time. According to the editors of the *Columbus Enquirer*, "Expedition—A theater, 70 feet long and 40 wide was commenced on Monday morning last by our enterprising fellow citizen, Asa Bates, and finished on Thursday afternoon in season, for the reception of Mr. Sol Smith's company on that evening. A great portion of the timber, on Monday morning waved to the breeze in its native forest; fourscore hours afterward its massive piles were shaken by the thunder of applause in the crowded assemblage of men." This house also witnessed Smith's version of *Pizarro* when he recruited twenty-four Creeks to play Incas. Whiskey served as part of their pay, and they were paid much too early. When the time for their performance arrived, they broke into their own chants and dances that lasted for thirty minutes and destroyed most of the props on the stage.[5]

Bates's construction survived the Creek dancing, and he continued to be an important local contractor. Given his construction skills and his position on the town commission in 1832, he was surely involved in selecting the bridge's builder. He himself might have submitted a bid; those records have not survived. Given his experience with heavy timber structures, Asa Bates and his slave crews must have helped Godwin and King with their bridge. He had definitely become a bridge builder a decade later.

Bates and other Columbus builders relied on African American carpen-

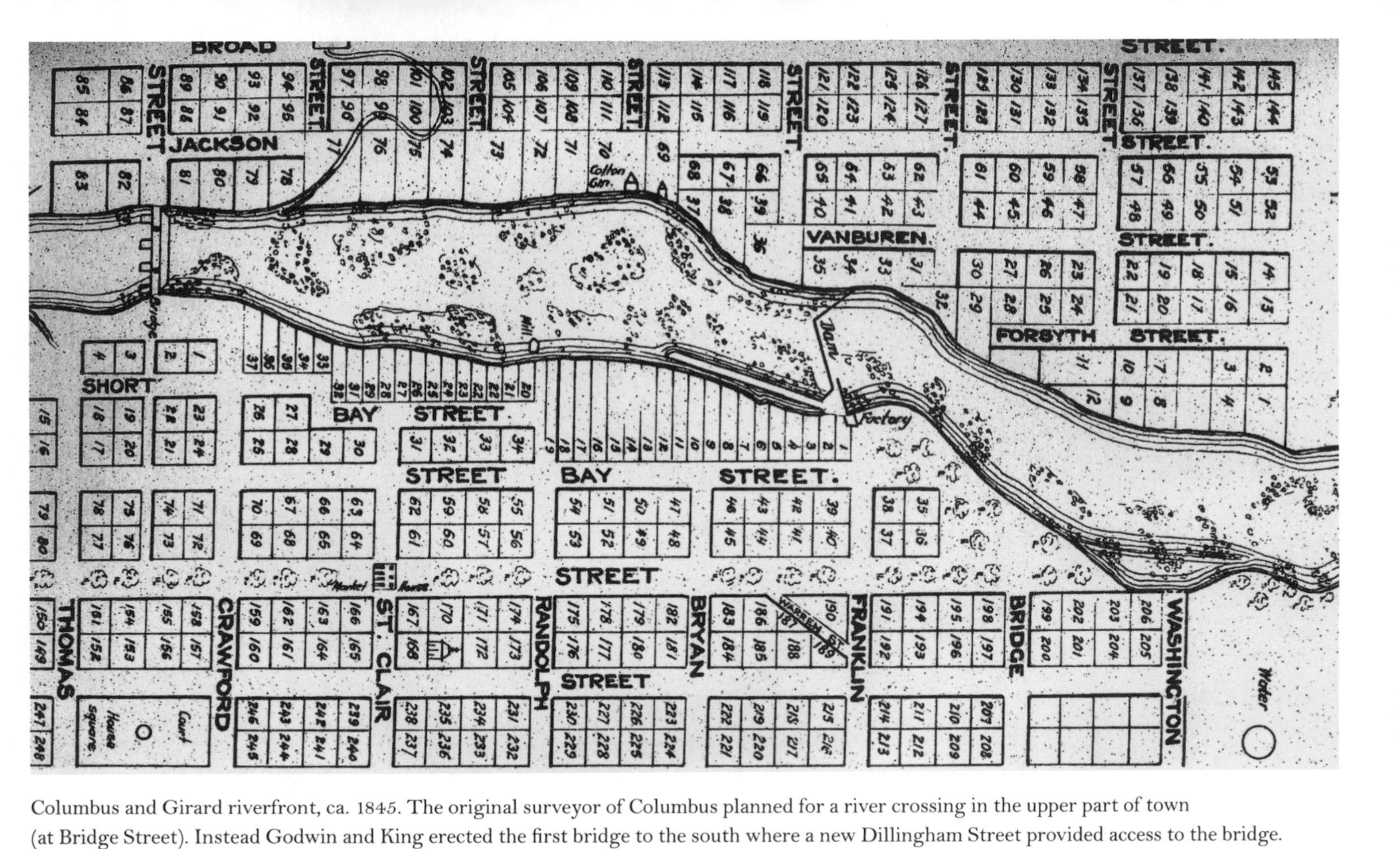

Columbus and Girard riverfront, ca. 1845. The original surveyor of Columbus planned for a river crossing in the upper part of town (at Bridge Street). Instead Godwin and King erected the first bridge to the south where a new Dillingham Street provided access to the bridge. The Muscogee County Courthouse built by Wells and John Godwin, along with King, is shown in the southeast corner of this map. Horace and John probably built the small bridge over Holland Creek on Broad Street in Girard. This map was originally produced to promote the sale of lots in Girard and Columbus's nineteen riverfront industrial lots. *A. J. F. Phelan and Calvin Stratton, G. and W. Endicott, Lith., New York City; reprinted in 1927 and 2000. Courtesy of Historic Columbus Foundation.*

ters—both slave and nonslave—who erected the majority of the built environment of Georgia and Alabama. African slaves made bricks, sawed lumber, hewed timbers, fashioned mortise-and-tenon joints, carved cornices, and assembled all the pieces in both modest and elaborate houses. Georgia lists of nonslave blacks show carpentry as a typical occupation, but few free blacks lived in Columbus during its first decade. No other African Americans possessed the skills or the degree of responsibility given to Horace by Godwin only two years after he acquired him.[6]

The span Godwin and King built across the Chattahoochee was unique in various aspects, and it underscores the continuing role of the state in promoting the commercial interests of its planned towns of Macon and Columbus. Although in 1805 the state legislature gave inferior courts the power to construct bridges in their counties, these were small bridges. The sponsors of large covered toll bridges petitioned the legislature for a state charter. Between 1790 and 1840 the Georgia legislature issued thirty such charters, of those twenty-three went to individuals, six to local governmental bodies, two to corporations, and one to a church congregation. Only two towns received state aid: Macon and Columbus.[7] In Augusta private speculators—Wade Hampton, James Gunn, and Henry Schultz—financed such activities, but in other towns, the state financed the bridges.[8] The legislature created the town of Macon in 1822, and two years later the governor appointed commissioners to contract a bridge, built by Daniel Pratt, across the Ocmulgee with state funds. In 1828 the city of Macon purchased the span from the state for twenty-five thousand dollars with payments spread over ten years, and then operated it as a municipal toll bridge. Specific tolls were enumerated in all the bridge charters. For most of them, the rates were similar, but not identical, and some acts just stated that the tolls be the same as those on existing ferries for that particular river.[9]

The legislature dictated an unusual toll structure for both the Macon and Columbus bridges. In the case of Macon, the General Assembly specified that Macon was not "permitted to collect toll for any waggon or other carriage loaded with cotton or corn." Perhaps, the legislators represented the interests of the planters and especially the cotton factors or merchants. The lawmakers viewed the prosperity of Georgia as being tied to cotton and corn. Even more than the span over the Ocmulgee, Columbus's bridge to Alabama was important to the state. Its early erection assured that Columbus, rather than a potential rival on the Alabama side would become the major trading town on the Chattahoochee. In the 1830s many Alabamians viewed their side of the river at Columbus as the site of a city that would equal Columbus in size and economic importance. Building a bridge before the Alabama side became organized would insure that Alabama cotton flowed into Georgia warehouses.

In 1831 the Georgia legislature loaned Columbus sixteen thousand dollars "for the purpose of building a Bridge across the Chattahoochee River, at the ledge of rocks terminating the falls of said river, between Crawford and Thomas streets." The specified location for this structure lay five and a half blocks south of the town's original Bridge Street in slack water, probably at the narrowest point across the river. The bridge did not align with any established street and over time both the bridge and the short street leading to it became known as Dillingham. The act required Columbus to pay 6 percent interest on the loan and to repay the full amount in ten equal annual payments beginning in January 1834.[10]

During the same session, the legislature authorized Columbus to lease lots for wharves below Thomas Street (now Ninth) for the "use and benefit of" the town. The bridge and wharves were both essential links in the cotton economy. The Columbus bridge law vested the power to set tolls in the Inferior Court of Muscogee County (which never exercised its authority over the span) and prohibited "any toll for any wagon or other vehicle carrying any agricultural productions to market in said town of Columbus, or in returning from the same."[11]

In January 1832 Columbus began implementing the state's plan for the bridge. The town commission advertised in the local *Columbus Enquirer*, the Milledgeville *Georgia Journal*, and the Washington, D.C., *National Intelligencer* for "sealed proposals . . . for the building of a Bridge across the Chattahoochee at this place; including the model and the price. . . . It will have to be about four hundred feet long, including the abutments and built hard and strong, of durable materials. One hundred dollars will be given for the most approved model." The ad seems to indicate that the contract and the model could have been separate items.[12]

On February 25, a few days before the due date for the bids, the *Enquirer* noted the city had received several plans "to throw a bridge across the Chattahoochee." Judging by where the town officers advertised, they wanted to cast a wide net in searching for a builder. Why did the town commission award the job to someone living three hundred miles away? Godwin apparently offered the lowest price, fourteen thousand dollars. Godwin might have been shrewd enough to realize how much the legislature loaned Columbus for the project, sixteen thousand dollars and he simply bid under that figure. Perhaps he assumed he could collect on any cost overruns, which he did. Certainly, Godwin's résumé, at age thirty-four, included earlier bridges built according to the Town plan. The mystery is that the council accepted Godwin's bid but paid the hundred dollar model premium to Daniel Pratt, one of Georgia's master builders.[13]

A native of New Hampshire, Pratt arrived in Milledgeville in 1819 and built some of its early mansions, which were usually characterized by unsupported circular staircases and graceful elliptical fanlights. Pratt also constructed boats and bridges, including the Macon span. In 1831 he partnered with Samuel Griswold, who manufactured cotton gins near Clinton, Georgia. In 1832 Pratt moved to Alabama to manufacture Griswold's gins there. By 1839 he had established his own industrial town, Prattville, Alabama, which became, along with Graniteville, South Carolina, the southern models for such enterprises. Horace King's building skills certainly equaled those of Pratt. Had it not been for the racism of American society, King's career might have followed a similar entrepreneurial path to that of Pratt's. Interestingly, Pratt served in the Alabama legislature from 1861 to 1863, just a few years before Horace King would win election to that body.[14]

In 1832 Pratt sought to earn some money from the Columbus bridge before he moved to Alabama. Unfortunately, the Columbus Town Commission minutes for that period were destroyed, so only the brief newspaper account illuminates the awarding of the contract. Pratt might have bid more than fourteen thousand dollars to build the bridge. Given his proximity to Milledgeville he certainly knew the amount of the loan provided to Columbus.

Pratt, through his wife's family, enjoyed Columbus connections that ran all the way into the council chambers. One of Mrs. Esther Ticknor Pratt's cousins was Francis Orray Ticknor, who later became a physician and a poet, best known for his Civil War piece "Little Giffen of Tennessee." This universal man also retailed Pratt's gins in Columbus. His older sister, Lucy Ticknor, married Dr. George W. Dillingham, who served on the Columbus council in 1832. Dillingham died in 1834, and the council eventually named the bridge and the street for him. Pratt probably stayed with his wife's people when he migrated to Alabama in 1833; perhaps he paused to consult with Godwin and King, who might have needed his advice given the slow pace of their construction.

Dillingham may have used his influence to encourage the council to award Pratt the prize for the best bridge model. The fact that the council accepted Godwin's bid and Pratt's model might indicate that Godwin did not submit a model. He might have felt a model or a specific plan was unnecessary, because Town's patent provided all the necessary details for the bridge's construction.

In 1832, as Pratt prepared for his move to Alabama, John and his brother Wells Godwin, along with Horace King, came to Columbus, threw their bridge across the Chattahoochee, and then built the first house for a white family in the Alabama suburb later known as Girard. When selecting a site to manufacture gins, Pratt leaped over east Alabama because it remained in Indian hands. Godwin, on the other hand, was a real pioneer, siting his new home in Indian or federal territory.

Moving with the first wave of settlers into this frontier region shaped the lives of Godwin and King. It provided them with ample work as bridge contractors. Horace built five major bridges across the Chattahoochee River within ten years. Their skills also garnered Godwin and King contracts for small bridges, houses, factories, and warehouses. Being on the frontier, where traditions and customs about slavery were not as rigid, also helped King to build his own metaphorical bridge out of slavery, to become a recognized craftsman as a slave and a respected citizen as a freedman. He began building that reputation by assembling and erecting the first Columbus bridge in 1832.

CHAPTER FIVE

TO THROW A BRIDGE ACROSS THE CHATTAHOOCHEE

Childhood! The Chattahoochee's banks,
My scene of summer playing;
The first to cross on the first planks
Of the first bridge—King's laying.

Francis Orray Ticknor, "Reminiscences: On Visiting the Eagle and Phoenix Mills at Columbus, Ga."

The smell of fresh pine engulfed Ticknor, the eleven-year-old poet-to-be, when he crossed the gleaming white bridge for the first time. The bridge impressed his neighbors, journalists, and travelers passing through Columbus. But it was more than just a pretty bridge, and its significance transcended both its engineering achievements and its commercial convenience. It occupied a central position in the early history of the region. It spawned an interstate conflict and brought desperados, impoverished Indians, and frightened settlers into Columbus. By providing easier access to Creek Indians in Alabama, it aided in their exploitation by white traders and land speculators.

Because of the history that unfolded around and through this bridge and because it was the first bridge built by King and Godwin in the Chattahoochee Valley, it, more than any other span, is identified with Horace King today. Its construction brought his master to the area, while the bridge's later destruction by flood and war, coupled with its constant need for maintenance, kept King busy and in the public eye. Because of its significance to King, it warrants a detailed account. Its history illustrates the frustrations of financing, building, maintaining, and collecting revenue for any large wooden span as well as the impact of politics and nature on covered bridges.[1]

The construction of this bridge consumed more than a year and a half, measuring from the contract date of March 1832 until its completion in September 1833. This protracted period suggests John, his brother Wells, and King were still somewhat inexperienced. It also shows the difficulties of assembling building materials for a large-scale project on the frontier.

The first step—sawing lumber—did not occur at the site but elsewhere along the river and its tributary streams. The bridge required approximately 300,000 board feet of lumber costing approximately three thousand dollars. Godwin could not simply order that lumber and have it appear on the riverbank the next day, the next week, even the next month.[2]

The Town lattice truss shifted bridge building from an emphasis on precise craftsmanship—for example, the woodworking skills necessary to fashion beams into arches—to mass-producing enough wood to construct timbered tunnels. The technology of sawing lumber and improving saws remained important for bridge builders. Later in his career King invented replaceable teeth for circular saw blades, but that technology did not exist in 1832.

The volume of sawn timber needed for the Columbus bridge sorely taxed the capacity of local sawmills. The portable steam engines associated with later southern sawmills were not developed until the late 1840s. Small water-powered mills with vertical saws cut Godwin's lumber. As a point of comparison, the maximum daily production of large commercial sawmills in Maine at the time was 6,000 board feet. One of those companies would have operated for sixty days to produce 360,000 board feet. Obviously, producing that much lumber in Columbus slowed the building process.[3]

In addition to lumber, Godwin assembled other materials: stones and bricks for piers and abutments, hewed timbers, nails, flat boats, and hundreds of miscellaneous items. The stone blocks for the piers came from small quarry sites on the riverbanks north of the bridge. Godwin's enterprises eventually included a blacksmith shop, and if it existed by 1832, it supplied the nails needed for attaching the shingles and siding. They used no other nails. Large white oak treenails (or round pegs) were formed by driving square pieces into a steel cylinder with an ax-sharp edge.[4] Soaking the two-foot-long treenails in oil extended their life.

Godwin's workers fell into three general categories: the superintendents or foremen who understood the mechanics of erecting piers and assembling trusses; the skilled carpenters, masons, blacksmiths, wheelwrights, and so on, who knew their craft but not necessarily how to construct a bridge; and the unskilled work gangs that provided only muscles and consisted mostly of slaves.

John's brother Wells, probably Asa Bates, and Horace King acted as the foremen, who gave orders to white craftsmen and slaves, verifying their work for Godwin who paid them. No records document the exact role of "Horace Godwin," as he was known at age twenty-five, but he acted as more than a manual laborer. Eight years later a newspaper ad listed Horace as a bridge builder along with John. By the early 1840s Horace served as the overall superintendent of a project in Columbus, Mississippi. The consistent oral tradition also attests to his important role, as noted by Ticknor's phrase "Of the first bridge—King's laying."[5]

The overwhelming part of the labor force consisted of slaves. John and Wells owned between six and twenty-four slaves for the project.[6] Judging from the later profile of John's slaves, most of these would have been carpenters or skilled workers and included Horace's family. Washington, Horace's eleven-year-old brother, assisted Horace and learned the craft of carpentry. Their mother, Susan, and their thirteen-year-old sister, Clarissa, cooked for the crews. Clarissa's future husband, Henry Murray, a wheelwright or blacksmith, may have forged the thousands of nails for the shingles and siding. Nails were not mass-produced until the 1840s, and they would have been hand forged in this frontier location. Godwin needed many more slaves, which he leased from their masters for about sixty cents per worker each day.[7]

Using slaves in a construction job netted substantial profits for owners. Robert Jemison Jr. of Tuscaloosa, who later employed Horace, and Dr. Alexander J. Robison, who acted as King's guardian after he was freed, both realized greater profit hiring out their slaves as carpenters and unskilled gangs for heavy timber jobs rather than for other work.

Horace would have taken Washington and a gang of slaves to cut virgin or first-growth pine trees for poles and hewed pieces.[8] The sawn chords and lattice diagonals for the trusses came from fresh pine lumber. Bridge builders used green rather than seasoned wood, since its shrinkage tightened the truss. The poles for the piers and scaffolding were simply debarked; other heart pine pieces with their dense, close grain were hewn into support beams for the substructure under the trusses. Skilled black workers dressed or trued the beams using axes and adzes (a tool with an axlike blade perpendicular to the handle). The workers rhythmic strokes left perpendicular marks about every three or four inches down the entire length of the squared beam. King did not hew these beams, even though he knew how and taught the process to others. His role resembled that of a black driver on a rice plantation, where slaves directed the daily labors of other chattel.

Construction began on the riverbank in May 1832 "with a large force" of workers, according to the *Columbus Enquirer.* Those men were preparing cofferdams, the first step in erecting the piers. They sunk log "cribs" filled with stones in the river, connected the cribs with logs, and then covered the log framework with boards. These log and board structures surrounded the pier site and diverted the river enough to create a dry area for the foundation. Slaves then dug through five or six feet of muck and mud to the bedrock. Godwin needed masons to raise the piers. In early July 1832 he advertised in the Columbus *Enquirer* and the Milledgeville *Georgia Journal* for "two or three stone masons, to work on the Piers of the Columbus bridge, to whom liberal wages will be paid."[9]

These craftsmen used hydraulic cement to lay the irregular stone blocks into a rock foundation about twenty to twenty-five feet high that supported wooden piers rising another eighteen feet. Iron bolts fastened the wooden members to the stone blocks. Planks covered the wooden piers down to the rock piers. The bridge required three of these piers in the river. On the low-lying western bank, a wooden pier without a stone base supported the final truss.[10]

After the pillars reached full height, workers built false work or light scaffolding between the piers. Apparently, they finished this process by October 20, 1832, when John Godwin and his family moved to Alabama as "the first white man to build a home there," in what became Russell County. Establishing their residence on the Alabama side of the river was more than likely a symbolic gesture to indicate Godwin's faith in his bridge. The white and the black Godwins lived in a simple residence near Fort Ingersoll, a trading post for Indians operated by Stephen M. Ingersoll.[11]

Godwin and Ingersoll, an entrepreneur, might have been partners in an Alabama sawmill. If they did establish a sawmill in Alabama, then Godwin conceivably built trusses on either side of the river. From the time that Godwin moved until the completion of the bridge, almost another year passed. Creating a sawmill on the far bank could account for some of that delay.

As the piers rose from the dank clay river bottom, carpenters on the riverbanks began work on the essential part of Town's design—the trusses, three of them, each extending about 186 feet in length. They approached the maximum length of a Town truss. The process involved laying out the lumber for the latticework in a jig formed by driving stakes into the ground. The truss consisted of numerous diagonal members and four horizontal stringers (two at the top and two at the bottom). Each stringer consisted of four boards with two boards spliced together with staggered joints on each side of the diagonals. Every intersection of the diagonals or where the diagonals intersected with the stringers was connected with two treenails. Placement of those holes was critical.[12]

Horace King probably used a sharp knife to mark the exact points where workers placed the tip of the auger bit as they began to drill. At the junctions of the diagonals they drilled through two pieces of lumber. At the junctions of the diagonals and the stringers, they drilled through six boards. The treenails for those holes measured more than eighteen inches.

Despite the extensive literature about the romance of covered bridges, the exact process for assembling the trusses remains open to conjecture. After the treenail holes had been drilled, two methods could have been used. One

The Fort Gaines, Georgia, bridge being rebuilt by Ernest King, Horace's grandson, ca. 1913. This photograph clearly shows the design of a Town truss with the diagonals and chords, the camber in the truss, and the false work under the truss that supported it during construction. Horace and his grandson used the same technology in two different centuries. *Courtesy of Georgia Division of Archives and History, Office of Secretary of State.*

method involved numbering the pieces in the jig, perhaps assembling a section of the truss while it was in the jig, removing the pieces or sections from the jig, and reassembling them over the false work. The second method involved assembling both sides of the truss on the riverbank, then raising the two sides and tying them together with cross beams, and rolling the entire truss over the false work and attaching it to the piers.

The essential part of the process involved placing camber in the spans. The middle of each truss needed a slight upward bow. A bridge that failed to maintain its camber would fail under a heavy load. Proper placement of the treenails insured the correct camber. The skill to place the treenails separated Horace King from ordinary carpenters and allowed his crew to throw a truss more than 180 feet in length. Once the trusses rested on the piers, the essential components of the bridge were in place.[13] The expensive Town truss only spanned the water. The approach bridges built over land lacked the latticework truss and remained uncovered since they could be easily replaced. The next step involved laying three-inch boards for flooring; people began crossing the bridge on foot at this point. The roof consisted of "good heart pine shingles," and the vertical siding of "dressed planks" came next. The boards on the sides started about a foot from the top, that gap creating a latticed strip of light inside the bridge. The bottom of these planks extended far enough down to cover the substructure under the bridge. Two windows, both set over the piers, graced each side.

The next to the last step was painting the siding with "two coats white lead and oil." The final step most likely occurred in September 1833, after sixteen months of construction, when the Godwins and King drove a fully loaded wagon across the bridge. The structure settled slightly and creaked as the treenails set. They wanted to be sure the settling stopped before it lost all its camber.

Godwin's role in "throwing" this bridge was to manage the financing. He needed to please the town commission since they provided the necessary funds. On March 30, 1833, after Godwin asked for a payment, the council ordered him to appear. The commissioners were impatient after a year of work, but Godwin explained his delays and left the April 3 meeting with a payment of $2,000. On July 27, 1833, even though the bridge remained uncompleted, the commissioners allowed Godwin to draw his final installment. Limited cash flows characterized much of his career but also the frontier in general. By the end of the project Godwin had accumulated $2,824 in extra expenses, which he managed to convince the council to absorb. Thus the total cost of the bridge exceeded the $16,000 loaned to Columbus by the state. On September 28, 1833, the council met at the bridge and accepted it as completed.[14]

Even before Godwin and King tested their white covered bridge, local pundits praised it as an essential improvement for the young town. "Well done, Columbus! four years ago a howling wilderness—now a handsome town, with a population of 1,800 souls, and three banks in successful operation," as well as three churches, a theatre, a bookstore, a circulating library, a reading room, and a public garden along the river. All these testified to the town's prosperity as would "a very handsome Bridge, which . . . will add very much to the beauty and convenience of the town." Its convenience proved to be short-lived. A conflict with the owners of the land where the Alabama end rested led Columbus to lock the bridge gates in late September 1833. Two months later, the commissioners ordered the removal of the flooring and began operating a ferry. They opened discussions with John Godwin about relocating the bridge.[15]

Even before the bridge existed, when the town ran a flat boat across the river, the Indian owners of the west bank claimed a half share of the ferry tolls. In 1830 a Lieutenant Clark of the U.S. Army garrison at Fort Mitchell interceded for the Indians and demanded that Columbus share its revenue. The town refused. Then as the town advertised for bids on their bridge in February 1832, the U.S. Congress passed an act authorizing Columbus to select two acres of federal property on the opposite bank of the Chattahoochee for the bridge's western abutment. The act, however, contained an important loophole that rendered it meaningless. This grant was to be "subject to the encumbrance of the Indian claim."[16]

Executing the provisions of the Treaty of Washington (1832), federal agents granted Chief Benjamin Marshall a section of land, 640 acres, directly across the river from Columbus. Marshall's selection of that site did not occur by chance. As a powerful mestizo leader, Marshall realized the economic potential of his grant. His title, or encumbrance, superseded Columbus's claim to the two acres in Alabama. Marshall then sold his section to Daniel McDougald, a white man then of Harris County and Robert Collins of Macon for $35,000 in June 1832.[17]

In December 1832 the Alabama legislature, intent on forcing the removal of the Creek Indians, established Russell County across the river from Columbus. The *Enquirer* noted in early 1833 that this was "the first movement towards the establishment of a rival town." Although that fear never became reality, Columbus fought Alabamians over control of the bridge for more than a decade. In January 1834 the Alabama legislature granted to the Alabama landowners the right to build a permanent bridge abutment, to pay Columbus one half the bridge's value (later set at $22,500), and to receive 50 percent of the tolls.[18]

McDougald and his associates forced the closing of the span because Columbus refused to pay the Alabama assessment. After five months of con-

Pont de Columbus, by Francis de la Porte, 1838. This painting shows the crucial ingredients in the Columbus economy: cotton, slaves, a riverboat, and the city bridge built by Horace King for John Godwin. The central purpose of that bridge was to bring cotton to Columbus warehouses for eventual shipment down the river to Apalachicola and then to the world market. The state of Georgia did not fund this bridge to aid pioneers moving westward—a recurring theme in local histories—but the city of Columbus did charge tolls to such travelers. *Collection of the Columbus Museum, Columbus, Georgia; Museum purchase.*

troversy, the Alabama landowners began advertising for the sale of five hundred lots including where the bridge landed. That threat most likely forced Columbus to act. The commissioners agreed to purchase one acre of land at the western abutment for $10,000. Even though that transaction would not terminate the controversy, it at least insured that the bridge stayed open for traffic.[19]

In December 1834 the *Enquirer* offered the following praise: "A Bridge across the majestic Chattahoochie, unequaled by any similar work in the South, adds to the artificial scenery of the place, while it facilitates the communication with the western country." While tourists might have noticed the beauty of the white bridge, they were probably more impressed by its convenience. Before the creation of Atlanta and the associated rail net, most people traveling from Savannah or Charleston to the west followed the fall line across Georgia and visited Columbus. Their writings illuminate the bridge, the roads, and the countryside.[20]

The modern, solid "very fine bridge" drew favorable comments because it stood in such stark contrast to the roads and inhabitants on the Alabama side. C. D. Arfwedson, a Swedish visitor, lamented, "No road in all America can be compared with that between Columbus and Fort Mitchell . . . the worst piece of ground in the Southern States." A half mile south of the Alabama end of the bridge, a collection of huts housed "a conglomerated mixture of gambler, black-leg, murderer, thief, and drunkard."[21] Since the federal government exercised little control there, "a certain number of loose persons . . . founded a village, for which their . . . atrocious misdeeds has procured the name of Sodom. Scarcely a day passed without some human blood being shed in its vicinity" very near the home of the Godwins and their slaves. Tyrone Power, an English traveler, noted how the Sodomites used the new bridge. "These bold outlaws, I was informed, occasionally assemble to enjoy an evening's frolic in Columbus, on which occasions they cross the dividing bridge in force, all armed to the teeth: the warrants in the hands of the U.S. Marshal are at such times necessarily suspended."[22] The lawlessness and rough frontier character of Sodom flowed across the bridge and affected the tone of life in Columbus for at least a decade.

Law and order, however, slowly emerged in both places. In June 1834, when McDougald and his associates began selling their building lots, they named the town Girard, apparently for Stephen Girard, a wealthy Philadelphia merchant and philanthropist, who had no connection with the area. Godwin purchased some of the first lots to be sold. The primary issue between Columbus and Girard was the Godwin-King bridge.[23]

The bridge represented a significant source of revenue for Columbus and a topic of constant debate. Unlike other spans on the Chattahoochee River that

Close-up of the illustration on a Bank of Columbus two-dollar note, 1850s. The image on the local banknote shows the bridge, the steamboat, and the new form of transportation—the Mobile and Girard Railroad. *Courtesy of C. Dexter Jordan Jr.*

Godwin and King later built for private companies, this municipal span had no precedents the commissioners could use for setting tolls. The council needed to balance the desire for revenue with the necessity of attracting business to Columbus. Also, the town's debt to the state dictated that ample moneys be collected from the bridge. Its management became a partisan issue. In a town with evenly matched parties—the Democrats and the Whigs—politics dominated every aspect of life.

The commissioners constantly tinkered with the toll structure, often issuing rulings that contradicted earlier ones. Alabamians who traded in or attended churches in Columbus were allowed free passage, as were Columbusites farming in Alabama. Legislation prohibiting the town from collecting tolls on agricultural products also engendered debates.[24]

The commission micromanaged the bridge by assessing annual rates for individuals, families, and companies, not at a flat rate but on a case by case basis, presumably based on their rate of usage or perhaps their connections to a council member. During many sessions the aldermen debated the fees, and members frequently called for the yeas and nays. In 1839, for example, the council set rates for 189 people. These ranged from $5 to $100 and netted $2,595.50 in income for the city. That figure did not include the charges made for stage lines, approximately a thousand dollars, nor the fees paid for single passage by hundreds of people. In 1840, voting four to three, the council assessed John Godwin an annual rate of $20, a typical amount for a moderate user. Family members and slaves, including Horace, also crossed as part of this package. Determining who was part of what family and which masters had paid for their slaves must have complicated the job of bridge keeper. He more than earned his annual salary of $600 the same as the city clerk and the treasurer.[25]

Given the bridge's financial importance to the city, the council worried in

1839 that the accumulated filth inside the bridge would rot the lower sills. Manure, refuse, and dirt constantly collected inside the bridge and its removal posed an ongoing problem. Despite the romantic image of covered bridges, the smell inside a bridge would offend contemporary nostrils. Perhaps because of the bridge's potential income, councilman John L. Lewis offered to purchase it on May 15, 1838, for $300,000 to be paid in fifteen annual installments. His scheme was never reported out of committee. The city apparently used those funds for general expenses, because in 1835 and 1839 Columbus begged the state for a postponement in paying against its bridge loan.[26] By 1839 the city's financial problems reflected the hard times resulting from the crash of 1837, which precipitated a major economic depression.

By the late 1830s the bridge, not yet a decade old, had witnessed a war—"our races rend apart" in the words of Ticknor. Remembering his childhood, the poet recalled sylvan pleasures, such as shooting, fishing, and snaring terrapin "in true Indian fashion." Few Creek Indians would have remembered those as the halcyon years of their life. Georgians and Alabamians had pushed the Creeks into a narrow strip of land by 1832, and the Indians experienced social dislocation and poverty. Not all were destitute, but Tyrone Power observed: "The condition of the majority of these poor people seemed wretched in the extreme: most of the families were living in wigwams, built of bark or green boughs, of the frailest and least comfortable construction; not an article of furniture, except a kettle, was in the possession of this class."[27]

The Godwins and their slaves lived about a mile west of the bridge on a hill adjacent to Fort Ingersoll among the Creeks, near the present location of the Girard Baptist Church. While some accounts portray the Godwins as moving into a raw frontier, the Creeks were hardly savages looking for scalps. After more than a century of contact with the English and their descendants, the Creeks had adopted most of the Europeans' habits and vices. Dr. Ingersoll catered to those tastes at his trading post. He had fought in the War of 1812 before arriving as a physician and Indian trader on the east side of the Chattahoochee River in 1826. He established a temporary home in Columbus in 1828 and operated Fort Ingersoll at a distance.

Some Creeks owned more than just a kettle. An elite group, including Benjamin Marshall, owned slaves who worked their cotton fields. Marshall's house, located a couple of hills to the north of the Godwins', was more elegant than their initial home near Fort Ingersoll. The Godwins and the Kings knew Marshall. Despite their shared Indian ancestry, King and Marshall probably had a limited relationship; they were divided by status in a slaveholding culture, with Marshall being the slaveholder and King the slave. The significance of King and Godwin to the Indians lay in their bridge, a span that carried the Creeks to Columbus.[28]

Many destitute Creeks, looking for something to put in that kettle, joined the "concourse of people, Christians and Indians" on the crowded streets of Columbus every morning. Arriving by ferry or boat before the bridge existed, Creek men, women, and children walked up the middle of Broad Street in a single file through the mud and slush, always looking straight ahead. Often when approaching a house to beg for food, they politely left their rifle at the gate. Sometimes they committed petty theft, but they posed less of a threat to civic order than did the Sodomites. Nevertheless local officials required all Indians to leave Columbus before dark, to recross the river back into federal territory.[29]

The bridge provided access for land speculators to seize Indian lands. Columbus began planning the bridge at the same time the Treaty of Washington defined the future Creek landholdings in Alabama. The treaty provided that every male Creek head of household would receive 320 acres of land, while chiefs obtained 640 acres. Most Creek leaders believed naively that they could retain their individual plots of land and become U.S. citizens. Some reformers in Washington also saw the Indians being integrated into white society. But racist white settlers bent on gaining more cotton land had no intention of respecting Indian-owned lands. The U.S. Army at Fort Mitchell attempted to enforce the treaty provision that protected Indian property and improvements. Other federal officials, however, encouraged the Creeks to move to Oklahoma. Benjamin Marshall himself moved and encouraged other Indians to do the same.[30] He had his money and was ready to leave, realizing he could not maintain his status within an integrated society. The Columbus Land Company and other speculative firms led by investors such as Daniel McDougald also worked to remove the Indians, once these capitalists secured titles to the native property.[31]

By 1836 Creek leaders called for an investigation of the massive land frauds. Secretary of War Lewis Cass might never have initiated the inquiry if an Indian uprising had not occurred in the spring of that year. Starving and homeless Indians had ample justification to attack white intruders, but land speculators may have paid some Indians as agent provocateurs to begin the conflict in order to insure the Creeks' removal. Attacks on stagecoaches, riverboats, and isolated white settlements brought people scurrying to the protection of Columbus. Later reminiscences may have exaggerated this flight. The King family tradition emphasized the importance of the Columbus bridge during this brief conflict. A local history of Chambers County, Alabama, noted that Horace King performed "heroic deeds" during this conflict but provided no further details. Most accounts stressed the importance of the bridge in describing the flight of settlers; so maybe building the bridge made Horace a hero.[32]

According to Thomas J. Jackson, the bridge "witnessed the wild rush of thousands of frightened Alabamians to Columbus at the breaking out of hostilities in the spring of '36." The city's first historian documented the hysteria: "Some neighborhoods hearing of the depredations of the Indians, would unite together and take such as they could of their most valuable effects and start for Columbus. . . . [A]nd for one day and night, the bridge at Columbus was crowded with the refugees from Alabama, coming in all sorts of style; some in wagons, some on horseback, some on foot—mothers calling for their children, husbands for their wives, and no response to their cries." The Godwins, including Horace and the other slaves, may have fled from their Girard home and crossed the bridge to the safety of Columbus, or they may have remained in Alabama among the two hundred "friendly" Indians encamped on Ben Marshall's land. The conflict quickly subsided; Gen. Winfield Scott arrived in Columbus with a small federal force augmented by state militia and quelled the uprising.[33]

The war served the ends of the speculators by extinguishing all Creek claims to the land. Approximately twenty thousand Creeks were removed, even though the U.S. Army only considered twenty-five hundred to have been hostile.[34] The forced migration of the Creeks opened more cotton land on the Alabama side of the Chattahoochee Valley and made the Columbus bridge, the city's warehouses, and the riverboat wharf that much more important.

Indian attacks also retarded the development of textile mills along the Chattahoochee River. Local businessmen began planning for Clapp's Factory in 1834. Located about three miles north of town, just above the site of today's Oliver Dam, it did not start operating until 1838, in part because of the Indian uprising. The primary investors included Charles Stewart and his son-in-law John Fontaine, wealthy cotton merchants.[35] By November 1840 Stewart and Fontaine had financed a bridge across the Chattahoochee River there. It consisted of two short spans that connected an island with the riverbanks. Since no steamboats operated there, these were low spans and probably not covered. Godwin and King may have worked on them, based on Horace's later role as a builder for John Fontaine. The county, in 1840, allowed Stewart and Fontaine "to charge the same rates for toll [for their bridge at Clapp's Factory] that are charged at the Columbus bridge."[36] As the expansion of riverside cotton cultivation began denuding the clay hills of the Piedmont, this small bridge would threaten the larger Godwin-King bridge in Columbus in a most unusual way.

The bridge survived an Indian war but succumbed to a flood named for an Indian fighter, William Henry Harrison, the victor of the Battle of Tippecanoe and the election of 1840. This flood became associated with Harrison's thirty-day term in office. "On Tuesday evening last [March 2] the clouds gathered up from every point of the horizon, looking watery and black, and soon overspread the heavens, a dense and portentous mass." In Columbus "[a]bout dusk the rain began to fall in torrents. For forty-eight hours it continued to descend with but little intermission, filling every nook and valley to overflowing, and threatening to deluge the whole face of the country."[37]

It certainly deluged the entire Southeast, and in Columbus the rapidly rising Chattahoochee came to resemble a lake as the falls disappeared beneath its raging waters. By Wednesday, March 10, the roaring, muddy river almost lapped at the bottom of the City Bridge. Water covered all but a few feet of its piers, when it was threatened by the Stewart-Fontaine bridge, which had floated from its moorings and was racing downstream. As the bridge tumbled over the falls, a group of men lassoed it and secured it to a tree before it could crash into the Columbus bridge. To prevent the Godwin-King bridge from acting as a dam, volunteers removed the siding from the bridge to allow the water to flow through the truss work. Their efforts proved futile: the force of the floodwaters bowed the trusses. Then, at daybreak on Thursday, March 11, just after the stagecoach reached the Girard side of the span, one end of the bridge lifted off its pier and "never was there a more majestic sight than the departure of that noble bridge on its remarkable voyage." It floated eight miles downstream to Woolfolk's Bend, where slaves secured it to a tree. As the water receded it came to rest in the middle of a muddy cotton field.[38]

The city council immediately made plans to replace the bridge. The next day, they reopened the city ferry, called for contracts to replace the span, and thanked the men who had tried to save the old bridge. Interestingly, the list of twenty-five men cited for their heroic efforts did not include John Godwin.[39] Only sixteen days later, on March 27, the council received bids for the reconstruction. These came from

> Joseph Davidson for $15,500, without any insurance, to be completed within twelve months;
>
> Asa Bates for $15,000, without any insurance, to be passable by August 1;
>
> John Bell for $14,800, without any insurance, to be completed within ten months;
>
> David Wright for $13,000, without any insurance (without a completion date);
>
> P. H. Nolan for $16,000, without any insurance (without a completion date); and
>
> John Godwin for $15,100, with insurance for five years, to be passable by July 20 and completed as soon thereafter as practical.[40]

The council's selection of Godwin as the builder surprised no one in Columbus. Everyone involved in the decision also realized that Horace would act as his cobuilder. The previous spring the owners of the new Florence bridge, south of Columbus, had advertised that their "superior Bridge . . . built by 'honest' John Godwin and Horace, is now ready for crossing." Godwin also used advertising to cultivate his image as a bridge builder. In January 1841 his newspaper ad began appearing; it identified his profession as bridge builder and his residence as Girard. The editors also endorsed Godwin, writing: "Every body hereabouts, knows that John Godwin heads all creation in building bridges." The city aldermen must have agreed with the ad, which was still running in April. The bid proposals suggest that Godwin knew the content of Bates's bid and simply set his completion date a week sooner than Bates. Given the short construction time, Bates and Godwin probably had agreed to work for each other before they submitted their price.[41]

Godwin's singular proposal to insure his work certainly influenced the council's decision, and the aldermen had required him to insure earlier work. In 1838 Godwin built a new rock pier for the bridge, "provided he will insure the bridge during the progress of the work and for ten years thereafter." The ten-year guarantee must have applied only to the pier, which apparently held during the flood. The council itself had purchased bridge insurance, which only covered destruction by fire. By the mid-1840s such coverage ranged from 1.5 to 2 percent per annum of the value insured. As part of the contract, the council required Godwin to post a bond of forty thousand dollars as a warranty that his work would be completed on time and insured for five years. Four prominent local merchants and planters—John Banks, Daniel McDougald, A. B. Davis, and John Fontaine—guaranteed Godwin's performance bond.[42]

The rebuilding of this span involved one of the most-cited achievements in the Godwin-King legend: retrieving the trusses from downstream, bringing them back up river, and replacing them on the piers. Apparently this procedure was not addressed during the bid process, because on April 8, 1841, the council dispatched Alderman J. L. Morton, a builder, to visit the beached bridge and "to make disposition of the old bridge." If Godwin purchased the structure, the transaction was never recorded in the council minutes, but he must have paid for the lumber, even at used prices. Maybe this reconstruction of the bridge represented a seminal moment in King's life. Some accounts cite his rapid rebuilding of this span as the reason why Godwin freed him, that Godwin promised King his freedom if he rebuilt the bridge by the prescribed date. While the speed of the construction must have pleased Godwin, given the later problems with the span, King might not have been eager to claim credit for its reconstruction. If Godwin freed him in 1841, then why did he wait until 1846 to have the state legislature ratify it? Whatever the stakes, King worked quickly to restore the bridge.[43]

King's crews partially disassembled the beached bridge for its trip back to

Dillingham Street. They numbered all the pieces and moved them in sections as large as possible. They most likely tried to minimize the number of treenails they removed and replaced. The trusses were returned to the piers in the same arrangement they were in before the flood. The ability to reuse the Town truss spoke to its durability, even though recycling it worked better in theory than in fact. By July 21, with the bridge passable, the council paid Godwin $7,500, with his advances deducted. On November 23, they agreed to pay him the remainder except for $500 since the span was not finished. Having collected the bulk of his money, Godwin's pace of construction slowed; another seven months passed before the job was completed. Considering that most of the old piers remained and that by using the old truss work he reduced the amount of lumber needed, Godwin should have profited from his contract, except that the reconstructed span continued to demand his attention.[44]

The two constants about this bridge were its frequent need for maintenance and the enduring conflict over tolls between Columbus and Girard. Just as the forces of nature continued to distort and stress the bridge, its potential income continued to create political controversy. The exact meaning of the state's prohibition on tolls for agricultural products continued to create debate. In 1841, possibly because of the reconstruction, the legislature restated the clause in more specific terms: "That cotton, corn, fodder, oats, wheat, rye, and potatoes, shall be [the] only products that shall be permitted to pass the Bridge at Columbus, free of tolls, any law to the contrary not withstanding." The lawmakers gave no break to the Scots Irish herders moving their pigs and cattle to market. Again, the purpose was to lure Alabama cotton all the way to Columbus warehouses and markets, just as it had been ten years earlier.[45]

The Columbus bridge continued to demand major maintenance: its recycled lumber probably required more attention than a new bridge. This interstate bridge also carried heavier loads than most spans. For example, in 1844 and 1845 a minimum of twenty-five tons, or ten thousand bales, of cotton crossed the bridge, free of charge but not without considerable stress on the structure. No other bridge over the Chattahoochee River moved a similar amount of cotton. Other spans must have had similar deficiencies. The problems of the Columbus bridge were universal, but its woes and solutions are the best documented. They serve here as examples of the tribulations of all covered bridges.[46]

In August 1845, less than three years after the bridge was rebuilt, the council declared it to be "in a dangerous and precarious condition." The aldermen, citing Godwin's pledge to maintain the span for five years, selected a committee of mechanics to inspect the bridge and recommend remedies while declaring they would "not accept any patching that may be put upon said bridge." The inspection committee consisted of two local builders, R. R. Goetchius and Alderman J. L. Morton, along with a rising architect, Stephen D. Button. A Philadelphia native, Button lived in Georgia and Florida in 1845–46 and won the architectural competition for the first Alabama state capitol in Montgomery. This distinguished panel found faults, seemingly small but measurable and apparently discernable to the untrained eye. The committee report noted that when approached from the Columbus side, the bridge appeared to lean and sag slightly to the left, while the lattice framework "swa[g]ged" in the same direction, toward the south, or downstream. The floodwaters bowed the lattice trusses enough that within a couple of years that swag reappeared, even though it was not obvious immediately after the reconstruction. The old trusses had been remounted in the same locations as before the freshet, and the memory of the pine lumber was stronger than the new braces applied by Horace's crews. John Godwin rather than King presumably supervised these repairs, since Horace was working in Mississippi for Robert Jemison Jr. during these months.[47]

The council approved Godwin's repairs and they never resurfaced as an issue. Nor did these sags and swags sully Godwin's or King's reputation with the aldermen. By December 1845 the council contracted with Godwin to replace the bridge flooring—a facet not guaranteed by his warranty. Almost simultaneously with the discovery of the leaning of the bridge, Russell County and the town of Girard launched another round of toll controversy. The Alabamians erected a tollgate and demanded payment from everyone crossing the bridge. Negotiations and recriminations followed. At one point, the Columbus council exempted every citizen of Muscogee County from tolls and charged every other user while adding fees for produce and cotton. At another point, Russell County agreed to pay an annual fee to Columbus to make the bridge free, but the Columbus councilmen could not relinquish that steady stream of revenue, and the conflict ended with an agreement to return to the status quo antebellum. Columbus kept its tolls but exempted certain items and continued to set annual rates. In later years, Columbus kept the peace by wooing Alabama lawmakers. Columbus allowed all the elected officials in Girard and their families to use the bridge without any charge.[48] Bridge builders were not exempt. In 1848 the free black Horace King paid five dollars a year to cross his bridge, while his former master John Godwin and Asa Bates contributed ten dollars to the Columbus coffers.[49]

In the same year the Columbus council noted the precarious nature of the bridge: "[A]t the first freshet in the River the eastern abutment will certainly give way & will be attended with great inconvenience to the traveling public & of great damage to the finances of the city." Inconvenience and income, rather than safety, ruled the council's actions in the 1840s. The bridge committee conferred with John Godwin, and he recommended replacing the open eastern land bridge with another 155 feet of covered, latticed truss work at a cost of $1,300. Godwin's plan would have stiffened the bridge by using a Town truss to connect it to a solid brick abutment on the Columbus side. The details of this repair are unknown, since the committee handled it without reporting to council, and Godwin's role is unclear. If he had been willing to build 155 feet of latticework for $1,300, then his prices had fallen precipitously since 1841. Whoever propped it up, the bridge endured, once again, despite its almost imminent collapse.[50]

By the 1850s Godwin had reached the half century mark, and his former slave was busy building new bridges in three states. Yet both the old master and his black protégé continued to earn money by repairing their first Chattahoochee bridge. In 1852 the bridge committee hired Horace King to repair three piers for $300. The committee informed the aldermen that they immediately contracted with King without seeking approval of the full council because "delaying the matter until the regular meeting of Council would endanger the safety of the Bridge."[51] In its haste, the committee ignored the fact that King, a free black with no legal status in Georgia, had no right to execute such a contract. But such formalities were often disregarded when local governments dealt with Horace King. One year later King's old master received $300 for replacing the sills under the bridge and in 1854 earned another $200 to build a shelter over the east end of the span. As Godwin grew older, the scale of his work got smaller and more geographically localized.[52]

The need for better illumination inside the dark tunnel may have led the council to adopt gas streetlights for the city in 1853. Apparently the aldermen began buying gas in order to light the bridge. They placed five gas fixtures within the span and another at the bridge keeper's house, while the twenty additional lights for the city's streets were an afterthought. The centrality of the bridge to the community is obvious in the city's list of property in 1856: the bridge accounted for $100,000 of the city's $257,825 possessions. The city council, therefore, continued to protect its most important asset.[53]

In 1856 King won the bid for a larger project—to re-side the bridge. This contract for $650 conformed to the legal restrictions against free blacks, because William E. Godwin, John's son, negotiated it for Horace. The next month, however, the council dealt directly with King. In May 1856 the bridge committee, noting problems with the piers, "conferred with Col. Asa Bates and Horace King." Bates proposed to make the repairs for $2,000, while King bid $800. The council accepted King's bid, which might have been lower since he was already working on the structure. Perhaps that price differential explains why King built so many bridges. The next year, King rebuilt the eastern approach bridge for $1,050; John Godwin negotiated that contract for his former slave, and both John and Horace signed the document.[54]

King's work withstood another mighty flood in 1862. The muddy waters rose within two or three feet of the 1841 mark and washed away the upper bridge built in 1858. On the night of February 18, "though the angry flood overtopped [the lower bridge's] stone piers and buffetted the wooded frame work between them and its floor, and though many a large tree and heavy beams battered against its weaker supports, [it] stood firm and unshaken, with its floor almost reached by the flood. It has proven itself a staunch and well knit structure and having breasted unharmed this mighty torrent, it will henceforth be relied on with more confidence than ever as secure from the utmost reach of the Chattahoochee." As a result of the flood, Horace repaired the well-knit structure in 1863. King remained identified with Godwin, even though Godwin had died in 1859. The bridge committee reported to the council that "Horace Godwin" had "thoroughly repaired" the bridge according to contract.[55]

While the bridge remained solid, the filth inside the tunnel in 1858 offended the editors of the *Enquirer* and belies the romantic image of covered bridges. "On that much-traveled structure, the dirt lies deep enough for a potato-patch and when two or three wagons are crossing at the same time the dusts rises into a cloud as impenetrable as the darkness of Egypt—at all events, it can be felt as well as seen. The condition of the bridge is really discreditable to our city."[56]

The bridge even played a role in local folklore. This tale rendered by Thomas J. Jackson in his 1890 reminiscence suggests one reason the council was so eager to put gaslights in the old bridge:

> Many superstitious persons firmly believed it [the Columbus bridge] to be haunted, and stoutly asserted that several victims had been murdered and thrown from the bridge windows. . . . . But the most indubitable evidence of all was the well defined outlines in blood of a human hand printed on one of the window frames, plainly showing that some miserable unfortunate had made one last desperate clutch to avoid being chugged into the river by a fiend incarnate. This symbol of foul play and bloody deeds was contemplated with awe and wonder from year to year by thousands of passers-by, and for a long time timid people were afraid to cross the bridge after dark.[57]

For thirty-three years the bridge was the most important enterprise of the city of Columbus. Bridge tolls and bridge maintenance occupied an inordinate amount of the city council's time. After another protracted controversy, the factory owners, in 1858, erected a bridge (at Franklin, later Fourteenth Street) to connect their mills with workers' homes in Alabama. At that point, the city finally stopped collecting tolls from its 1833 bridge. In the span of twenty-four years, the city probably collected about $70,000 in tolls, more than twice the amount expended on construction, maintenance, and insurance. It had served as an excellent source of revenue for the frontier city on the Chattahoochee.

The bridge had also been a constant source of revenue and frustration in the life of Horace King. He had built, rebuilt, repaired, repiered, resilled, refloored, and re-sided this span that had played such a central role in the economic and political history of the city. At the same time he kept the Columbus bridge functioning and built other spans, Horace also developed his unique relationship with the Godwins and established his own family on the Alabama side of the Chattahoochee River.

CHAPTER SIX

FAMILY TIES

[I]n consideration of the natural love & affection which he [John Godwin] bears toward the said parties [William C. Wright in trust for Ann H. Godwin] . . . as well as for the consideration of one peppercorn in hand paid . . . doth give . . . the following described negro slaves, that is to say, Ben a boy about nineteen years old, Lucky, a woman about forty-five years old and her three children, Clarissa, Washington, & Horace.

May 5, 1837, Russell County Deed Book B, 65

In 1832, while still building the first Columbus bridge, Horace King moved with the Godwins from a temporary residence in Columbus to what would become the town of Girard, Alabama. For the next fourteen years King lived there as a slave. Most of his life during that period was consumed by his profession, working with John Godwin and his brother Wells to construct bridges, courthouses, warehouses, and homes throughout the Chattahoochee Valley. That construction can be described because of surviving contracts and newspaper accounts. Other aspects, such as the family life of the Kings and Godwins, Horace's relationship with John and Ann Godwin, and the daily life of slaves on the Godwin Place—important themes of this chapter—can only be documented through the lives of the masters. From 1832 until 1846 the biography of Horace must be viewed through the actions of John Godwin.

Most of the historic records left by John Godwin, however, fail to illuminate his family life in any detail. Instead they involve his land purchases and his debts, as he became one of the largest landowners or speculators in the new town of Girard. Although he helped start several institutions within the town, he never played an active role in either politics or a church. Work, land speculation, and family dominated his life. At the same time John and Wells expanded their families in Girard, Horace married and began his own family. The relationship between these families of masters and slaves was unusual. The widely held assumption that Godwin must be King's father is false. Their relationship was about making money, which is the major theme of this chapter. Godwin's speculating affected Horace's life.

Maybe Godwin settled on the western side of his bridge to show his faith in his new bridge or because he saw the unsettled area as a fertile one for land speculation. The Godwin family were speculators and they were attracted to upstart rival towns. John and Wells's move to Girard paralleled their father's venture in Sneedsboro, North Carolina, in the 1790s. Another factor influencing the move to Alabama could have been the presence of Ann Godwin's family in Montgomery. Godwin's unorthodox, liberal treatment of his skilled slaves may also have prompted him not to settle in Columbus, where municipal regulations were stricter on slave owners and slaves. Whatever the motive for his pioneering move across the river, John Godwin played an important role in the early history of Girard. Ironically Godwin's choice of residence shaped the later political career of his slave Horace King, who became an Alabamian and served in its legislature. Undoubtedly such an eventuality never entered Godwin's mind when he moved in the fall of 1832 to the hills of Alabama, just west of the Godwin-King bridge.

As one of its pioneer citizens, John Godwin shared in the birth of the county's government by allowing his original house to serve as the first county courthouse. The first session of the court convened in his carpentry or blacksmith shop, but Godwin never sought public office. Considering that one of Wells's sons was named Andrew Jackson, the Godwins probably voted the Democratic ticket, but John never caught the political fever. Even so, people recognized Godwin as a man of influence. The first history of Columbus, written by journalist John Martin, consists primarily of entries from the newspaper with brief biographies of about two dozen men; one of these was John Godwin. The bridge builder associated himself with prominent individuals, such as Ingersoll and the Abercrombies in Girard, Robert Jemison Jr. of Tuscaloosa, and the entrepreneurs who guaranteed his performance bond for the 1841 Columbus bridge. Horace King learned the importance of such associations from Godwin, and his success as a free black craftsman rested on his relationships with numerous influential men.[1]

Godwin's venture in land speculation—America's principal business—came even before he finished the Columbus bridge. Soon after they moved to Alabama, John and Wells Godwin partnered with D. K. Dodge, an Apalachicola, Florida, cotton factor and commission merchant, who eventually became one of the richest men in that town. During the scramble for Creek Indian lands, they purchased three half sections, or 960 acres, from Ho-lar-se-he-nar, No-cose-Yo-ho-lo, and Fixico. Wells served as a commissioner appointed to certify the legality of the transfers, and he personally certified the title to the land he owned with his brother and Dodge. The grants do not reveal the amount of money or rum they paid for this land, but they lay to the west of Girard. Their more speculative purchases lay along the river on the eastern edge of the new town organized by Daniel McDougald.[2]

John and Wells planned to make money from the development of this town, which they hoped would eventually rival Columbus. Daniel McDougald and his associates purchased a square mile from Chief Benjamin Marshall in 1833 and created Girard. They delineated the property into 393 half-acre business lots on the eastern side as well as another 72 settlement lots of four acres along the western edge. Extensive building began in Girard in 1834 and 1835, which probably involved the Godwins and King. Even so, the sale of property moved slowly. In 1836 McDougald's consortium sold three-quarters of their property for "about $70,000—a sum," that showed "the hope of building up a commercial rival to Columbus was then strong."[3]

The Godwins and Dodge bought at least nine business lots near or adjacent to the river. Appropriately, they were located where today's Thirteenth Street in Phenix City connects to the Horace King Friendship Bridge. In the heat of the initial Girard boom, Godwin and Dodge paid from $200 to $600 per acre. They shared the motives of the other entrepreneurs of Columbus and Girard, who sought both to capture a portion of the river trade and to harness the power of the river for manufacturing. John and Wells built a warehouse north of Holland Creek, possibly as a component of Dodge's cotton marketing business.[4]

Apalachicola played a crucial role in the river trade. Until the 1850s steamboats provided the only means of moving cotton from Georgia and Alabama to Apalachicola, on the coast, for export to the world market. The Godwin-Dodge warehouse never evolved into a full-scale operation because Girard never had its own steamboat dock. On one hand, Georgians denied Alabamians the right to use the river. On the other, the bridge built by the Godwins and Kings served its purpose. It drew Alabama cotton to Columbus, and where cotton went, so did commerce.

Given the location of two of their waterfront lots, the Godwins must have intended to use them for manufacturing, if for no more than a saw- or gristmill. Everyone realized the enormous waterpower potential of the falls of the Chattahoochee, which the U.S. Army Corps of Engineers had surveyed in 1827 as a possible site for a federal arsenal. Perhaps, the Godwins established a sawmill on this property, but they quickly sold these lots. The Panic of 1837 reversed the financial fortunes of everyone and dashed any hopes of realizing quick profits. At about that time, Dodge also withdrew from the partnership.[5]

In January 1838 the Godwins sold their interest in these nine lots, for about what they paid for them, to Stephen M. Ingersoll—the prominent, if slightly eccentric entrepreneur of Girard. Ingersoll, who later claimed to have invented the telegraph and shared its secrets with Samuel Morse when he visited Girard, definitely had plans to industrialize the western bank of the river in 1838. Maybe the Godwins simply stepped aside and allowed Ingersoll to pursue that goal while they chased building contracts. A circa-1845 map shows two mills in the lots (69 and 70) earlier owned by the Godwins.[6]

Ultimately Ingersoll's Girard failed to become a true rival to Columbus. The dispute over whether Alabamians had the right to use the waterpower or to build wharfs into the river reached the U.S. Supreme Court in the early 1850s. Georgians argued that when their state ceded what became Alabama and Mississippi to the federal government in 1802, Georgia drew its western boundary at the high water mark on the western bank of the Chattahoochee. The Court affirmed that only the Georgia riverbank owners enjoyed riparian rights. Unable to use the river, Girard never became a commercial or manufacturing center; it remained a suburb or a bedroom community rather than a rival to Columbus.[7]

Because of the Court's ruling, no speculator ever built a major steamboat wharf on the Alabama shore and the Godwin-Dodge warehouse did not flourish. Instead it became a house of worship, less than a mile north of the former site of Sodom. As early as 1838 the Godwins allowed a mission of the Methodist Church to meet in their building. Five years later, with Dodge apparently out of the picture, John donated his interest in the structure to the church, whereas Wells charged them a hundred dollars for his. The newly organized Girard Methodist Church made the old warehouse its home until 1851.[8] Although John Godwin helped establish the Methodist Church in Girard, he is not recorded as a member of the congregation. None of his brief biographies link him with a church in Girard, even though he associated with Baptists in Cheraw.

By the early 1840s John and Wells stopped buying land together, although they continued to sell what they owned. By 1850 Wells lived outside of town and began identifying himself as a farmer, while John continued to be a mechanic—the generic term for anyone working with his hands. In 1842 John and Ann sold their half interest in a 361-acre rural tract to Wells, maybe to raise capital to cover John's recent purchase of fifteen and a half settler's lots or about eighty-eight acres in the northwest corner of Girard.[9]

A portion of this property became the site of the Godwin Place where John, Ann, and their slaves lived. It was located adjacent to Holland Creek, which meanders through Girard in the shape of a backward capital S. On its upper reaches, Holland Creek was known as Marshall's Mill Creek and there John and his slaves built the Godwin Place, a complex of buildings including a main house, slave quarters, workshops, and a sawmill. The latter structure, old Marshall's Mill, appeared next to the creek on lot 69 of the circa-1845 map.

Some of Godwin's earlier acquisitions outside of the town appear to be contiguous to his Girard lots, including the family cemetery where Horace would later honor his master.[10]

This land next to Holland Creek was not bought for speculation but to make a home for his family and slaves. By the time John and Ann moved to their new house in 1843, seven of their children had already been born. Their first four—William E., Mary Ann, Napoleon, and Messina—had migrated from South Carolina with them, and three more—John Dill, Sarah Ashurst, and Thomas Metternick—were Russell County natives. Only Susan Albertha was born at the Godwin Place. In addition to naming their children after family members—Ann's uncle William; her mother, Mary Ann; and John—they appeared to favor European heroes: Napoleon, Messina, and Metternich. André Masséna served as Napoleon's best field marshal, and Metternick dominated European diplomacy after 1815.[11]

The Godwin's familial ties reached beyond their nuclear family. Wells and his wife, Malinda, had nine children, and although the business and land-buying partnership between John and Wells appeared to be less intense after 1842, they certainly gave each other support. In 1838 John sold one of his Girard lots to John C. Twitty, who may have been Ann Godwin's maternal grandfather. She also had kinship connections with the Abercrombies, a prominent Russell County family. Even more important were Ann's relatives in Montgomery.[12] The entrée into Montgomery and Alabama provided by the Wrights was even more important for Horace King. William Carney Wright may have introduced Godwin and King to Robert Jemison Jr., who became, after Godwin, the most important agent in Horace's professional life.

Ann Wright Godwin's parents and her uncle William Carney Wright lived in Montgomery County, Alabama, after 1824. William Wright played a major role in Ann's life. He and her father, Joseph, were both born in 1775 and were probably twins. Joseph and his wife, Mary Ann Twitty, had thirteen children; William never married. For a bachelor uncle to shower attention on his brother's oldest child is typical. William seems to have "adopted" Ann, acting as her legal trustee and giving her money.

Ann and her uncle became the owner of Godwin's most valuable slaves, including Horace, in 1837. The panic that year created an immediate depression. By March 1837 the price of cotton in New Orleans had fallen by 50 percent and unemployed workers had rioted in New York City. Godwin obviously had outstanding debts, and thus his property, including his slaves, would have been subject to liens by his creditors. According to tradition, Godwin had rejected an offer of six thousand dollars for Horace; thus the control of this skilled craftsman represented an attractive target for Godwin's creditors. Therefore in May 1837 Godwin passed the title for "Ben a boy about nineteen years old, Lucky, a woman about forty-five years old and her three children, Clarissa, Washington, & Horace" to Ann and Wright. This transaction may have been more than simply protecting John's assets; it increases the likelihood that Ann Wright's family money bought Horace and his family in 1830. If there were a kinship relationship between Horace and the Godwins, it came from the Kings through the Wrights, but the proof of such a link is tenuous.[13]

Ann's uncle William continued to provide financial support for John, Wells, and Ann. Wright loaned money to John in 1842 and to Wells in 1849. In his 1854 will, John did not leave his wife any money because she had received $2,000 from her uncle's estate in 1851.

Beyond his family, John Godwin's associates appeared to be his fellow builders and sawmill men. John bought what became the Godwin Place from two lumbermen, William D. Lucas and William Brooks. The construction men within the area seemed to form a fraternity of sorts. Brooks, in a manner similar to Asa Bates, remained a friend of the family. Also included within this circle of contractors was Horace King.[14]

The slaves on the Godwin Place lived exceptional lives when compared to other slaves. Slavery, as practiced in the antebellum South, was an intensely cruel institution. A slave owner could inflict any punishment on his or her property, even death. Given the basic premise of this system of forced labor, the master was obligated to discipline his slaves in a regular and stern manner. White southerners believed that the owner or overseer who spared the whip spoiled the workers. Frederick Douglass, the most famous escaped slave and African American abolitionist, accurately defined the dilemma of the master: "Beat and cuff your slave, keep him hungry and spiritless, and he will follow the chain of this master like a dog. But feed and clothe him well, work him moderately, surround him with physical comfort, and dreams of freedom intrude. Give him a bad master, and he aspires to a good master. Give him a good master and he wishes to become his own master." The need for social control and the constant reminders of the slave's inferiority permeated all customs and legislation dealing with the South's peculiar institution.[15]

Urban slavery strained the system of control, because masters frequently rented out their slaves. An 1832 Columbus ad, "A Negro Woman Wanted," sought a "cleanly, neat, and industrious" slave to cook, wash, and iron for seven dollars a month, paid to the owner. Such slave workers performed a wide variety of services from menial labor to sophisticated craftsmanship. The records of Columbus slave traders show that hiring out as well as selling slaves was routine.[16] In larger cities such as Charleston and Savannah, some slaves were allowed to live as virtually free, renting out their own services. But in Columbus, the city council prohibited slaves from "hiring their own time" and required a slave to live on the lot "where his or her owner lives." In order to capture runaways, groups of white southerners, both nonslaveholders and owners, rode slave patrol at night. The Columbus aldermen used the bridge to control slaves. They instructed the bridge keeper "to lock the small gate to the bridge and let no slave pass or repass after the hour of 10 O'Clock at night without a written permit on business." However, no bridge keeper ever asked Horace Godwin (or Horace King) for a pass. Given his frequent work on the bridge, he was too well known to need a permit.[17]

The laws on the books regarding slavery rarely reflected the actual relationship of an owner such as Godwin to a skilled slave such as Horace King. Despite the oppression, the racism, and the specific regulations, slavery also rested on personal relationships between the master and the enslaved. Godwin's treatment of his slaves remains a matter of conjecture, but he certainly did not whip King or any of his other skilled slaves. He did, however, sell them, a typical practice of entrepreneurial slave owners. But he never relinquished his five master carpenters or his wheelwright. He owned eighteen slaves in both 1840 and 1850 but probably not the same ones. In 1850 he had twelve males and six females—four of childbearing age. Considering that these women almost certainly had children, Godwin probably sold some slaves between 1840 and 1850. He freed Horace in 1846, and his slaveholdings continued to decrease during the 1850s. By the time of his death in 1859, he owned twelve adults and four children.

Godwin's practice of recording his debts in deed books provides another glimpse of his slaves. Godwin's chattel became collateral for loans, much in the way twenty-first-century homeowners use the equity in their houses for a second mortgage. In 1842 Godwin mortgaged thirteen slaves—but not those owned by Ann—as security for a $2,744 loan from William Wright. Even though the debt was due the next year, the deed book shows the note was not satisfied until 1853, after Wright's death. In the meantime, Godwin used some of the same African Americans to guarantee a $2,300 loan in 1851.[18]

An unusual feature of Godwin's slaves was that some of them, such as the wheelwright Henry Murray and the entire King family, were listed by their own family name. Obviously, the Kings formed a very special family of slaves within Godwin's household: Horace (born in 1807), his mother, known as Lucky or Susan (born about 1791), his sister, Clarissa (born 1821), and his brother, Washington (born 1823). Godwin relinquished his title to the Kings and a slave named Ben in 1837, turning the chattel over to his wife, Ann, and her uncle. After Godwin's death, his sons bought the Kings (all except Horace, who was free). Ironically, Horace, who had accumulated wealth as a free black for thirteen years by 1859, did not buy freedom for any of his family members. Instead, he purchased a monument for Godwin's grave.

Godwin was a benevolent owner, one of the kindest in the spectrum of slaveholders, but his treatment of Horace may have been more or less dictated by Horace's skills. A comparison with Robert Jemison Jr., a much larger slaveholder than Godwin, illustrates King's significance. In discussing a possible bridge project in Jackson, Mississippi, Jemison wrote: "I have several carpenters who with an experienced Bridge builder" to direct them could make more money "than they now make me." Godwin was in a unique situation among slaveholders; he, or after 1837 his wife, owned the slave superintendent who could direct other slaves and greatly increase the earning power of a master.[19]

John Godwin, or more legalistically Ann acting through William Wright, allowed Horace King to marry a free woman and thus to control his own family while still a slave. In 1838 or 1839 while building the West Point bridge, King met Frances Gould Thomas, a fourteen-year-old girl. Her African, European, and Native American ethnic background paralleled that of Horace's; her Indian roots came from her Creek ancestors. Her mother, Rachael Gould, had been freed before 1829 in Milledgeville, Georgia, and according to tradition, she received an 1832 allotment near the Chattahoochee River before being removed to Oklahoma. Her daughter stayed in the West Point area. Her father might have been a white master, since during the Civil War Frances claimed Masonic membership for her father—a highly improbable circumstance for a mulatto in central and west Georgia before 1865. The brief courtship between this free woman of color and the slave Horace King culminated in marriage on April 28, 1839.[20]

Frances moved to Girard, where she and Horace lived in their own home on the Godwin Place. Living in a separate house, or half of a double-pen house, constituted a minor indulgence on the part of Horace's masters, compared to their major concession in allowing him to marry a free woman. The Godwins relinquished their ownership of Horace and Frances's children. Again, the economic potential of Horace and his special relationship with John and Ann produced this extremely unusual situation. Even the most benign slaveholders rarely relinquished the "natural" increase of their capital.

Frances and Horace were married two years before they had children. Frances then gave birth to five children in seven years: Washington W. (1843–1910), Marshal Ney (1844–79), John Thomas (1846–1926), Annie Elizabeth (1848–1919), and George H. (1850–99). While slaveholders often selected Greek or Roman names to reinforce the illusion that southern slavery replicated these noble civilizations, Horace rejected that practice. John Thomas might have been named for John Godwin or for Frances's family. In selecting the name Marshal Ney, Horace followed John Godwin's pattern and named his son for one of Napoleon's generals.[21]

King family tradition has associated the name Marshall with the Creek leader, but actually Michel Ney was Napoleon's field marshal. A legend, well known in the land of King's birth, surrounded the death of Marshal Ney. When Napoleon broke out of exile from Malta, Ney, who had declared his allegiance to Louis XVIII, pledged to stop Napoleon's advance on Paris. Instead Ney switched sides, joined his old commander, and helped swell the ranks of a new Napoleonic army, which Lord Wellington destroyed at Waterloo. After that defeat, the anti-Bonaparte French leaders executed Ney. According to legend, however, Ney was not executed: he escaped, moved to the Charlotte, North Carolina, area, and taught school under the pseudonym N. S. Ney. Allegedly Ney's final grave was decorated with symbols appropriate for a French field marshal. Marshal Ney was an appropriate appellation for a slave's son whose master's sons were named Napoleon, Messina, and Metternick.[22]

Horace's sister was part of another family that lived on the Godwin Place. Clarissa King married Henry Murray, Godwin's wheelwright (or blacksmith), another skilled and valuable slave who would have merited his own dwelling. Also a mulatto, Henry, who was fifteen years older than Clarissa, was about the same age as Horace. In 1859 Henry and Clarissa had an infant daughter also named Clarissa, but she died before she was ten. Clarissa and Horace's mother lived with the Murrays after the Civil War, and she may have been living in their household before 1865. According to another local myth Clarissa was a Creek Indian who married Godwin's slave before the removal of the Creeks in 1836. However, Clarissa's Indian blood, like her brother's, came from a Virginia tribe or the Catawbas of Carolina rather than the Chattahoochee Creeks.[23]

Henry Murray's relationship with John Godwin inadvertently became an Alabama Supreme Court case. Using his skills as a wheelwright, Henry found opportunities to earn his own money. On at least one occasion, the master borrowed money from Murray, and he accepted a promissory note from Godwin for fifty-eight dollars. Then as was customary, Henry passed, or sold, the note, probably at a discount, to a white man, W. B. Martin, who then sold it to John M. C. Reed. When Martin refused to pay the note, Reed took the matter to court. The case eventually reached the Alabama Supreme Court, which ruled the note was valueless because a slave had no rights to make contracts or promissory notes.[24] Such restrictive laws limiting the rights of slaves did not apply to skilled craftsmen on the Godwin Place.

Another example of the restrictions on slaves involved their possessions. After the death of John Godwin in 1859, his son William E. bought a wagon from his father's estate for eleven dollars for Henry Murray; more than likely Henry had always used that wagon. The slave Henry could "own" that wagon as a personal possession, but he would have to use it as yard decoration or borrow a mule from a white man since Alabama law prohibited slaves from owning mules or horses. Such anecdotes illustrate the absurdity of slavery as well

Frances Gould Thomas King, ca. 1855. Horace met Frances while he was building the West Point, Georgia, bridge. They were married in April 1839; she was fourteen years old, thirteen years younger than King. They shared a similar triracial background. Frances's dress and jewelry, which were not provided by the studio, is indicative of the Kings' affluence as free blacks. *Collection of Thomas L. French Jr.*

as how such regulations did not apply on the Godwin Place. Ironically, Henry Murray, in 1872, became the owner of the Godwin Place.[25] Such a transition from slave to owner of the master's place was unusual but perhaps appropriate in the case of the Godwins, given their treatment of the King family, the Murrays, and their other slaves.

Because of the manner in which Godwin used his slaves as master builders and carpenters, he had to give them a significant degree of freedom and responsibility. Unlike a gang of slaves growing cotton, who were constantly supervised, Godwin's slaves worked on tasks that were crucial to the integrity

Horace King, ca. 1855. This view of King apparently dates from the same session that recorded his wife Frances. Photographs of African American families in the South before the Civil War are extremely rare. Horace appears confident and relaxed, indications of his status and position as a public figure within the Chattahoochee Valley. Its residents were not surprised to see him supervising major construction jobs. *Collection of the Columbus Museum, Columbus, Georgia; Museum purchase.*

of a bridge or a courthouse. Since they had so much responsibility on the job, in comparison with other slaves, they enjoyed a great deal of freedom on the Godwin Place.

Even though Horace had the luxury of maintaining his family while still a slave, he spent much of his time away from home. Working as a heavy contractor meant Godwin netted larger profits than a house carpenter, but it also required him to work in a broader geographical area. Especially from July through October when the flow of rivers was at the lowest, King tended to be working away from Girard.

Some writers who have perpetuated the myth of Horace King seem to imply that Godwin's treatment of King proves that slavery was a benign institution. The actions of the Godwins as slaveholders were not typical of most owners. John and Ann's kindnesses toward King cannot be used to make positive generalizations about southern slavery. Horace, because of his extraordinary skills and his ability to make money, received very special treatment. In turn, he worked with John and Wells Godwin to build the major heavy timber structures in the Chattahoochee Valley from 1832 until 1842.

CHAPTER SEVEN

"HONEST" JOHN GODWIN AND HORACE

The superior Bridge across the Chattahoochee River at Florence, built by "honest" John Godwin and Horace, is now ready for crossing.
Columbus Enquirer, August 5, 1840

From 1832 to 1842, operating from their base in Girard, John and Wells Godwin along with Horace King built bridges, warehouses, courthouses, and homes throughout the lower Chattahoochee Valley. The King family tradition makes Horace a partner of John, implying they worked as equals. Given the fact that by 1837 Ann rather than John owned Horace, the money he earned may have gone to her, and Horace might well have kept a share of his earnings. Horace's role as a superintendent on various jobs was obvious to the public. By 1840 a newspaper ad billed John and Horace as cobuilders, an extraordinary status for a slave. While their projects brought profits to the Godwins, economic uncertainty haunted this period. The Panic of 1837 affected John's finances, especially as he moved from building bridges to owning them.

The Godwins and Horace constructed some houses, but the Rev. Francis L. Cherry exaggerated Godwin's role by claiming he "built most of the houses in Columbus and Girard during the first fifteen years of their growth." Those erected by Godwin and his slaves in Girard were modest structures. Godwin has never been associated with any of the mansions in Columbus or its eastern elite suburb of Wynnton. Godwin established his reputation as a contractor for massive structures rather than as a house builder. The exact extent of Godwin's work remains unknown, but enough of his bridges and buildings are documented to show that Horace King gained a variety of experience working with his master from 1832 until 1842. The real question is how they executed all their overlapping contracts.[1]

From 1834 until 1841 they launched at least eight major projects, any one of which could have insured the reputation of its contractor. In 1834 they probably began building warehouses in Apalachicola, a project that consumed more than a year. By February 1838 their bridge stood at West Point, Georgia; from April 1838 until October 1840 they erected the Muscogee County

Courthouse in Columbus. During the same period they fabricated bridges in Eufaula (1838–39) and Florence, Georgia (completed by August 1840); they rebuilt the Columbus bridge from April to July 1841; in September 1841 Godwin agreed to create a breakwater in the river at Columbus; and from 1839 until 1841 Godwin and King constructed a brick courthouse for Russell County in Crawford, Alabama. In 1842 Godwin contracted with Muscogee County for two small spans across Bull Creek. The geographical dispersion of these jobs meant that John Godwin cooperated with other contractors—his brother Wells, Asa Bates, or Simon H. Williams. He also relied on his other partner, Horace King, for without his supervisory skills, Godwin could never have completed this range of projects.

Godwin and Horace were probably involved in a large project that created forty cotton warehouses in Apalachicola, Florida, during the 1830s. The records illuminating this construction remain sketchy, but the town's oral history tradition identifies John Godwin as having built eleven of them. The Apalachicola Land Company, which developed the town, required cotton merchants who purchased riverfront lots to build standardized storage facilities. Apparently designed by a New York architect, these storehouses mimicked the features of extant New York warehouses, and the exterior construction materials—granite and bricks—for those in Florida also came from New York, probably as ballast for ships arriving to transport the port's cotton.

In 1834 cotton merchants B. F. Norse and H. W. Brooks hired John Godwin to erect one of the earliest warehouses. Then he contracted for ten more facilities. While only one source notes Godwin's participation, other factors reinforce the possibility of his involvement. He already enjoyed a business partnership with D. K. Dodge, a commission merchant who had offices in Columbus and Apalachicola. The lack of a permanent population in that tiny seaport probably necessitated the use of outside contractors. Many of the merchants who profited from handling the cotton that came down the Chattahoochee and Flint Rivers came only for the cotton season and never became Apalachicola residents. The town probably had no heavy-timber contractor. Also, the assembling of these warehouses employed beams larger than those used for houses and such heavy-timber-frame construction paralleled bridge-building techniques.[2]

Sources in Tuscaloosa, Alabama, have Horace King fabricating a massive bridge over the Black Warrior River there in 1834. Horace did not build that span. Seth, not Horace, King constructed it. Even though for other bridges—Cheraw, Wetumpka, and Columbus, Mississippi—Seth only provided capital, he designed the original bridge in Tuscaloosa. This design—a Town double lattice—differed from those Godwin and King were using. It employed two sets of crossed diagonal braces separated by another horizontal chord. Viewed

Apalachicola waterfront, 1838. Godwin and King may have built as many as eleven antebellum cotton warehouses in Apalachicola, Florida. Merchants who purchased waterfront lots were required to erect a warehouse based on a standard plan. The demand for skilled contractors makes Godwin and King's involvement extremely likely. *H. A. Norris and N. Calyo, P. A. Mesir and Co. Lith. Courtesy of Lynn Willoughby.*

from the end, the truss consisted of ten boards that measured slightly more than thirty inches across, held together by three-foot treenails. Its construction consumed 40 percent more lumber than the Columbus, Mississippi, bridge of about the same length and, at $30,000, cost approximately twice as much money.[3]

Both Seth King, the largest stockholder in the Cheraw bridge, and Robert Jemison Jr., who became an important client of Godwin and King, were major investors in the Tuscaloosa span. Jemison, an entrepreneurial slaveholder with diversified business interests first ventured into bridge building with this span. By the 1850s Jemison and Seth King were bitter enemies, mired in litigation against each other. Seth's work as the builder of the 1834 span is documented by Jemison's complaint to his friend Horace King about Seth's charging "the Tuskaloosa Bridge Co. with one thousand dollars for his services in rebuilding the Bridge and $2770.00 for two Piers, one built 1838, the other in 1842, the necessity for both which was occasioned by his [Seth] making the two center reaches too long[,] originally one 250 ft.[,] the other 220 ft, and both required to save his guaranty of sufficiency of plan[,] material and workmanship."[4]

Since Jemison assigned the insufficiencies to Seth King and then described its center spans to Horace, clearly Horace did not build this bridge. Earlier accounts in Tuscaloosa probably attributed the span to "Mr. King," meaning Seth. Given the appeal of the Horace King legend, Tuscaloosans later assumed Horace King built the bridge, having long forgotten the white man with the same name. Rather than venturing across Alabama to Tuscaloosa in 1834, Godwin and his skilled slave remained in the Chattahoochee Valley.

While Godwin and King became known historically for their bridges, in the 1830s and early 1840s Wells and John contracted for large structures. Wells acted as the lead contractor on the most impressive structure in antebellum Columbus. In April 1838 the Muscogee County Inferior Court and the Columbus City Council acting in concert hired Wells and John to build a 16,200-square-foot county courthouse–city hall. It replaced the initial wooden one, which an arsonist, according to local tradition, had burned. The county paid $20,000 and the city $10,000 to the Godwins for this handsome two-story building with massive Doric columns on the back and the front. The contract specified the construction details.[5]

This courthouse played an important role in Horace's education: he applied lessons learned here to the Alabama state capitol in the next decade and to the Lee County, Alabama, Courthouse after the Civil War. The exterior arrangement of the steps of the Columbus courthouse, which were not specified by the contract, could have been influenced by Robert Mills's "open-arms" courthouses in South Carolina, which were characterized by the circular exterior stairs. In 1867 King built what was apparently a miniature version of the Godwins' large Columbus courthouse for Lee County, Alabama, in Opelika.[6]

Financing the construction of this impressive structure presented problems for the Godwin brothers. The contract called for five equal payments when they completed various phases of their work. Granted a contract on April 28, 1838, the Godwins, in less than two months, beseeched the council for money. Their plea received little encouragement from the aldermen. "A petition from John & Wells Godwin was read and laid on the table forever." They had received several payments and at least one advance from the council by January 25, 1840, when another plea, slightly more oblique, again raised the ire of council. "A communication was received from George Smith during the Council to accept a note on John Godwin which on motion was ordered to be laid on the table."[7] A paucity of capital and currency plagued the entire society in the antebellum period. The Republic and especially the South floated on a sea of personal notes, and the Godwins were no exceptions.

Muscogee County Courthouse–Columbus City Hall, ca. 1890. In 1838 Wells and John Godwin along with Horace King built this public building in the latest style, Greek revival. King learned techniques from this work that he applied on the Alabama state capitol (1851) and the Lee County (Alabama) Courthouse (1867–68). *Courtesy of Harding B. Givens Jr.*

The Godwins finished the courthouse ten months after the specified completion date and incurred extra expenses, amounting to $7,800 by their calculations. Arbitrators reduced it to $6,000. The city councilors apparently paid their share of the initial and extra costs, perhaps because they enjoyed revenue from Godwin's bridge. The county, on the other hand, failed to pay the builders. In January 1841 the Godwins filed a lien for $12,184 plus interest against both governments.[8]

The county issued a note to the Godwins for $3,737.63, but they did not wait for the county to redeem it. They transferred it to Hampton Smith, at a reduced value, and as late as March 1848 he was still attempting to collect from the county. Obviously, the Godwins never realized the income they expected from the massive courthouse. Despite the litigation the two governments paid the Godwins $200 to plant trees on the "Courthouse Square."[9]

Despite the problems raised by John's delay and his extra expenses, both governments continued to hire John Godwin and Horace. They rebuilt the bridge in April 1841 and in September of the same year began erecting a breakwater to prevent the city's riverbank from eroding. This two-hundred-foot structure, most likely constructed just north of the present Fourteenth Street Bridge, angled downstream in a southwesterly direction and shifted the current from the bank to the middle of the stream. The city paid Godwin $4 a foot, but again his work proceeded slowly. In December 1841 the council urged the completion of Godwin's breakwater as a means of countering the effect of S. M. Ingersoll's illegal dam extending from the Alabama shore. In June 1842 the council reported having to force the completion of Godwin's contract. Ironically, Ingersoll's offending dam might have extended from land he purchased from the Godwins. Working on this breakwater also served to train King for what he later considered an onerous task—erecting obstacles during the Civil War to deter the passage of Yankee boats up the Apalachicola and Alabama Rivers.

In 1842 John Godwin contracted with Muscogee County to build two small bridges over Bull Creek, a major stream to the south of the city. He finished one in January at Moore's Mill for $500; another at Tom's Mill for $790 was completed in May. Godwin was experiencing financial problems. He did not wait for payment from the county, but "assigned his claim upon the county through John Warren to Edward Carry."[10] These men either represented his suppliers or Godwin accepted a reduced amount from Warren or Carry, who in turn tried to collect from the county.

At the same time Wells and John undertook the Muscogee–Columbus Courthouse, John contracted for a similar facility in Russell County. The first county court met at John Godwin's house on Ingersoll Hill in 1833. By 1839 county officials had moved the seat ten miles west of Girard to Crockettsville (known as Crawford after 1843). By 1841 or 1842 Godwin and perhaps King built a brick courthouse there. One account asserted that Godwin finished his courthouse and King rebuilt the Columbus bridge in 1841. Given what must have been disparity in potential income from the two projects, both men likely concentrated on the bridge and neglected Russell County.[11]

In spite of their work on government projects, Godwin and King became renowned for their bridges. During their first decade in Girard they constructed three major spans across the Chattahoochee River and many smaller undocumented ones. Having established their credentials with the Columbus bridge, they were able to raise their prices. The cost of their Town lattice trusses rose sharply even as the economy slumped. The total expenditure for the spans at West Point, Irwinton (Eufaula), and Florence reached approximately $60,000, of which Godwin may have grossed $20,000 in cash or stock. From that figure he absorbed all the labor costs. If he were paid for supplying the timbers and lumber, his gross would have been even larger. Private investors or companies financed all three of these crossings. Two of the companies created banks that circulated their notes and change bills. Godwin himself became a speculator as he moved from building bridges to owning them. The fates of these three spans typify the history of such endeavors: Interstate litigation forced one into municipal ownership; two suffered from floodwaters; and two ultimately failed—a tragic end for such expensive enterprises but not surprising for a frontier setting.

Godwin and King's second major span across the Chattahoochee came at West Point, Georgia, approximately forty miles north of Columbus. Unlike Columbus or Eufaula, the West Point bridge avoided interstate conflicts because the town existed on both sides of the river. At this point the Georgia border stopped following the riverbank and became a straight line as the river turned eastward. West Point's location marks the upper end of the rapids that extend northward from the fall line at Columbus. North of West Point, the river became calmer, and smaller, nonmotorized box boats operated in the slack water.

At the same time the state organized Columbus, private speculators created the town of West Point. Indians for centuries and English traders since the 1690s had forded the Chattahoochee near West Point. Planters and merchants incorporated a small town there in 1831–32. They named it Franklin and then realized a Franklin existed in the adjacent county; thus West Point became its name. A modern covered bridge represented a logical progression and a necessity for these investors bent on making money from their property and from the commerce of West Point. In 1835 a state charter authorized eight men "to build a bridge across the Chattahoochie river, at any point within the limits of said incorporation they may deem expedient, and that they be entitled to such tolls as are received by the existing ferry at that place." Three years later nine men, who controlled most of the property in the town, organized the West Point Company, a real estate venture; six of them already owned the bridge.[12]

To erect their span, they hired the builders of the Columbus bridge. Their new bridge reached 652 feet, probably had three piers in the water, and cost $22,000, approximately 17 percent more per foot than the Columbus bridge. Godwin and King finished their work in February 1838. Horace gained more than income from this job. While in West Point he met Frances Gould Thomas, who became his wife in April 1839.[13]

The new span must have been profitable because someone challenged its ownership. In 1850 the state legislature restated its entitlement to the same men and permitted the company "to charge sixty cents for each wagon crossing said bridge, and drawn by six horses, mules or oxen, and otherwise the rates of toll to remain as before established." The act prohibited the building of a bridge or ferry "within one mile and a half of said bridge . . . without the consent of the [bridge] owners." Prophetically, the act waived that restriction for a future railroad bridge. When the major rail line between Atlanta and Montgomery came through West Point in the 1850s, a railroad rather than a wagon bridge determined the future of the town.[14]

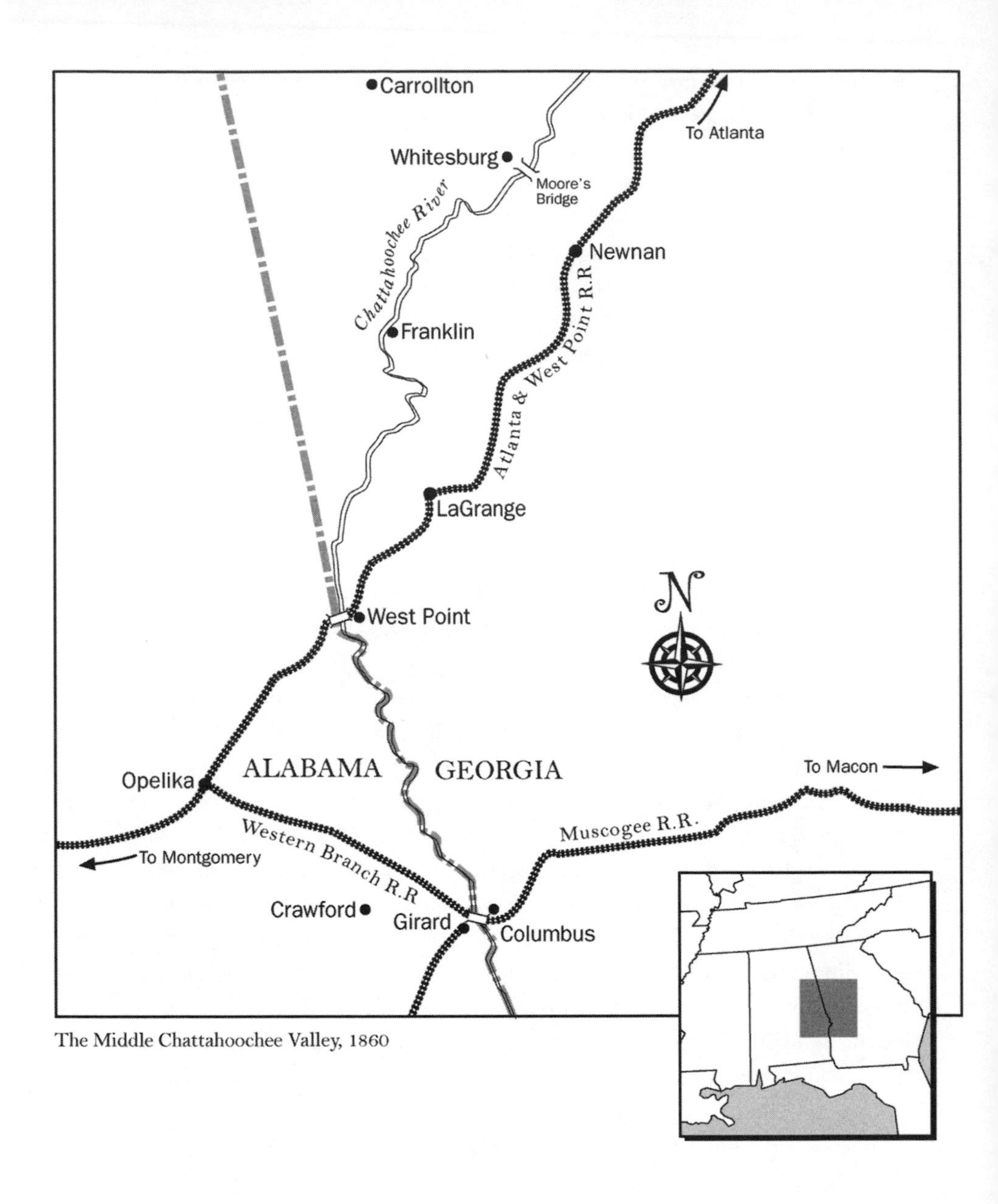

The Middle Chattahoochee Valley, 1860

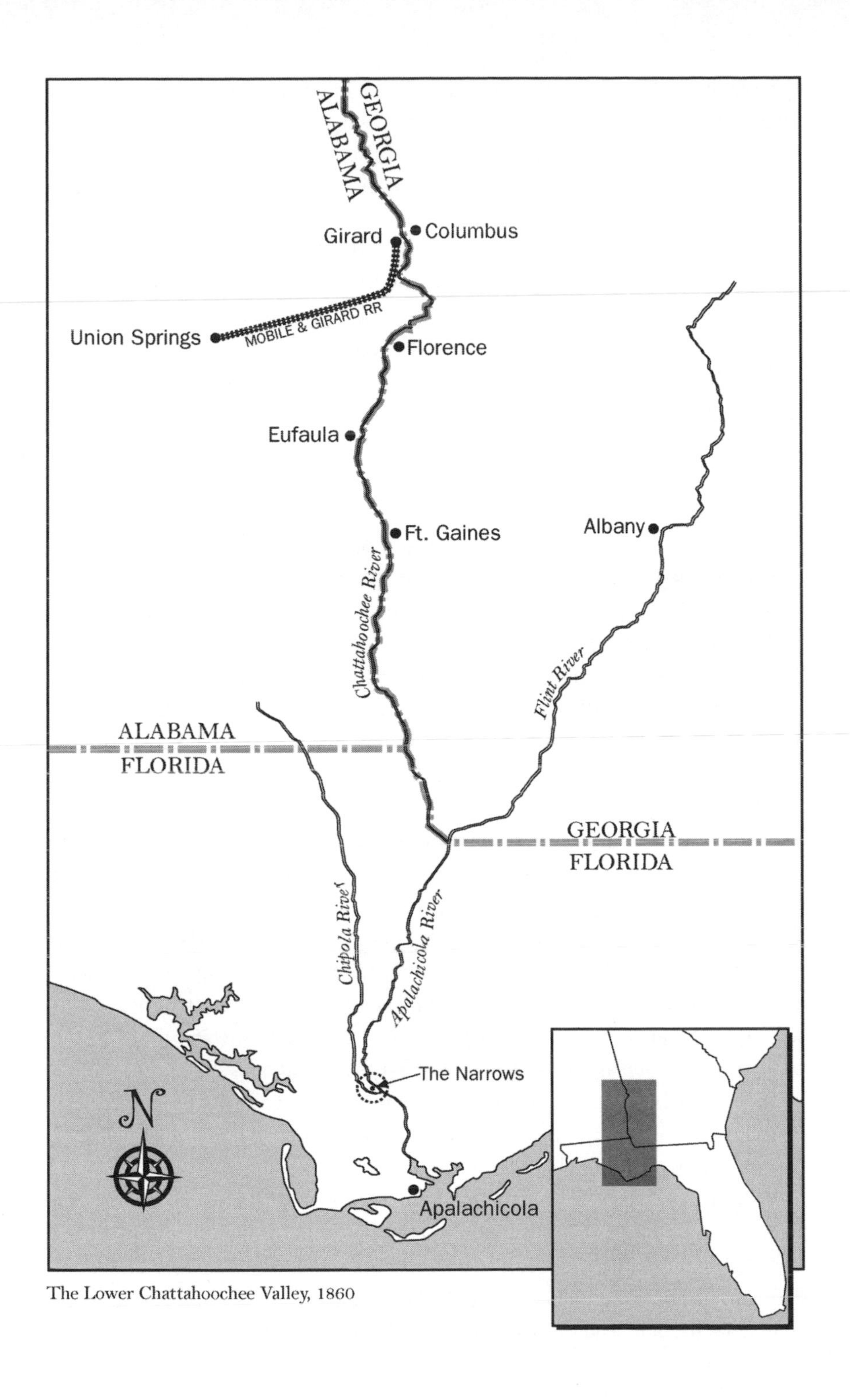

The Lower Chattahoochee Valley, 1860

Godwin and King built their next major Chattahoochee bridge approximately sixty miles downriver of Girard at Irwinton (later known as Eufaula). In 1830 the Alabama legislature chartered the town situated on a high bluff. The first settlers exploited the Creek Indians and their property. While the land was still under federal control, whites began building houses, taverns, and even a cotton warehouse in violation of the Treaty of 1832. Federal officials protecting Indian lands sent troops who burned this illegal settlement. According to the federal marshal in charge of the operation, these white squatters were "some of the most lawless and uncouth men" he had ever seen.[15]

The settlers almost immediately moved back in and began resurrecting the town. By 1837 a bridge across the Chattahoochee became a necessity if Irwinton hoped to rival Columbus as a center of the river-borne trade. Rich cotton lands extended from both sides of the river at Irwinton; only a bridge could lure the precious staple to Alabama warehouses. Without it, Georgia planters tended to ship from local river landings to merchants in Apalachicola or Columbus. In 1837 seventeen men petitioned the Georgia legislature to charter the Irwinton Bridge Company. The Georgia border extended to the high-water mark on the Alabama side; thus Georgia had to grant the charter. The state empowered the company, which could issue stock up to $75,000, to construct a bridge within three years. Unlike the Columbus bridge, no state moneys financed this crossing, and the coffers of nascent Irwinton could not provide any municipal support at least initially.[16]

The company moved deliberately. In November 1838 it contracted with John Godwin and Simon H. Williams to build a 540-foot "Town Patent Lattice" bridge resting on wooden abutments and 84-foot-high timber piers. Godwin and Williams, a fellow contractor, agreed to complete the span within a year for $22,500. The cost exceeded the price of the Columbus bridge by 40 percent, even though the Irwinton span rested on wooden rather than stone piers. To avoid the problem of raising capital for the entire construction, Godwin and Williams received four payments of $5,000 at specific times in the process. They even agreed to accept payments as small as $200 as those funds became available to the company. The company borrowed so much money from Edward B. Young, the town's most important cotton merchant, that he became the primary stockholder.[17]

With Young's capital and the skills of Godwin, Williams, and King, the span rose out of the Chattahoochee. Horace in all probability applied his knowledge to every phase of the construction, including the piers. The foundation of the original piers consisted of wooden cribs, large open crates constructed with mortised timbers, that resembled the walls of log cabins. Workers built these bases without constructing a cofferdam, and crews floated the boxes, which measured twenty-two by fifty feet, to the proper places in the

Eufaula Bridge, early 1920s. Captions for this image have frequently cited Horace King as its builder. This photograph documents the bridge before its final repairs. When this picture was taken, the only surviving portions built by King were the log foundations of the piers, buried in the river. *From Hardaway Construction Company,* 20 Years along the Rivers of the Southeastern United States *(1931). Courtesy of Columbus State University Archives.*

river. Rocks were added to these timber structures, which then sank to the bottom, and more rocks were piled around the cribs to keep the current from undermining them. These bases provided support for the first wooden piers and the later brick ones.[18]

An often-published twentieth-century aerial photograph of the bridge at this location usually carries a caption saying that Horace King built the span. But King's bridge did not survive; the only extant portions of his work at this site were the log cribs at the bottom of the river. In March 1841 the Harrison Freshet brought a raging wall of water down the Chattahoochee. "Considerable damage has been done to the abutment of the Irwinton bridge on the Georgia side—the foundation having been washed from under the pier, when it was forced to give way." A local, overly optimistic journalist expected its restoration within a matter of days. Perhaps the repair involved sinking another crib on the Georgia side. Godwin and King had trouble finding the time to work on the piers for the Irwinton bridge, because they needed to replace the entire Columbus bridge, as well as replacing piers under the Florence bridge.[19]

Godwin exhibited confidence in his repairs. In January 1842 he and Wells accepted one hundred shares of stock as collateral for a loan of $2,300 to John A. Norton of Muscogee County. This debt was never canceled, so maybe the Godwins acquired this stock when Norton failed to satisfy his indebtedness.[20] By 1842—the nadir of the worst pre–Civil War depression—the bottom had fallen out of the local economy, and few individuals had money to loan. Despite the financial uncertainty after 1837, the value of Irwinton bridge stock remained high; the original cost of one hundred shares equaled $2,500. On the other hand, the Godwins may have been unduly optimistic about the economic potential of the bridge and overvalued the stock. Unfortunately for the Godwins, the company's value declined precipitously, eventually because of the condition of the span, but more immediately due to the general economic climate and the interstate squabble over compensation for the Georgia bridge landing.

By 1844 the Irwinton Bridge Company faced financial difficulties. The company, following the standard practices for the era, created a bank that issued notes and change bills in small denominations. Such local banking operations floated bank notes far in excess of their assets. By 1844, when the economy began to improve, most original banks within the Chattahoochee Valley had failed. In that year, the notes for Irwinton Bridge Company traded at only 50 percent of their face value and the firm's officials decided to close the doors of its bank permanently. The company appointed Edward B. Young "to wind up the affairs of the institution," and he also "assume[d] all its liabilities." He eventually purchased the entire corporation for $10,000.[21]

In 1845, at the request of the company, the legislature restated the tolls and exempted all agricultural products, but the new rates applied for no more than two years. By 1847 "the bridge had fallen down." The exact circumstances surrounding the collapse are unknown. Local sources mention a flood. An 1847 court case instituted by Young, stated he "had purchased the bridge, . . . that the bridge had fallen down, and that the complainant [Young] was desirous of rebuilding it." The dispute reached the Georgia Supreme Court on at least four occasions and the litigation referred to the collapse of the bridge near the western bank. How many portions fell is unclear.[22]

This litigation involved the Georgia owners of three acres of sandy riverbank; they disputed the amount of compensation paid to them by the company. The primary Georgia litigant, James Harrison, began as a company official but challenged the company or Young by 1847. The issues involved a convoluted title and a technicality about the arbitrators who assessed the property. The arbitrators set the compensation for the Georgia owners at $500 and only considered the value of the land. Harrison and his fellow owners refused that offer; perhaps they compared it with the $10,000 paid by Columbus to the Alabama owners of the western abutment there. The supreme court initially supported Harrison's litigation and stopped the reconstruction of the span; Harrison then convinced the Georgia legislature to withdraw the original charter for the Irwinton Bridge Company.[23]

By that time, Young had sold the company and its bridge, which had already "fallen down," to the city of Eufaula for $10,000, while he kept its liabilities. The Georgia Supreme Court reversed the Georgia legislature and reconfirmed the original charter vested by then in the city of Eufaula. The decisions of two more juries in Georgia superior courts reached the supreme court on appeals. The higher court set aside the jury's decision of $13,007 for Harrison. Young, Eufaula, or both then paid $1,500 to Harrison, but the supreme court still remanded the case back to the lower court in 1857.[24]

Apparently Young and the city reconstructed the entire span over time. In the 1849 case Young referred to his rebuilding as "putting of timbers together at a point near the middle of the Chattahoochee River," rather than just replacing the western truss. By the mid-1850s Eufaula businessmen had raised $6,000 to complete the rebuilding, with John Barnett as architect and George Whipple, a local builder, as contractor.[25]

To the north of old Irwinton, "'honest' John Godwin and Horace" built the Florence bridge, the only crossing over the Chattahoochee between Eufaula and Columbus until the last decade of the twentieth century. In typical Godwin fashion, John became involved in the frontier boomtown of Florence, not simply as a builder but eventually as an owner of the bridge. As in the case in Eufaula, this span proved to be short-lived, and its demise contributed to Godwin's financial woes.

Erecting a new pier foundation, Eufaula Bridge, 1923–25. Under the direction of J. L. Land, workers built a coffer dam and excavated around the old pier foundation in preparation for pouring larger concrete piers around the old ones. These supports carried the new, metal McDowell Bridge at the site of Horace's old bridge. *Courtesy of Alabama Department of Archives and History, Montgomery, Alabama.*

The surviving wooden crib foundations at Eufaula in 1923, originally sunk by Horace King's crew in 1838. Workers excavating the foundation for a new pier discovered this wooden structure, a rare example of King's original work. The close fit of these hand-hewn beams demonstrates excellent craftsmanship. *Courtesy of Alabama Department of Archives and History, Montgomery, Alabama.*

On the eastern bank of the Chattahoochee, approximately sixteen miles west of Lumpkin, Georgia, and about three miles north of Roanoke—a community burned by Creek Indians in 1836—a group of planters and merchants created Florence in 1837. Henry W. Jernigan, a local militia officer during the Creek War, and his associates purchased six hundred acres. Jernigan and his partners paralleled the actions of the West Point Company. The twenty-eight investors in the Florence Company envisioned more than a small trading center and a riverboat landing. They saw Florence as a rival to Eufaula, if not to Columbus. Some of the early settlers toyed with calling it Liverpool after the largest cotton-trading center in the world. The rich lands on both sides of the river and its crop served as the economic impetus for this urban venture.[26]

In 1837 the Georgia legislature chartered Florence, its academy, and a ferry. Jernigan, as an agent for the Florence Company, and Matthew Everett, the owner of the Alabama land, gained the right to operate the ferry. This crossing must have attracted Alabama cotton because in April 1838 the Florence *Georgia Mirror* boasted of shipping five thousand bales of cotton harvested in the fall of 1837 from the new town.[27]

The presence of a newspaper in this nascent town underscores its founders' aspirations. The *Georgia Mirror* documented the town's other institutions: the Florence Bank, the Phoenix Hotel, two cotton warehouses, two churches, a male and a female academy, a tanning vat, an early cotton seed mill, and thirteen physicians, as well as the usual stores and small businesses.[28] Jernigan, who acted as the town's first intendant, or mayor, and one of its cotton merchants, oversaw the creation of Florence's Independent and Female Academy in 1838. His ad "To Contractors" provided detailed specifications for a two-story building, the upper for a girl's academy and the lower for an independent church. He closed his description with an entreaty echoed by many public officials: "Intending it [the school and church] not only as a convenience, but as an ornament to our town, the Trustees are opposed to having it botched." Perhaps Godwin responded to Jernigan's plea and that building began their relationship.[29]

Godwin and King built the essential ingredient for this booming cotton town—its bridge to the fertile Alabama land, including Jernigan's plantation. On April 2, 1838, the *Georgia Mirror* emphasized the need for such a span: "It is folly to slumber over this matter." A large volume of Alabama cotton "will, if a bridge is completed, undoubtedly come up to Florence and with it will come an immense trade, which will greatly advance the interest of our young and flourishing town." Without this crossing "this cotton and trade will be taken, almost from under our very eyes, either to Columbus or Irwinton." The editors also noted that Florence lay on the most direct route between Milledgeville and Macon to the east and Montgomery to the west. The daily mail between those places would cross the Chattahoochee at the Florence bridge. "Will we shut our eyes to the light that shines so forcibly? We think not. We believe we have men at the head of affairs here, who know their duties and will promptly discharge them."[30]

Actually these men failed to act promptly. Chartered in 1838, the Florence Bridge Company failed to organize within the thirty days prescribed by the law and repetitioned the legislature the next year. Their charter resembled the Irwinton one, with capital stock up to $75,000 and the toll rate as in Columbus. The legislation also provided for a punishment, from four to ten years imprisonment, of any person found guilty of destroying or damaging the bridge. Sometime during late 1839 or early 1840 Godwin and King began constructing the bridge. Given his subsequent purchase of this span, Godwin more than likely accepted one-third of the company's stock as his payment for building it. Since the price of this bridge equaled those in West Point ($22,000) and Eufaula ($22,500), Godwin probably received about $7,000 worth of stock in payment.

In tune with its competition in Irwinton and resonating with the speculative fever of the 1840s, the Florence company officers, under the liberal provisions of Georgia law, also created a bank that issued notes and change bills. When travelers paid in gold or silver, they received their change in the form of paper bills issued by the company. If someone collected five dollars worth of change bills and presented them to the company, then it was obliged to exchange them for five dollars in specie or hard money. Such transactions rarely occurred, however; the bridge keeper never kept specie at the tollhouse. More often, when a bridge user tried to pass these pieces of paper in a store or inn, especially outside of the area, the merchant greatly discounted the value of the note, and the amount of discount increased proportionately with the distance from the bank or bridge. Realistically, bearers of these change bills could only use them to pay their tolls. One of the surviving notes, worth two bits, depicts a bridge—two short through-trusses supported by stone piers in the midst of a New England village. This generic scene, used on similar currency throughout the nation, provides no information about the actual bridge at Florence.[31]

The Florence bridge resembled the other Godwin-King spans over the Chattahoochee River, except for the configuration of its approach, or land bridges. Although the exact location of the Florence bridge remains unknown, neither riverbank has high bluffs in that general area. In order to allow river navigation, the trusses must have rested on piers higher than the riverbanks; therefore, both approach bridges climbed a long incline. By August 1840 Jernigan advertised, "The superior Bridge across the Chattahoochee River at Florence, built by 'honest' John Godwin and Horace, is now ready for crossing." Obviously, King must have constructed this bridge, while Godwin attended to other matters. Jernigan lived near the bridge and probably developed a friendship with Horace as he watched him construct the span. Thus Jernigan recognized the role of this African American contractor in the bridge's construction.[32] Certainly the citing of a slave craftsman by name in a newspaper ad was extremely rare in the antebellum South. King's skills had become well known throughout the Chattahoochee Valley.

Despite the unique status of its builder, the Godwin-King bridge in Florence barely survived the Harrison Freshet in 1841. The waters rose so high that "[t]he Steamer *Siren*, in her passage up[river] came around the Florence Bridge, over an adjacent field, and pursued the boisterous tenor of her way the greater part of the distance over submerged plantations." As with the Irwinton bridge, the river washed out the foundation under one pier. Godwin and King rebuilt the failed pier but the town of Florence languished in a manner similar to the earlier Sneedsboro, North Carolina, where Godwin's father had invested in urban lots.[33]

The merchants and planters of western Stewart County had picked an inauspicious time to begin a new town. Conceived as the nation plummeted into the Panic of 1837, Florence was almost stillborn. Owners of many of the houses being built in 1837 and 1838 never finished constructing them. In 1840 and 1841 members of the Florence Company filed several lawsuits against purchasers who had failed to pay.[34] While many owners lost faith in the place, the Godwins remained convinced of its potential and continued to invest in its bridge. Or perhaps all nineteenth-century bridge builders dreamed of eventually owning a bridge and retiring on the profits from a steady flow of tolls.

In 1843 Wells and John Godwin purchased the entire Florence Bridge Company for $17,000 from Matthew Everett, the firm's president and presumably the same man who owned the initial ferry with Jernigan. The Godwins might have already controlled some of its stock, since the purchase price was $5,000 less than the cost of contemporary bridges at Eufaula and West Point.[35] This action was one of Godwin's most significant economic decisions, one that may have haunted him until his death in 1859.

The years from 1832 until 1842 had been an economic roller-coaster ride for most planters and merchants in the Chattahoochee Valley, and the Godwins had experienced the same vicissitudes. Their fortunes had fluctuated since 1832. During the Panic of 1837 John transferred his most valuable asset, Horace King, to his wife and her uncle. Perhaps debtors were knocking on the door at that time, and on other occasions the Godwins certainly appeared to be short of working capital. In 1842 John borrowed money from his wife's uncle, but in the same year he bought extensive property in Girard. Presumably, 1843 represented a prosperous year for John and Wells. By 1843, with the economy rebounding and with payments from numerous projects, John and Wells felt confident enough to stake their economic future on their covered bridge in what they hoped would be a frontier boomtown.

The amount of traffic over the Florence bridge never met the expectations of its promoters or of Godwin. At age forty-five, John seemed to be slowing down. He contracted for fewer major projects. The construction of new bridges required geographical mobility, and he lacked the inclination or the energy to pursue them. Rather than traveling to new sites, John sent King to work on projects in other areas. Perhaps John planned to retire on his share of Horace's earnings and the profits from Florence. Then in 1846 a flood washed the Florence bridge away, dashing John's hope of easy income.

During the period from 1832 until 1842 Horace honed his skills as a builder by constructing five major bridges (Columbus twice, West Point, Irwinton [Eufaula], and Florence) over the Chattahoochee River, a large and a small courthouse, warehouses in Apalachicola, a breakwater, and probably numerous houses. At the same time he gained a public reputation as being an excellent bridge engineer and contractor. While Godwin's level of activity declined after 1842, Horace was eager to work in a larger geographical area. From 1843 until 1845, in partnership with Robert Jemison Jr., this slave applied his skills to projects in northeast Mississippi.

CHAPTER EIGHT

ROBERT JEMISON JR. AND HORACE

I have also engaged the services of Horace the Boy who was the chief architect in the building of the Bridge at Columbus & who has been the most extensive and successful Bridge Builder in the South.
Robert Jemison Jr. to Jas. Smith, 27 September 1845

From the late summer of 1843 through December 1845, while John Godwin stayed in Girard, Horace King worked as his own boss, raising two major bridges at Columbus, Mississippi, and Wetumpka, Alabama, and small spans near Luxapalila, Mississippi. Horace, whose legend places him firmly within the Old South, was actually a man of the frontier, riding the wave of westward expansion and building new bridges that were essential for the cotton economy.

The manner in which Godwin employed King during these years was explained at the end of Horace's life: "Mr. Godwin, during slavery, owned a number of colored mechanics—all bridge builders. Among them Horace, who was his foreman, and who drew all the plans and specifications, and made the contracts for building bridges. Mr. Godwin had the utmost confidence in Horace's judgment and ability. So much so, that he would send him with the hands to any part of the south to make contracts for building bridges." By 1843 Horace's skills had reached such a level that Asa Bates, who acted as a partner to Godwin and King on the Columbus, Mississippi, bridge, took orders from Horace on that project.[1]

The most important man in King's life during this period, however, was Robert Jemison Jr., the Tuscaloosa entrepreneur. Because of the surviving correspondence between Jemison and his contractors, Horace King and John Godwin, this two-and-a-half-year period is the most documented phase of Horace's life. These writings illustrate the unique status of King as a slave. Jemison's letters document Horace's level of skill, his value as a craftsman, his freedom of movement, and his warm personal relationship with Jemison. What began with Jemison's need for a skilled architect-builder blossomed over time into a true friendship. Jemison provided the political influence necessary to emancipate Horace in 1846, and Horace's income from Jemison's projects probably allowed King to buy his freedom.

Robert Jemison moved from central Georgia to Tuscaloosa with his family in 1822. At the death of his father in 1826, Jemison inherited considerable wealth, much of which had been acquired through land speculation, a practice continued by the younger Jemison. He lived for ten years on his father's Garden Plantation in Pickens County west of Tuscaloosa. In 1836 he moved to Cherokee Place, a four-thousand-acre plantation three miles west of Tuscaloosa on the Black Warrior River. Eventually he owned six plantations with more than two hundred slaves, eleven thousand acres of land, and timber rights for thousands of additional acres. While Jemison devoted much of his land to cotton cultivation, he also acted as an investor and entrepreneur in a myriad of economic ventures in the region around Tuscaloosa and northeast Mississippi. In 1860 he built his urban Italianate mansion in Tuscaloosa, which various writers still insist on calling his plantation house.[2]

Many of Jemison's investments involved making money from various forms of transportation. The slaveholding elite often owned ferries, but Jemison took his involvement to a higher level. In 1834 his sawmill supplied $5,000 worth of lumber for the construction of the Tuscaloosa bridge. The next year, in the process of collecting a debt, Jemison fell heir to a stagecoach, its slave driver, and its team of six horses. Consequently, he entered this business. A shrewd businessman with deep pockets, Jemison forced his competitors to combine as the stage line of Jemison, Powell, and Ficklin, with Jemison as the dominant player. Within a short period, this stage company carried the largest volume of mail and passengers in Alabama and ranked among the largest in the South. Jemison also bought and built plank roads, ferries, and, of course, bridges. In the prerailroad economy, the horrid condition of most roads allowed an entrepreneur to reap sizable profits from tolls on improved roads and bridges.[3]

In 1837, a year after Jemison moved back to Tuscaloosa County, his neighbors elected him to the Alabama legislature as a Whig, the party of many elite planters. The Whigs had formed in opposition to Andrew Jackson and his more democratic supporters, who attacked the elite for using governmental power to gain monopolistic economic favors. While Jemison never enjoyed the power to obtain monopolies, he used his legislative position to aid his stage lines, obtain waterpower rights, and acquire lucrative lumber contracts. Jemison's correspondence always dealt more with economic issues than political controversies, especially after the Panic of 1837. The ensuing depressions lasted until 1844, and cotton prices, the primary barometer of the southern economy, remained low until 1846. As a legislator during this period, Jemi-

Robert Jemison Jr. (1802–71). A wealthy entrepreneur in Tuscaloosa, Alabama, Jemison pursued diversified economic activities including agriculture, mining, milling, textile manufacturing, stagecoaches, ferries, and bridges. His political career as a moderate Whig reached from the Alabama legislature to the Confederate Senate. King and Jemison's symbiotic relationship began in the arena of bridge building and grew into a warm friendship. Jemison pushed Horace's emancipation through the Alabama legislature in 1846. *Courtesy of Alabama Department of Archives and History, Montgomery, Alabama.*

Robert Jemison's Italianate mansion (built ca. 1860), Tuscaloosa, Alabama, 2003. Many authors identify this as a plantation house even though it stands in the middle of the city. King visited Jemison's home on several occasions. If their correspondence is any indication, Jemison probably greeted King more as a colleague than as a black man. In the early 1850s Jemison tried unsuccessfully to convince King to move to Tuscaloosa. *Photograph by John Lupold.*

son sought banking reform to improve the reputation of Alabama banks since many of them had collapsed after the crash.[4]

During the 1840s Jemison broadened his investments and moved even more of his capital from planting cotton to other enterprises; he also encouraged others in the area to make similar changes. His actions belie the popular stereotype of southern planters—blindly refusing to stray from their dependence on growing cotton, buying slaves, and acquiring more land. In fact, scholarly studies of other wealthy entrepreneurial slaveholders testify to their willingness to shift their capital, especially during this decade when cotton prices plummeted. The first boom in building southern cotton mills came during the 1840s. Some wealthy planters, like Farish Carter of Milledgeville, Georgia, moved their money and their slaves to textile mills during that decade.[5]

Jemison continued to diversify his holdings even after the economy recovered and cotton prices rose in the late 1840s. His holdings eventually included a Tuscaloosa hotel and, after he foreclosed on an editor, a local newspaper, though his role as a publisher lasted for a only brief period. His enterprises also produced pig iron near Talladega and coal in Brookwood. One of his plank roads linked the latter with Tuscaloosa in the early 1850s. Within the city he controlled a foundry and the city's only textile mill, which eventually failed. The production of lumber, flour, and cornmeal occupied a central position in his economic ventures. He continually worked to improve his three major grist- and sawmill complexes near Tuscaloosa, Pickensville, and Columbus, Mississippi.

Jemison purchased large tracts of the first lots sold in Pickensville, and Columbus. He sought to provide access to those towns and to supply some of their basic necessities. When moving westward from Tuscaloosa toward these towns, travelers rode on his turnpikes and perhaps on his stagecoaches. The turnpikes were short affairs that spanned swampy areas and creeks like Coalfire, Sipsie, and Lubub, but they could be very profitable. The receipts from one of these ranged from $1,000 to $2,000 per year.[6]

The grist- and sawmills of Jemison played an important role in settling these areas. One of his largest grinding and sawing complexes operated at what is now Steens, Mississippi, on the Luxapalila River (actually more of a creek) about nine miles northeast of Columbus. There, in 1838, Jemison paid $15,000 for 1,020 acres of land and probably expanded those holdings in later years. By 1843 his three-story brick gristmill at Steens contained two sets of stones for corn and the same number for wheat. He toyed with establishing a textile mill there. Horace would later play an important role in bridging the streams on either side of these mills.[7]

In 1845 Jemison wrote H. B. Gevatheney, a friend in Richmond: "I have

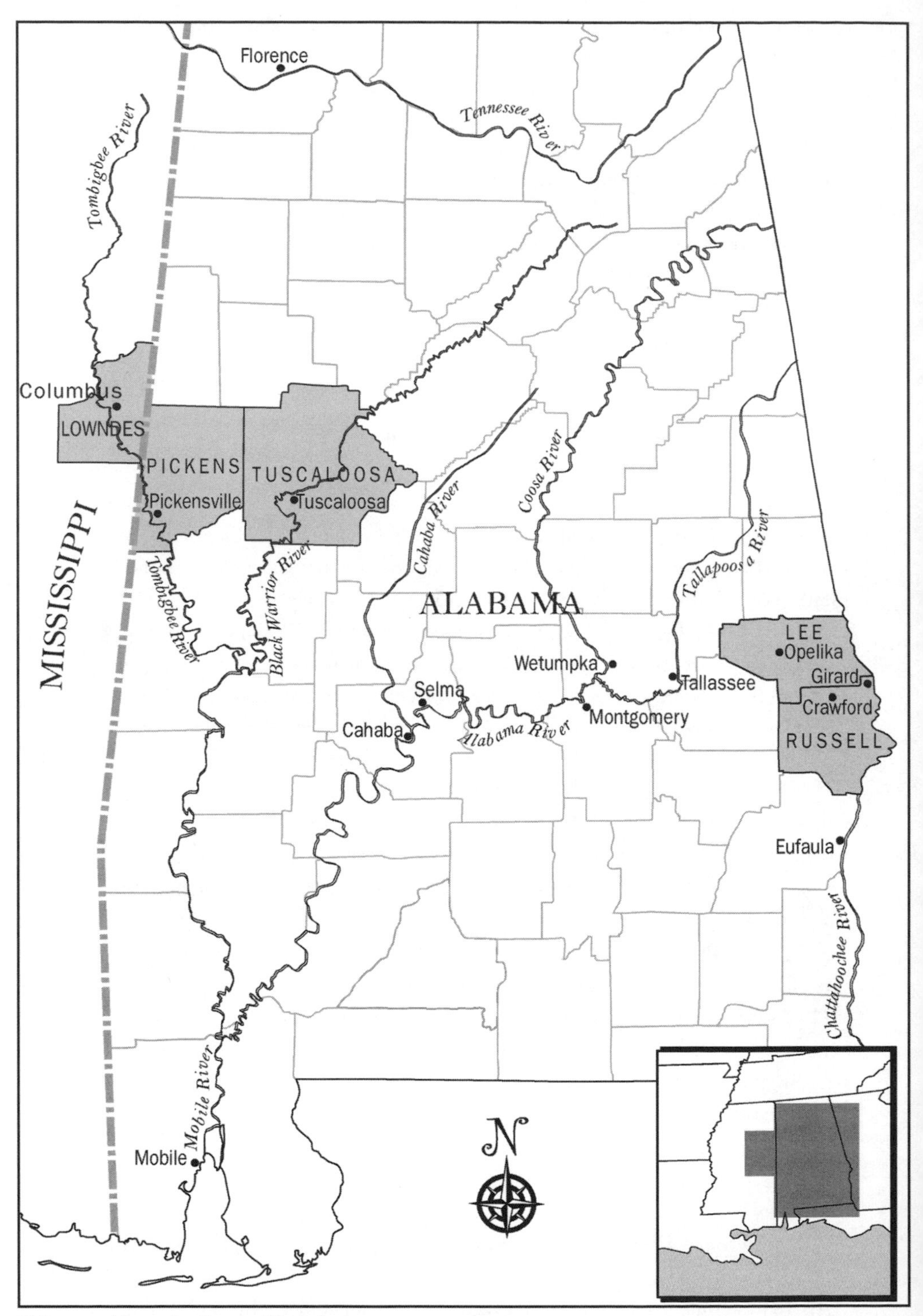

Alabama and Lowndes County, Mississippi

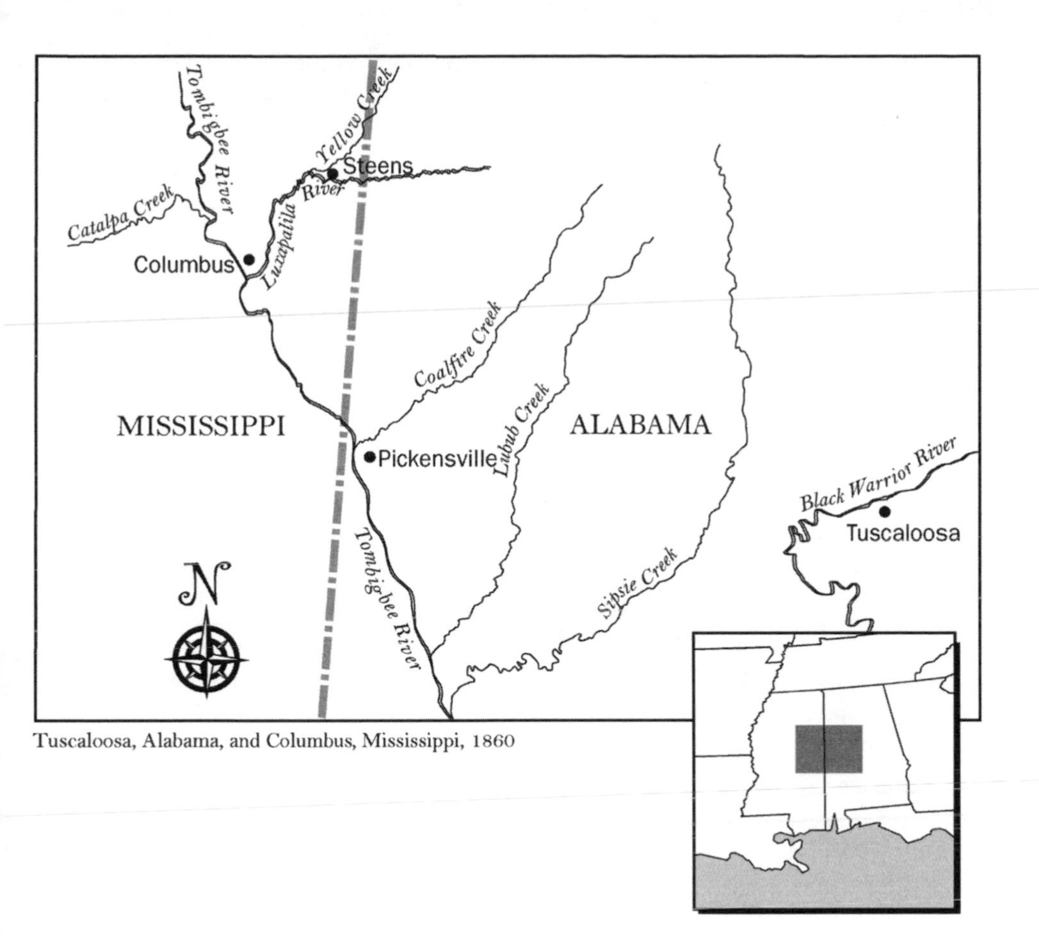

Tuscaloosa, Alabama, and Columbus, Mississippi, 1860

lately been directing my attention to growing wheat and building mills which I am confident will prove a much better business than cotton. At one set of my mills near Columbus [Mississippi] I shall manufacture this year which is my second year near 4000 Bls Flour & every prospect of a rapid increase. My greatest difficulty is the want of an experienced & practical miller." Jemison then proceeded to ask Gevatheney a series of questions about milling flour and whether he could find him a qualified miller. Notwithstanding the fact that Gevatheney had written Jemison to inquire about the cotton crop and that Gevatheney appeared to have little knowledge of producing flour, Jemison assumed his presence in Richmond, the South's largest flour milling center, would allow Gevatheney to find the answers to his queries.[8]

The problems faced by Jemison in attempting to mill flour in rural northeast Mississippi reveal the South's general technological deficiencies and hint as to why Jemison found Horace's skills so appealing. Stated in nineteenth-century terms, the South lacked adequately trained mechanics—millers, millwrights, foundry men, steam engineers, sawmill men, machine tool builders, and bridge designers. Jemison imported Horace from almost three hundred miles away to build relatively small bridges. Southern society acknowledged its lack of mechanics by accepting and even preferring northern men for these professions. Advertisements in southern newspaper placed by mechanics often mentioned their northern roots. No other group of Yankees in the South ever touted their heritage.

The geographical isolation of a rural society also produced problems in finding technological expertise. Jemison was trying to mill wheat where few if any skilled millers could provide informed advice. Jemison, even though a planter and a fine southern gentleman, was not a Luddite who hated technology, nor did he object to dirtying his hands fiddling with machinery. Jemison seemed to enjoy devising ways to modify the turbines for his waterpower mills. His correspondence was replete with references to his new wheels and his improved saws. Just as he sought information about milling from his Virginia acquaintance, many of Jemison's later letters to Horace asked advice about sawmill equipment. Jemison lacked readily available experts—a problem shared by other southerners who tried to employ new technologies.

Despite the paucity of skilled millers, Jemison envisioned his Luxapalila operation as a large merchant mill. In June 1846 Jemison wrote his commission merchants in Mobile, who apparently handled most of his cotton: "Never saw times so dull here. Corn crops promise to be good. Cotton sorry. The wheat crop has never been so good. In a very few years the manifest of Boats for this Section of country will present amongst the articles of Freight the article of Flour. I am not sure I will not send you some next season." While his Luxapalila flour never became the favorite brand of Mobile cooks, Jemison's sawmill on the Luxapalila played a crucial role in his integrated empire. The same millpond that fed his flour mill also supplied falling water for the turbines turning his saws. They produced some of the lumber for the Columbus bridge, in which Jemison became the major stockholder.[9]

The Columbus bridge project was begun ten years after the Columbus, Georgia, span, even though settlement began in Columbus, Mississippi, in 1816, twelve years before Georgia established its Columbus. This chronology illustrates the uneven spread of the white frontier in the Gulf South. The first occupants of the Mississippi town occupied land recently ceded by Choctaws and, like its Georgia cousin, this Columbus originally stood on the east bank of a river while Indians still controlled the land on the western side. Columbus, Mississippi, was more of a frontier town than its Georgia namesake and grew more slowly. Its bridge came later because the far bank of the Tombigbee was not open for settlement until 1836–37. Then, cotton planters flooded into the fertile area known as the prairie and created a need for a bridge.[10]

Unlike Georgia, the state of Mississippi did not loan bridge money to its nascent Columbus. As with other major bridges in this period, the capital for it came from private investors. Being the closest large town east of Columbus, Tuscaloosa or more specifically its financial capital played a prominent role in the development of Columbus. Jemison, realizing the possible stream of money he could earn from the bridge, joined or perhaps initiated the formation of the Columbus Bridge Company. Its February 1842 charter provided for a company capitalized at $50,000 to bridge the Tombigbee River within twenty-four months.[11]

The company's nine original stockholders included Jemison and his nephew Green T. Hill, who oversaw Jemison's stage line and his other business interests in Mississippi. The actual financing and construction of the bridge involved Jemison, who supplied the lumber; E. F. Calhoun, who kept the books and supplied goods and laborers as one of the major stockholders; Horace King, who, working for Godwin, built the bridge; and Seth King, who advanced the operating capital, just as he had for earlier bridges in Cheraw, Tuscaloosa, and a later one in Wetumpka.[12]

Godwin offered to construct the bridge for $6,000. He and Horace would have provided the design, the superintendent, the skilled mechanics or carpenters, and the unskilled gang laborers—slaves—who would have erected the bridge. Godwin's offer did not include any building supplies but presumably would have provided lodging and food for his workers. The stockholders, especially Jemison and Calhoun, decided instead to act as contractors. They hired Horace as a superintendent, the skilled carpenters, and the slaves or unskilled laborers and paid them, or their masters, wages by the day or the month. They finally realized the expense of this practice and before the end of the project began paying by the job. Jemison later admitted that the company spent between $3,000 and $6,000 beyond the $6,000 they would have paid Godwin.

With Jemison and Calhoun acting as their own contractors, Godwin sent King to design and supervise the bridge's construction. Jemison referred to Horace as both the architect and superintendent of the bridge. Asa Bates also worked on the bridge, apparently as a subcontractor for Horace. In a later discussion about the internal finances of the project, Jemison wrote Godwin to delineate the role of Seth King and E. P. Calhoun in the project: "Was it not your understanding that he [Seth King] was to advance funds for Mr. Calhoun. On this Branch of the Subject Horace & Mr. Bates I expect will know more than your self."[13]

This brief sentence illustrates several points. Godwin was not in Columbus, Mississippi, during the construction and had little knowledge of the company's financial dealings. Horace King, on the other hand, acted as more than a master carpenter and supervisor of work crews. He shared in the internal workings of the operation and understood the arrangement between the various investors or understood them as well as anyone did on this mismanaged project. Judging by the order their names appear in the sentence and by Jemison's later comments, King and not Bates was ultimately in charge of building this bridge, almost three hundred miles from their home base. King not only directed laborers but also managed subcontractors. Jemison never corresponded with Bates about future jobs. The King family tradition recorded a partnership between Bates, Godwin, and King that existed while King was still a slave; this venture represents one of their tripartite enterprises.[14]

Construction of the Columbus bridge proceeded slowly. The company formed in 1842; building apparently took place in the late summer and fall of 1843; and the span was passable in 1844. Management problems undoubtedly retarded the work, and Jemison might have had trouble supplying lumber in a timely manner. Rather than paying Jemison in cash, the company probably advanced him stock for the lumber, a typical practice in both the antebellum and postbellum South where cash money was always scarce. Perhaps Jemison and Calhoun sought to swell the company's debt to them, which expanded their share of the stock and increased their percentage of the annual profits. Thus the proceeds from Jemison's lumber continued to reap profits in bridge tolls for years.

Jemison's Luxapalila sawmill lacked the capacity to supply all the sawn boards for the Columbus bridge. For a later project Jemison estimated he could saw 2,000 to 2,500 board feet a day at Luxapalila. His figures for the Columbus bridge indicate that the lumber cost $5,000 at $17 per 1,000 board feet. Thus he supplied almost 300,000 board feet of lumber; that quantity would have required 120 days of sawing at Luxapalila. He must have transported lumber from Pickensville and maybe even Tuscaloosa. That additional transportation cost explains his charge of $17 per 1,000 board feet, when a more typical price was $10 per 1,000.

Once Horace received the lumber, poles, and timbers, he had no problem constructing the bridge. The interior and exterior dimensions of this timbered tunnel resembled the single Town lattice truss bridges he had built in Columbus, Georgia, ten years earlier. There both sides of the river have high banks, whereas at the Tombigbee the topography consists of a high bank on the town side and a much lower one on the west side. Consequently, the eastern end of this span started on a bluff with little or any land bridge, while the western side had a long sloping land bridge.[15]

The Columbus abutment and four wooden piers supported the four spans of the 420-foot bridge. The eastern truss extended from the abutment to a pier, about 40 or 50 feet high, located on the flat between the bluff and the

Columbus, Mississippi, Bridge, 1842 by Samuel H. Kaye. While still a slave, Horace King served as the architect and contractor for this bridge, almost three hundred miles away from his master. The topography of a high riverbank on the east and a low one on the west produced a high bridge (the river piers stood sixty feet high) with a long, steep land bridge on the west. This bridge failed by 1856. Seth King, the primary investor in the Cheraw, S.C., Bridge Company, and Robert Jemison Jr., who became one of Horace's best friends, provided the capital for this span. *Courtesy of Samuel H. Kaye.*

river. Two 60-foot wooden piers, without any stone or brickwork foundations, stood in the river and carried the covered sections to a fourth pier. It rose about 20 to 25 feet from the top of the west bank. From there, a sloping land bridge, from 160 to 175 feet long, connected the bridge to the ground.[16]

The bridge cost between $15,000 and $18,000. The uncertainty about the total figure stemmed from the chaotic or slipshod record keeping of the company. Unfortunately no records similar to those for the Cheraw bridge survived for the Tombigbee span, perhaps because they never existed. E. F. Calhoun, who had died by April 1845, kept the books "very loosely." He died in debt to the bridge company and his estate was insolvent. Jemison believed that Seth King, who claimed a one-third share in the bridge, still owed money to the company, but Jemison could not produce the account books to prove that point.[17] Consequently, Jemison wrote Godwin:

> Matters touching Columbus Bridge between Mr. [Seth] King & myself remain unsettled I am in consequence most awkwardly situated. After having unexpectedly had the chief burden of the cost thrown on me I find I have still to bear the brunt of all further difficulties. . . . Will you be pleased to write me at yr. earliest convenience fully the understanding between yr. self and Mr. King. Was he to have furnished all necessary fund, you to have done the work & I to have furnished the sawed lumber. Was there ever at any time an agreement between yr. self & Mr. King that neither of us should be interested in building any other Bridge across the [Sittle? Bigbie] river without the consent of the other two. When the first disagreement as to the price of building the Bridge occurred between yourself and Mr. King in Tuscaloosa on what terms did you propose to build the Bridge was there at that time any conditional contract made between you and myself by which I was to take your interest in the stock & pay you a certain price for building.[18]

After a series of questions, Jemison concluded by asking Godwin to solicit the opinions of Horace and Mr. Bates about these matters. The conflict between Seth King and Jemison continued for years. It affected the success of the Columbus bridge, Jemison's future involvement in bridge building, and, indirectly, the career of Horace King. The internecine warfare between its major stockholders represented only one of the problems faced by the company.

When the bridge opened in 1844, the company, over the objection of Jemison, adopted a yearly toll structure, because of local pressure and competition from the ferries. Under this scheme a family paid $5 per year, which included no carriage or wagon. The use of those vehicles cost another $10. This annual fee yielded the company a significantly smaller income. "The plan of taking subscriptions by the year is most unique & unjust & one repugnant to all my views & feelings," vented Jemison. "[C]ould I have foreseen that I should have had to resort to such an expedient as yearly subscriptions for all crossing to say nothing of other difficulties that have unexpectedly arose, I would have sought some other agreeable if not more profitable investment for my capitol [*sic*]."[19]

Competition with two ferries aggravated the toll problem. In December 1844 one of the ferries began charging a dollar per year as their fee. Acting in a manner that came to typify the transportation industry, the bridge stockholders tried to buy out their competition, but it backfired. They purchased one half of the ferry below the bridge from one owner, only to have the Mississippi legislature lodge the entire ownership of the conveyance in the other partner, the brother of the seller. Perhaps the lawmakers sought to block what some Columbusites saw as a grasping outsider, Jemison, and his bid to monopolize the river crossing. E. F. Calhoun, before his death purchased the upper ferry, but without the lower ferry, ownership of the upper remained irrelevant. Jemison claimed Calhoun and not the company contracted to purchase the upper ferry and refused to pay for it. Litigation followed. And the bridge tolls remained a nuisance to Columbusites. The editor of the *Primitive Republican* in 1851 lamented the paucity of cotton arriving in Columbus and attributed the lack of trade to "those pests of travelers, toll bridges!"[20]

High water also limited the use of the bridge. The span provided 50 feet of clearance in ordinary high water, but in those times of peak stream flow—most likely late winter and early spring—shippers from Aberdeen, thirty miles to the north complained that their steamboats could not pass under the bridge. During a flood in 1847 the river rose so high that one boat passed to the west of the bridge.[21]

With debts—including a $9,000 bank note to the Agricultural Bank—and controversy—internal squabbles, cutthroat competition, and public attacks on its tolls—the legacy of the Columbus bridge continued to be a negative one. By 1856 the bridge had failed for unknown reasons, and the antipathy of the company officers prevented any immediate effort to reconstruct the span.[22] Despite all the problems, Jemison enjoyed at least one positive association with this project, the work of its architect-superintendent, Horace King. When, in 1845, Jemison turned his attention to building smaller bridges in the same county, he once again called for Horace.

During the intervening building season, in the late summer and fall of 1844, Horace King built the Wetumpka bridge, again without the supervision of his master. Wetumpka, like both Columbuses, began as an urban center on the edge of Indian territory. It evolved at the head of navigation on the Coosa River below the Devil's Staircase, the tumultuous rapids at the fall line. Located on the western edge of the Creek Indians' domain, its early history was a mirror image to Columbus, Georgia. The Chattahoochee marked the eastern and the Coosa the western borders of the last Creek lands in the eastern United States. Even before Wetumpka became incorporated and named, a group of frontier structures—a gristmill, taverns, bars, hotels, blacksmith shop, post office, and so on—stood on the west bank below the falls. On the eastern side, on the fringe of the Creek nation, "Chicken Row" harbored similar desperados as those thriving in Sodom at the same time. The extension of state authority within the Indian territory excited Wetumpkans, who envisioned an important city booming on the Coosa, maybe even a site for the new state capital.

As in other frontier riverside towns, bridging the river became an essential task. By 1835 the Wetumpka Bridge Company, a private concern, provided easy passage to Chicken Row and the newly opened lands to the east. Their span survived high water in 1840 and the Harrison Freshet of 1841 but succumbed to a flood in the spring of 1844. At that point, Godwin sent King to supervise the construction of a bridge of similar "length & dimensions" as the Columbus, Mississippi, span. This 600-foot structure with four trussed sections and another 200 feet of land bridges cost the stockholders about $15,000, a considerably lower price than Godwin's earlier spans over the Chattahoochee. Perhaps King and Godwin secured the contract because King worked for a lower fee. The project involved the same investors as on the Columbus bridge the previous year. Seth King apparently financed the construction and Robert Jemison owned a sizable share of the stock. Horace's sturdy span, erected just below the Devil's Staircase, withstood the current of the thunderous river.[23]

The barely lit interior of the Wetumpka bridge spawned local folk legends. Illuminated by only three "dim and flickering" lanterns that failed to penetrate the shadows of the night, it became the scene of a criminal intent on retaliation. Under the cover of darkness, a former convict, seeking revenge on the judge who had sentenced him, removed two planks from the floor, hoping the jurist would fall through the hole. The bridge keeper, however, discovered the gap and alerted the judge. Since King usually laid double flooring in his bridges, the criminal must have expended a great effort in hopes of disposing of the judge.[24]

Also typical of other bridges, the Wetumpka span engendered disputes over tolls. Unlike the two Columbuses, settlement in Wetumpka moved—after Indian removal—to the eastern side of the river beneath the hill. Most business activities occurred on the east side and western residents paid the company's tolls. They urged the town to purchase the bridge and make it free. Apparently King's bridge stood until a flood in 1886. At that time, the company rebuilt the bridge of iron and steel, and maintained tolls on it until 1908, when the city forced the investors to sell the bridge. By that date the fate of

Wetumpka Bridge, Alabama, 1847, by Adrian E. Thompson. Horace King built this six-hundred-foot bridge in 1844 for fifteen thousand dollars. It stands at the bottom of the Devil's Staircase, the violent rapids where the Coosa plunges across the fall line. Seth King and Robert Jemison financed this span. Their conflicts over this bridge and the one in Columbus, Mississippi, which they also financed, limited Jemison's bridge building. This may have affected Horace's career. This bridge apparently stood until 1886.
Oil on canvas, 108.27 × 147.95 cm. Museum of Fine Arts, Boston. Gift of Maxim Karolik for the M. and M. Karolik Collection of American Paintings, 1815–1865. 48.482. Photograph © 2003 Museum of Fine Arts, Boston.

Wetumpka failed to match the aspirations of the original town boosters who lost their bid to become the state capital. To add insult to injury, over time steamboats traveling upstream tended to come only to Montgomery, the new capital, and not steam all the way to Wetumpka. Another frontier town with large expectations, Wetumpka failed to achieve her early dreams.[25]

While investors encouraged by urban boosters tended to build large bridges over major rivers, smaller bridges came within the purview of county governments. Rather than being joint stock ventures, these small toll-free bridges were financed by county taxes often in concert with large local landowners. Jemison's actions in 1844 show the role of the entrepreneur slaveholder in the southern economy. He used his slaves to build two small Lowndes County spans for the county, which also served his economic interests. A freshet in May 1844 swept away the two bridges leading to Jemison's Luxapalila mills. "As one of the Bridges lies between me and the only market for my timber [Columbus] and the other between my Mills and a large body of my timber, I am deeply interested in having them built on some permanent plan." At the same time Jemison petitioned county officials for the contract to rebuild these bridges, he wrote Godwin to secure the services of Horace King and to seek advice about the price of the construction.[26]

Jemison and the Board of Police of Lowndes County, the body that controlled roads and bridges, negotiated about several issues: whether to build one or two bridges and whether Jemison would construct them for a set price or just charge expenses. The board's primary concern was the larger bridge over the Luxapalila, which carried more public traffic and linked Jemison's mills to Columbus. Jemison sought some county support for the Yellow Creek bridge, north of his complex, that served to bring timber to his mill. Jemison proposed different types of spans, either a full-height covered truss or an uncovered half-truss. He offered to build both bridges for expenses only or for $3,000. In listing expenses for the board, he quoted the cost of lumber at $16 per thousand board foot, while in his letter to Godwin, where he suggested that they might go halves on the bridges, he cited a cost of $10 per thousand. The board members should have realized how he inflated that figure, since the Luxapalila bridge was adjacent to his sawmill and the Yellow Creek crossing less than a mile away.

The question of insuring the bridges also engendered debate. Jemison offered to guarantee the bridge, if it were a covered Town truss, for an unspecified number of years against water and wind damage "and everything else so far as the sufficiency of the plan and faithfull execution of the work" at 2 percent per annum. In other words, Jemison would insure the two bridges, if the county paid him that percentage of cost of the bridges every year. The county then wanted it protected against "fire, Insurrection, Civil War, or the

Yellow Creek Bridge, near Steens, Mississippi, 1936. In the 1930s this bridge was identified by local sources as being almost a hundred years old. However, King's bridge built in the 1840s at this site was not covered. *Photograph by James Butter. Library of Congress, Prints and Photographs Division,* HABS, MISS, *44-*STEN. *V, 1-2.*

acts of evil disposed persons." Jemison balked at that proposal. And the board refused to fund the smaller Yellow Creek bridge that served Jemison's interests exclusively.[27]

Since Jemison's bridge-building skills consisted of sawing lumber and raising capital, he needed a capable superintendent. He concluded his formal proposal to the board as follows: "All the foregoing propositions is made with this provision that I am enabled to obtain the services of Horace (the Boy who Superintended the erection of the Columbus Bridge) or some other practical Bridge Builder on Town's plan competent to superintend the work." King's competence, their personal relationship, and the lack of skilled craftsman in the area led Jemison to insist on bringing Horace three hundred miles to oversee the construction of a 170-foot bridge. Only a week after he submitted his bid, Jemison, anxious to employ King, implored the board chairman to give him a quick decision since Godwin had written that he could use King, if Jemison let Godwin know immediately.[28]

Jemison floated the idea of Godwin's taking "a half interest in the Contract." Godwin would supply Horace; in return, Jemison would provide five carpenters; and they would share all other labor and expenses. But Jemison, after making the offer, began hedging on the idea, noting that he might sell his mill or either one of them might die, a fear that often appeared in Jemison's correspondence. Apparently, Godwin only sent King and had no other financial interest in Jemison's bridges.[29]

Godwin offered King's services at $5.00 a day and that price included a young man, perhaps his brother, Washington, who accompanied King. That was an exorbitant wage for a slave. By comparison, a white, master carpenter in Savannah charged $3.00 a day for his services; $2.75 for his best slave carpenters (as compared to $1.98, the nominal rate for white artisans); and $.75 a day for semiskilled slaves (as compared to $.83 for unskilled white labor). For the Luxapalila bridge, Jemison planned to supply five carpenters and to Godwin he quoted their wages as one (probably Tom, who appears in other correspondence as his most skilled slave) for $2.00 per day, one for $1.50, and three for $1.00. In Jemison's proposal to the county, he included among the expenses the carpenter rates as one for $50.00 per month, the others for $25.00 (or about $1.00 per day), and $15.00 per month (or about $.60 per day) for his "own hands" or unskilled slaves. In 1855 Jemison rented two slaves at $13.50 and $6.50 per month with their master being paid at the end of the year.[30]

If the young man Godwin sent with Horace earned about $1.00 per day, then Horace's wages equaled $4.00 per day, $100 per month, potentially $1,200 per year, and realistically at least $800 per annum. These figures clearly indicate that Horace King served as a contractor, a bridge designer, and a superintendent while still a slave. According to King family tradition, someone offered Godwin $6,000 for Horace. That price appeared to be consistent with King's earning power; an annual income of $1,200 would bring his owner a 20 percent return on his investment. King's working in Mississippi garnered money for Godwin in Girard and perhaps earned Horace the money to buy his own freedom.

Part of the agreement between Jemison and Godwin for the hiring of King involved Jemison "paying their [Horace and the young man] stage fare from Montgomery & back to Montgomery when returning." Jemison and Godwin simply shared the cost of transporting King to Mississippi. But King and his slave assistant riding a stage from Columbus, Georgia, to Columbus, Mississippi, raised two interesting issues: first, their traveling alone as slaves across two southern states, and, second, their riding with white passengers. For a master to rent out his skilled slave was routine, but hiring out a slave at that distance was unusual. Because of the fear of insurrections, white southerners controlled the movement of slave and nonslave blacks. In this case, Horace was known and occupied a unique status, because Jemison owned the stage line.[31]

Other African Americans also rode Jemison's stages. In July 1853 he wrote a company official in Montgomery, "If a negro belonging to Daniel Stewart of this County applies for passage to be paid here [Tuscaloosa] you'll so enter him." The issue here was not the passage of a slave, but who would pay his fare. The question is where did Stewart's slave, Horace, and his companion ride—with the other passengers or outside with the driver? Any answer would be conjecture, but one of Jemison's exchanges with his sisters indicates that Africans did at least occasionally ride inside the coach.[32]

After his sister's visit to the Crystal Palace exposition in New York in 1853, she commented on the presence of Africans at the event. Jemison replied that he had observed the organizers were making "no distinction of color. That is upon the good old Christian maxim 'It matters not an Indian or an African, the Sun may have burned upon a pilgrim.'" He reminded her of "her refusal to permit the stage drivers negro mistress to get into the coach when you were but a child. I suppose if you have not learned to dislike or condemn such associations and amalgamations less, you have learned to suppress your feelings and opinions." Jemison seems less bothered by racial contacts than his sister and presumably that attitude was shaped by his involvement with stagecoaches. Perhaps as a businessman, he tended to ignore color and frequently used that Christian maxim to calm irate passengers.[33]

In point of fact the antebellum South did not practice racial segregation where the two races were rigidly separated in any spatial sense. In the cotton South, blacks and whites were constantly in contact with one another on a daily basis during their entire lives. While blacks did not maintain a separate existence, they were forced to display a subservient manner whereby their behavior indicated their inferiority to whites. Even King, the skilled builder, deferred to white passengers, if he rode inside the coach. Somehow King did it while maintaining his dignity. His light skin tone also made him more acceptable to his fellow travelers.[34]

The baggage carried by King and his assistant on their way to the Luxapalila certainly differed from that of the other passengers. While Jemison contracted to build bridges, he lacked the tools and the knowledge to construct them. As requested by Jemison, Godwin's men brought "a Trunnill machine & auger." The trunnel or treenail machine was probably a steel cylinder with an ax-sharp blade. Workers drove square pieces of wood into the trunnel machine to produce the round wooden pegs. The auger was probably an oversized one used for drilling trunnel holes through the truss members.[35]

Even before Horace left Girard, Jemison started sawing timber into lumber. He asked Godwin for a "Bill" or a list of the required boards for the floor, the roof framing, the stringers, and the weatherboarding. He did not ask about the crossed braces, since they were standard in a Town truss. Having sawn them for the Columbus bridge, he knew their dimensions and their approximate number. The uniformity of latticed Xs illustrates the standardization of the Town truss, which made it so popular.[36]

King apparently arrived at the Luxapalila by the middle of August, before all the boards were ready. He stayed there at least through December 6, 1845, perhaps even later. According to local tradition, Jemison built seven bridges in Lowndes County. During the fall of 1845 Horace probably supervised the bridging of the Luxapalila River, Yellow Creek, the Catalpa Creek on the road from Columbus to Starkville, and, perhaps, the span at "Kirks old Bridge."[37]

Jemison traveled from Tuscaloosa to his Luxapalila mills to meet with Horace and plan the various bridges. They conferred with William Reynolds, who oversaw Jemison's grist- and sawmills. Jemison certainly returned home, leaving Horace with full responsibility for building the bridges, a job he completed in cooperation with Reynolds, to whom fell the task of insuring an ample supply of timber. Ironically, the best time of the year to build bridges was the fall when rivers and creeks were low, which was the hardest time to saw lumber at water-powered mills. In August 1845 "the water is lower than ever known." Thus Jemison worried about producing enough boards in a timely manner, especially when his gristmills needed the same falling water. Reynolds, who had to balance the needs of the saw- and gristmills, also acquired the other bridge materials and supplies such as the shingles, which his mill did not produce, and the hewed timber. Jemison preferred cypress for both these items. Reynolds found the source of the poles and timbers, either on Jemison's or a neighbor's land. He then sent his crews, perhaps directed by Horace, to cut, debark, and hew these pieces.[38]

Jemison worried about the timing of all the interrelated aspects of the project and wanted to make sure that Horace's talents were properly utilized when he arrived. "If we have not enough sawed for him to commence framing," Jemison informed Reynolds, "we can take the hands & go to getting the hewed timber which by the by will be wanted as early as any other for the Piers ought to be got in as early as practicable." This passage confirms that King built his trusses on the land and then moved them, probably rolling them over falsework onto the abutments and piers. Had he assembled them over the water, securing the poles and hewed timbers would have preceded rather than paralleled the framing.

Unfortunately, once King arrived in Mississippi to begin framing, the correspondence with Jemison ceased and the details become less clear. King clearly built a 125- to 130-foot covered Town lattice truss bridge over the Luxapalila. It evidently consisted of two truss sections. The end of one truss rested on an abutment against a "soapstone bank" and the other end on a 30- to 35-foot pier in midstream. The second truss span reached from the midstream pier to another pier on a low-lying gravel bank. That pier supported the covered section and a 50-foot land bridge that descended to the road. Brick piers eventually replaced the original wooden piers, because more substantial structures were needed to guarantee the survival of the bridge as specified in the county contract.

Less certain is the origin of the Yellow Creek bridge. The county declined to fund its construction as part of the package that Jemison wanted, even though he guaranteed the county its cost only equaled half of the actual expense. Jemison proposed to build it for $500, and when the county refused to provide the funds he turned to neighboring landowners to raise the money as a subscription bridge. Even if they refused, he apparently built it because of his economic interests. This span only cost $500 because it was not covered and was a half-truss. The sides of a full Town lattice truss rose about 18 feet above the floor of the bridge. On this bridge the trusses only rose about 9 feet, so it had no covering. Instead, its open truss work more than likely consisted of a series of king or queen posts rather than Town's crossed Xs. It likely had only two spans of 30 to 35 feet each. The topography of the Yellow Creek site and its bridge plan paralleled that of the Luxapalila span, with a firm soapstone rock bank anchoring one side, a 30- to 35-foot pier in the stream, and another similar support rising from a low gravel bank on the other side. Again, a 30- to 40-foot land bridge finished the structure on the side of the lower bank. Because of the short length of the spans over the water, rebuilding the bridge, if it failed, was cheaper than incurring the expense of a covered span, which would have survived for more years. Undoubtedly, Horace built it.

On December 6, 1845, Jemison completed a Lowndes County–funded bridge across Catalpa Creek on the road from Columbus to Starkville, Mississippi. He posted a $2,240 bond that warranted the bridge for five years except "against fire, Tornadoes, Insurrections, the acts of evil disposed persons & the like." Judging by the bond, this span may have been similar to either the Luxapalila or the Yellow Creek bridges. Horace undoubtedly built it because he was still in Lowndes County at the time. On the same day as the bond was executed, Jemison instructed Eli Abbott of Columbus, who had cosigned the bond, that "Mr. Reynolds & Horace are authorised to make a contract for me" at the site of "Kirks old Bridge." Note again the role of King as a contractor while still a slave. If that contract were made, King probably stayed to complete the span, making a total of at least one major and four smaller bridges he constructed in Lowndes County, three hundred miles from home, during a three-year period from 1843 to 1845.

King left more than new bridges behind when he returned to Girard. He left a group of skilled slave carpenters who knew how to build small bridges. On Christmas Day 1845 Jemison wrote Reynolds to have Tom, presumably his most skilled carpenter, and the "Boy's to plank up the pier on this side of the river at Columbus Bridge." When that was "well done two or three feet above high water mark . . . they can all come over." They must have been strengthening the piers. These workers were probably the five carpenters Jemison supplied to build the Luxapalila and other bridges. They then returned to Cherokee Place near Tuscaloosa, Jemison's residence, where he apparently kept his most skilled craftsmen. During the previous four months they had greatly increased their skill level by working with Horace King. While they missed Christmas with their families, as skilled slaves they enjoyed mobility that most southern blacks never had. They could return home at their own pace when they finished their work.

These craftsmen most likely continued to construct minor spans within the county. Local tradition credits Green T. Hill with building several bridges in the area. The accounts make Hill appear to be a builder, while in fact he was Jemison's nephew and oversaw Jemison's interest in Columbus, especially his stage operation. Hill's acting as a contractor for bridges is consistent with the family's business interests. In 1860, perhaps using Tom or other carpenters who earlier worked with King, Hill acquired the contract to rebuild the bridge at Jemison's Mill over the Luxapalila for $1,500; thus their span superceded the one built by Horace.[39]

Another covered bridge replaced Hill's in 1885. Yet in 1925 Charles Wood in his brief history of the county noted that two of Jemison's bridges, presumably at Luxapalila and Yellow Creek "are still standing and giving excellent service." One was two generations removed from Jemison's bridge, and his bridge over Yellow Creek was not "housed-in" as Wood noted. Similarly in 1936 the Memphis *Commercial Appeal* declared the Luxapalila bridge to be ninety-four years old and the oldest in Mississippi. Thus is the legend of covered bridges spun; any covered bridge is thought to be antebellum, even though most of the ones that survived into the 1920s and 1930s dated from after 1880 and often from after 1900.[40]

While King's bridges did not last forever, his stint as a bridge builder in Mississippi played a pivotal role in shaping his life. While there, he acted as his own master, planning construction, supervising white and slave labor, deciding the terms of contracts, and handling a myriad of details. He also became a closer friend to Robert Jemison Jr. In the next session of the general assembly, after King had spent almost four months working for him, Jemison steered the emancipation legislation through the Alabama House and Senate that freed King on February 3, 1846. While Godwin made the decision to manumit King, Jemison provided an essential service. The law made it difficult for a southern slave owner to free a slave. The Alabama legislature only manumitted two slaves during its 1846 session. Godwin's relationship with Jemison made it relatively easy to emancipate King.

CHAPTER NINE

FREEDOM

Please let me hear from you as early as possible as others want my Services and I wish to accommodate You before any other man.
Postscript, Horace King to Robert Jemison Jr., April 6, 1871

Nothing about Horace King's life is more obscure than his manumission. The following account untangles myth from fact, but in the end exactly how or why he was freed remains unknown. The primary characters in this script were his master John Godwin, his de jure owner, Ann Godwin, William C. Wright, Horace himself, and his friend Robert Jemison. Horace was an active participant in his bid for freedom; he did not simply wait for the "privilege" to be handed him. The freeing of Horace was probably motivated by economic concerns rather than kindliness. None of the individuals involved, including Horace, opposed the institution of slavery. Once Horace was free and had joined the small group of Alabama nonslave blacks, his life changed very little for a decade. He continued to live in Girard and contract for the same type of work.

The freeing of Horace King by the Alabama general assembly on February 3, 1846, was not a routine legislative act. John Godwin, Ann H. Godwin, and William C. Wright requested the action so King could remain in Alabama. According to its codified laws, free blacks could not enter the state, and any black freed within the state must leave within a year unless the state legislature exempted him or her from that requirement. White Alabamians restricted the freeing of slaves because they feared nonslave blacks. Both northern and southern whites saw free African Americans as the most dangerous class of Americans.[1]

In the North racism motivated actions against free blacks. In the South those prejudices were reinforced by fears of slave revolts led by free blacks. White southerners made no distinctions between slaves and nonslave blacks. Everyone of African descent represented a potential insurrectionist. Hence, while a slave master could inflict almost any imaginable punishment on a slave—even death—and usually not be punished by the law, a master could not emancipate a slave because freedmen threatened a society floating on a cauldron of simmering racial animosity. At any time that hostility, especially if fanned by northern abolitionists, could erupt into a race war such as occurred in Haiti in the 1790s. The prejudice against free blacks was reflected in the fact that their population growth was only one half that for whites and slaves during the decades preceding the Civil War.

Given the racial fears of whites, the freeing of any slave, even Horace King, was complicated and required a legislative act, if the slave wished to remain in the state legally. But legal codes never took precedence over personal relationships in the Old South, even when African Americans were involved. Local customs, family reputations, and codes of honor rather than law dictated behavior in antebellum Alabama. At that time the law required a bond to be posted to insure that the freed slave did not become a burden on the state. But practice rarely conformed to those regulations. The historian Gary B. Mills, after a detailed study of free blacks in Alabama, concluded, "that almost 50 percent of . . . Alabama's free Negroes were either illegally manumitted or illegally residing within the state." Half the free blacks in Alabama could never produce valid freedom papers.[2]

Godwin could have freed King by merely recording the transaction in the county court records. Why go to the trouble of petitioning the legislature? Why manumit Horace when the life of a free—or more accurately—a nonslave black remained almost as circumscribed as that of a slave? Furthermore, Horace already enjoyed de facto freedom so why go to the trouble of securing de jure freedom? He spent much of the three years before 1846 in the area of Columbus, Mississippi, and Wetumpka, Alabama, working independently.

Many local historians and family members of the King and Godwin families have concocted their own story of Horace's manumission, but it remains the most misunderstood part of his life. One fragment of the Horace King legend and the King family tradition has him traveling to Ohio before 1829 to receive his freedom. Then, he migrated to Alabama and the 1846 act was passed to allow him to make contracts within the state. Certainly Alabama would never have recognized King's emancipation in Ohio.

John Godwin plays the predominant role in most of these emancipation scenarios. Godwin's motives have been ascribed to rewarding King for a specific heroic act. Godwin's debts or his declining health has also been viewed as the root cause. J. M. Glenn of Midway, Alabama, who claimed kinship with Godwin's family, asserted that King's freedom came because of the rapidity with which he rebuilt the Columbus City Bridge after the Harrison Freshet in 1841. If that were true, King should have been emancipated in the summer of 1841 rather than the winter of 1846. According to the Rev. Francis L. Cherry, Godwin "became so involved that he lost nearly or quite all his property, except Horace," and to protect King from being seized by creditors, Godwin freed his skilled slave—but only after he had rejected a six thousand dollar offer to purchase Horace. Cherry exaggerated Godwin's losses; he never relinquished his homestead or his most skilled slaves.[3]

Technically, the 1837 transfer of Horace and his family to the ownership of Ann H. Godwin and her uncle William C. Wright should have protected King from John's creditors. Maybe the advancing age of William Wright, who died in 1851, necessitated a new arrangement to safeguard Horace. John Godwin may well have experienced economic problems in 1846 with the destruction of his bridge at Florence, Georgia, but not of the magnitude suggested by Cherry. And the Florence calamity occurred after he decided to free King. Ann and William C. Wright, who controlled Horace after 1837, are the forgotten actors in this play. By 1839 Ann and her uncle allowed King to enjoy an essential freedom—the right to marry a free woman and to have free children. That privilege probably meant more to Horace than his own freedom in 1846, or perhaps the children's freedom encouraged him to change his freedom from de facto to de jure. Ann obviously had to agree to his obtaining his freedom, and she probably helped to initiate it.

Local Wright family tradition points to the involvement of William C. Wright in financing a bridge in Tallassee, Alabama, about 1845, which could be linked to Horace's freedom. Located at the fall line of the Tallapoosa River, the next major river west of the Chattahoochee, Tallassee had two antebellum bridges. About a mile south of the falls, at what was called Old Tallassee, J. H. McKenzie, an early settler and planter, operated a river ferry, a gristmill, and a sawmill. About 1845 he acted to bridge the river and apparently entered into a partnership with Wright to help create this span. Wright continued to have an interest in the bridge and sent his nephew Joseph James Wright Jr., Ann's brother, to manage the bridge operation. The younger Wright married Mary, J. H. McKenzie's daughter. Certainly Horace built this bridge, given the Wrights' involvement. He may well have built other small spans in Alabama under the sponsorship of Wright. The timing of the Tallassee construction meshes perfectly with King's return from Mississippi. Also, Godwin appears to have stopped working on major spans by this period. Perhaps, King's pay for this work came in the form of manumission.[4]

Most accounts of King's manumission have him playing a passive role. But in truth, Horace King decided he wanted freedom, and he caused it to happen. After the Civil War, as part of the 1878 hearing regarding his wartime loyalty to the Union, King gave a brief autobiographical sketch, which sounded similar to the stories told by most freed slaves. "I learned the trade of bridge building, worked at leisure time[,] made money and bought my freedom in 1848." Urban and plantation slaves made baskets, furniture, boats, and countless other small items on their own time and then sold them. With hard work and frugality a very small percentage of slaves bought their way out of slavery. However, building bridges during one's leisure time hardly seems feasible. By some arrangement, King retained a certain percentage of his earnings, perhaps during the period from 1843 to 1845 when he worked for Jemison. Or maybe King bought his freedom from working on the Old Tallassee bridge for William C. Wright. The fact that he earned money as a slave was well known. In 1859 the *Union Springs Gazette* noted that Horace had "accumulated a comfortable competency" both "during his servitude" and after. Like Horace, as many as a third of the free blacks in Ohio owed their emancipation to self-purchase, as did many free blacks in Alabama.[5]

The masters of some other well-known nonslave blacks simply freed their chattel. The barber of Natchez, William Johnson, and William Ellison, the South Carolina cotton gin maker, acquired their freedom in that manner. Ellison changed his name from April, a typical slave appellation, to William to increase his level of respect within the community as he started his career as a mechanic. Horace kept his first name, because it already identified him as the best bridge builder from the Chattahoochee to the Tombigbee. Horace probably pushed to be recognized as King rather than Godwin. In April 1846 he received mail at the Columbus post office addressed to Horace Godwin, but not from his friend Jemison, who always addressed him as King.[6]

Robert Jemison played a pivotal role in King's freedom, hence the debt King noted in his 1871 correspondence: "I wish to accommodate You before any other man." Senator Jemison, who had already served in the House, certainly guided the manumission bill through the hurdles of both bodies. The *Union Springs Gazette* in 1859 mentions that Godwin "at considerable trouble and expense, obtained the passage of a bill granting him [King] his freedom."[7] In actuality, Jemison's position meant Godwin expended neither effort nor money to free him. King's Tuscaloosa friend probably encouraged the emancipation, so he could better employ King on his various projects spread across two states.

Jemison took a role in freeing King because of his business needs and his personal relationship with Horace, not because of any reservations he had about the institution of slavery. Jemison had no qualms about African American servitude. A year after Horace was freed, Jemison wrote J. D. Watson, who was handling the sale of some of his slaves: "The boy Washington if not yet sold, I think I can sell here for five Hundred Dollars," and Jemison thought he could get $3,000 "for Jupiter and family and Madline."[8] On another occasion, when his sister needed money for her wedding, he complained of the difficulty in obtaining cash money because the Black Warrior River in Tuscaloosa "is now at summer even low summer stage. I cannot send off either cotton or lumber." Thus Jemison instructed his business associate in

Columbus, Mississippi, to sell Jess and send the proceeds ($1,000) to his sister. Later, in 1860 Jemison received $1,400 for "a negro man slave Tom aged about twenty eight years." Tom committed an unspecified offense and Jemison sold him to Texas with the stipulation that Tom could never return to Alabama or Mississippi. Despite all the kindness Jemison showed to King, he disciplined Tom in the harshest way possible, short of death—he separated him from his family by exiling him to Texas.[9]

Jemison allowed at least one of his slaves to purchase his freedom. Rube Jemison, a blacksmith at the Luxapalila mills, bought his freedom for $5,000. While either a slave or a freedman, Rube must have worked with King on various projects in the area. According to a 1930s account, Rube continued to work for his master until the former slave died. Jemison also allowed his skilled slaves freedom of movement, especially between Tuscaloosa and Columbus, Mississippi.[10]

His relationship with Horace transcended that of slave and master, however. Jemison treated him as a colleague as well as a fellow saw miller and builder. Exactly a month after the legislature insured King's freedom, Jemison sent Horace a "certified copy of the Act . . . made out on Parchment, thinking it most suitable and that it would be most agreeable to you in this form." After apologizing for taking so long to forward this document, Jemison discussed future business at some length and then closed in a chattier fashion than characterized most of his letters.

> Tell Yr. Master I recd. his letter a day or two ago and give to him and Napoleon my respects. Tell the old man [John Godwin] I may write him soon if I can find leisure. My health about as when you left. I think a little improved. My new saw continues to perform well. Bealle [a mutual friend] has started his on same plan. She [the new saw] also does well.
> In hast[e] yr friend
> R. Jemison Jr.[11]

Jemison, the stereotypical southern planter, respected this gentleman of color who shared his affection for sawmill technology. Jemison closed very few of his letters with "your friend." The real frustration in documenting King's interpersonal relationships is the lack of surviving correspondence between Godwin and King. Presumably, such letters would have sounded similar to this missive from Jemison.

A casual glance makes King's freedom seem unimportant, even to him, because his circumstances changed very little. Later in life, the exact date of his manumission seemed to elude him. He told the commissioners of claims in 1878 and the Reverend Cherry in 1883 that his freedom came in 1848 rather than the actual date of 1846. Once he was free, however, King stretched the limits of his free status. He purchased his own land adjacent to his former master in 1852, and later moved his family to Carroll County, Georgia.

With his freedom, King joined a small group of free blacks in antebellum Alabama who sought to maintain their status as nonslaves in a society where whites feared them, dismissed them as lazy, and viewed even mulattos as racially inferior.

Some historians suggest that former slaves became slaveholders in order to convince their white neighbors that they did not plan to threaten the social order. Their ownership of chattel confirmed their acceptance of the racial status quo. King never bought any slaves nor did he free his relatives, as many former slaves did. Instead, King worked his brother, Washington, who was still owned by Godwin, as well as Godwin's other chattel and the slaves of various other entrepreneurs. Horace King never attacked the institution of slavery. His livelihood depended on working with the white elite as part of the existing system.

Though they could own property, including slaves, the actions of free blacks were restricted under law. The Georgia Supreme Court justice Joseph Henry Lumpkin emphatically stated the inferior legal position of African Americans.

> [T]he African in Georgia, whether bond or free, is such that he has no civil, social or political rights whatever, except such as are bestowed on him by statute; that he can neither contract, nor be contracted with; that the free Negro can act only by and through his guardian; and that he is in a condition of perpetual pupilage or wardship; and that this condition can never change by his own volition. It can only be done by Legislation . . . the act of manumission confers no other right but that of freedom of locomotion . . . to be civilly and politically free, to be the peer and equal of the white man, to enjoy the offices, trusts, and privileges our institutions confer[red] on the white man, is not now, never had been, and never will be, the condition of this degraded race.[12]

Race defined slavery and limited the freedom of nonslave blacks. Lumpkin's strictures, which were the same in Alabama, technically applied to King when he later moved his family to Georgia. An oral tradition has King's friends sponsoring a law that made him a citizen of Georgia. However, the name Horace King never appeared in any Georgia legislative *Acts and Resolution* during the nineteenth century.

Alabamians, like other southerners, ignored many of the legal restrictions on free blacks and even on slaves. The writers of popular histories often emphasize the taboo against educating a slave along with the requirement of free blacks to carry proof of their freedom in a waterproof container. In actuality these laws operated at the pleasure of local elites. If a master needed a skilled slave who could read, then no one interfered with that slave's education, and it did not take place under the cover of darkness. Similarly, a well-known free black such as Horace seldom needed to verify his status. And in King's case, his travel occurred on the stage lines of his friend Jemison and later on the same train route between Newnan and Columbus.

Contemporary stereotypical images of free blacks presented them as "'lazy,' 'ignorant,' and more 'depraved' than slaves." This argument portrayed freed slaves as incapable of providing for themselves and as always being wards of the community. In Alabama the law required the emancipator to post a bond of one thousand dollars, with the judge of the Russell County court in the case of Horace King, to guarantee that the former slave "shall never become a charge to this State, or any county or town therein." Apparently, neither the Godwins nor Wright ever posted this bond, displaying a typical disregard of the law as it applied to free blacks.[13]

Given Horace King's income-earning potential, no one worried about his becoming a ward of the county. The economic success and prestige of free blacks depended on the reputation of their former master, their patrons, or both. Godwin's standing, as well as that of Jemison and the other men who employed King, insured his acceptance and guaranteed his freedom of movement. As a freedman King continued working for those old masters and for other prominent men. Prosperity characterized the South during the 1850s, and free blacks, including King, shared in this economic boom. The extent and exact details of his contracts are even sketchier than those for Godwin, but in the span of fifteen years, from the time he gained his freedom until secession, King became one of the wealthiest men of color in Alabama or Georgia.

According to state law, King, as a free black, needed a trustee or guardian for any legal transaction, such as a contract. While he seemed to ignore many of the restrictive laws on free blacks, he did use a guardian for his larger projects. In Alabama, Jemison represented him, and by the 1850s Dr. Alexander Irwin Robison signed for King in Georgia. Robison lived in Columbus by 1840 where he served as a health officer and an alderman in 1852–53. If he practiced medicine, he invested his profits in slaves. By 1860 he owned ninety-five slaves, making him the third largest slaveholder in Muscogee County.[14]

Robison and John Godwin were close friends and business associates. John's will named his son Messina and "my friend Alexander Robison" as the executors of his estate. Godwin must have hired Robison's slaves on various construction jobs, including bridges. Robison's situation resembled that of Jemison: he could make more money with his carpenters and skilled slaves building bridges than in other endeavors. He just needed a superintendent, such as King. As both a slave and a free black, Horace managed Robison's slaves and probably taught them the art of bridge building. Robison earned profits from King's contracts, and he most likely found him new jobs. Robison continued to use his slaves for bridge building. In 1862 he joined with Asa Bates and his son Thomas Jefferson Bates to build a county bridge at Woolfolk's on Upatoi Creek.[15]

Horace, as a freedman, also joined with a Columbus carpenter, James Meeler, on at least one project before the war, building a warehouse for John Fontaine and J. R. Clapp, probably to be used by Clapp's Factory. He and Meeler appeared to be equal partners. They worked together after the war to rebuild Fontaine's cotton warehouse. They may well have worked on other undocumented buildings in the intervening years. King probably had similar relationships with other builders.

King's role on specific projects after buying his freedom probably varied from job to job. He always acted as a designer and superintendent; on some projects that constituted his only role, the same function he had performed as Horace Godwin. His business relationship with Godwin after 1846 is unclear. The short biographies of King portray them as partners before and after Horace's freedom. King might simply have used Godwin's crews and lumber and continued in the same role as before 1846, but probably only on local jobs. King's projects with Jemison and Robison did not include Godwin.

King may have continued working with and for Godwin in the late 1840s, and then as the "old man" became less active, King contracted for himself with Robison. By the 1850s King apparently sought larger projects in a wider geographical area than had Godwin or Asa Bates. These older bridge builders concentrated on small county bridges.

In 1849 Godwin secured a $4,050 contract for two bridges over Bull Creek in Muscogee County but never acted on it. The county abrogated the agreement, apparently the last dealings Godwin had with that county. Bates built at least eight bridges in Muscogee County between 1847 and 1854, and starting in 1852 and continuing at least through 1864, he received an annual salary, usually $1,500, to keep all the county's spans in good order.[16]

Both Godwin and Bates continued to work for Russell County, especially on Uchee Creek. John repaired or built a small span over that stream in 1850; Asa built a $1,600 wooden structure for another road, while John erected an $1,800 one at a different location the next year. John and Asa remained friends; Asa, along with Stephen M. Ingersoll, secured John's bond as the executor of his wife's estate. John retained influence within the local area. He constructed the Russell County jail in 1851. He also continued to saw lumber. In 1853 the county granted him permission to erect a dam, presumably for a sawmill on Mill (Holland) Creek to the west of his house outside of Girard. In 1858, a year before his death, Godwin earned $450 for another span over Uchee Creek.[17]

Red Oak Creek Bridge, Meriwether County, Georgia, ca. 1980. The county historical society asserts that Horace King built this extant span during the 1840s. If he built it in the late 1840s, it may be one of his first projects as a freedman. He and his sons built more than a hundred bridges of this scale (127 feet of truss work) and smaller. *Photograph by Thomas L. French Jr.*

In keeping with the family tradition, John's oldest son, William E. Godwin, continued to build bridges, but except for signing a contract for King to reside the Columbus bridge in 1856, Horace's and William's names do not appear together in the surviving documents. Perhaps William operated the sawmill for his father and supplied the boards and battens for that particular job. King, rather than William, succeeded John as the major bridge builder in the region. While William, John, and Asa Bates remained close to home, King was more mobile and answered calls to build bridges throughout a three state area.

When King served as the designer and builder of major bridges, his involvement in the other aspects of contracting—supplying timbers, lumber, and other laborers—can only be inferred. At various times, King appeared to control a small crew who cut and hewed the timbers.[18] These men included some skilled workers and, by the mid-1850s, two or three of his sons who began learning the business. In 1851, when King's schedule did not permit him to work for Jemison, he wrote Horace, "Can't you send me some man who may be relied on to Frame and raise the Piers at Columbus? The timbers are all on the ground. I can start a force of eight or ten carpenters [to go from

Tuscaloosa to Columbus, Mississippi]. How long should they be in framing and raising the three piers?"[19] Obviously, King's crew included competent superintendents, one of whom could have directed Jemison's slaves. Later in the 1850s, when King transferred hewed timbers from Milledgeville to Albany, his crews prepared those items. The degree to which King participated in the ancillary activities made a significant difference in the level of his profit. However, he probably did not serve as overall contractor, where he supplied all the elements for a bridge, before 1865.

Another business arrangement involving neither Godwin nor King's formal guardian might have been made between King and a local entrepreneurial slaveholder. With slave carpenters and lumber from a local sawmill, King could erect a short covered span. Whether working with his former master or Robison or a ferry owner who wanted a small bridge, his partners could always trust Horace King. Jemison wrote to one of King's potential clients that he had "never dealt or settled with a more correct and honest man of any colour."[20] While King was personally honest, his status forced him to be more scrupulous than a white contractor. His honor, his integrity, and his workmanship needed to be above reproach. Handicapped by his color, Horace's freedom to do business rested on his stellar reputation, even more so than for a white businessman, who could drive a shrewder bargain.

The fraternity of skilled bridge builders in the Deep South remained small, thereby guaranteeing King employment. The limited number of contractors capable of sinking cribs, raising piers, and building camber into trusses necessitated that they work together. In 1851 King wrote Jemison to seek the whereabouts of a Mr. Williams, perhaps Simon Williams who had contracted with Godwin at Irwinton. King needed Williams's help for an unspecified project. Later that same year Williams tried to employ Horace for the piers and cribs in Columbus, Mississippi. Because of other obligations King could not undertake the project, and Jemison finally begged King to send a skilled man who could.[21]

While King obviously worked on many spans in a variety of locations, the exact extent of his work remains unknown. Many of his associations with specific bridges are only supported by oral histories, some of which first appeared long after the building of the bridge. For example, local history in Meriwether County, Georgia, attributes the 1840s construction of today's Red Oak Creek Bridge (north of the present town of Woodbury) to Horace as a slave.[22] His mention in the Florence bridge advertisement of 1840 indicates that King might have been erecting spans without Godwin by that decade. And he certainly could have built this bridge, but the date seems too early for that bridge. More than likely, Horace or one of his sons built it after 1846 and probably after 1865.

The most detailed discussion of Horace's jobs or, in some cases, his possible jobs comes from Jemison's letterbooks. The correspondence between the two men reveals more than perfunctory instructions about their joint ventures. Jemison solicited advice from King about bridge building, carpentry, and sawmilling machinery, and he shared the details of his business dealings with Horace. Jemison's letters to him, though limited in number, reveal King to be a bright, knowledgeable colleague and friend. On several occasions, King visited Jemison in Tuscaloosa, and when Jemison traveled to the home of his sister Helen Plane in Columbus, Georgia, he spent time with King. The real question, not revealed in their letters, is the social relationship between the two, especially in Tuscaloosa. Did Horace come to the front door? Did he sleep in the slave quarters? Did they climb together to the belvedere on Jemison's mansion in order to survey the town? Even though they were friends and Horace was very light skinned, how far did Jemison care to bend social customs?

Jemison's correspondence with his friend Horace provides a glimpse into the multitude of projects that Jemison envisioned and hints at many of Horace's activities. Jemison's letter of March 3, 1846, discussed their mutual business opportunities. The "Bridge across North River in this county" while chartered would not be built that season but apparently remained a possibility in the future. Jemison had no prospects to report from Columbus, Mississippi, since he had not been there since Horace and had "heard little from there, nothing from Jackson, [Mississippi] as to the Bridge there." Unfortunately, their joint bridge building projects probably ended about this time and were suspended until after the war or more accurately until after the passing of Seth King, this unknown constant in Horace's life. Jemison, the lumber dealer, and Seth King, the financier, agreed in the early 1840s not to build a bridge without the permission of the other man. They apparently cooperated on the Tuscaloosa, Columbus (Miss.), and Wetumpka spans. By late 1844 they had fallen out and an extended dispute ensued. In 1858 they entered into arbitration, which Jemison won, but Seth King never paid Jemison the amount awarded by the court. These problems prevented Jemison from financing future bridges.[23]

When Jemison listed their possible projects in his letter of March 1846, he devoted the most ink to the prospects for a Tuscaloosa textile mill. "We are about commencing a Cotton Factory here. . . . The contract for the Building will be let this Spring or Summer. . . . I think it likely I may have some interest beyond the Lumber in putting up our Factory building here." Jemison and his friend and fellow entrepreneur Dr. John R. Drish had "proposed to do the buildings and take pay in the stock. Whether the Dr. should . . . desire me to do any thing more than furnish the Lumber, I do not know." But whatever

Jemison's level of involvement, he wanted Horace to participate in the undertaking. "If I should be any further interested, I would like to have your services, and I think whether I am or not you can get employment from whoever may be. If you desire it & I can aid you in it, I will do so."[24]

A small boom in building textile mills occurred in the Deep South in the 1840s with the downturn in cotton prices. That trend, linked with Tuscaloosa's losing the state capital to Montgomery, led local elites to launch this enterprise. George Daniels, Jemison's biographer, has Horace building the Warrior Manufacturing Company, but the correspondence only suggests that possibility. Jemison's confidence in Horace shows the breadth of this freedman's skills. If King did construct it, he was one of the few to profit from the enterprise. Jemison and apparently Drish acquired significant stock holdings, and Jemison probably sold the company its coal, but dividends never flowed from this operation. Chartered in 1843, with capital between sixty thousand and a hundred thousand dollars, and modeled after a New Jersey mill, the firm floundered from the beginning.[25]

Jemison blamed labor problems, but other deficiencies were related to marketing (the limited local demand coupled with transportation problems) and technology (specifically problems with its steam engine). Across the South many early mills begun in the 1840s failed during the 1850s. By 1860 Jemison sought to sell the closed factory. He urged his sister's husband, William F. Plane, to consider buying it. Plane as a former Columbus, Georgia, resident realized textile mills could survive in the South. Jemison also contacted Daniel Pratt, another successful industrialist.[26] King probably knew Pratt longer and better than Jemison, since they worked together on the Alabama state capitol.

While no correspondence between Jemison and King survives for the period between 1846 and early 1851, Jemison's letters definitely place King in Montgomery in March, July, and September 1851 while Horace worked on rebuilding the state capitol. In 1846 Jemison and other northwest Alabama legislators lost the fight to keep the capital in Tuscaloosa. Stephen D. Button, the Philadelphia architect who briefly lived in Columbus, Georgia, and inspected the Godwin-King bridge in 1845, is credited with designing the first capitol in Montgomery. Completed in December 1847, the building burned in 1849.

Daniel Pratt played a major role in rebuilding the capitol, the design of which is usually credited to the architect Barahias Holt, along with Montgomery residents John P. Figh and B. F. Randolph as the contractors. However, King definitely contributed to the massive Greek revival structure. Jemison, in a later recommendation for Horace, credited him with "the Carpenters work of our present State House" and closed a September 1851 letter with the question, "Have you finished the capitol?" King's role on this project would have involved building the heavy timber framework of the massive building. The mystery surrounding King's involvement here concerns whether he built the double spiral staircases that grace the central entrance hall of the building. This feature resembles the early work of Daniel Pratt in Milledgeville, Georgia. Using mortise-and-tenon joints and treenails, King would have cantilevered the beams supporting the floating stairs. He may have done the finish work on the banisters and balustrades. The popular history of the capitol suggests King built the steps, but the conclusive evidence has been lost.[27]

By 1850 the Kings lived in a separate house, not necessarily in close proximity to the Godwins. In that year's census Horace's family appears after James Abercrombie, who lived on the river south of Girard. The Kings were probably not neighbors of the Abercrombies. The enumerator might have realized the Kings needed to be listed as free and just added them to the bottom of the page. This nuclear family consisted of: Horace, a forty-five-year-old mulatto mechanic; Frances, a twenty-six-year-old mulatto; and four children (Washington, ten; Marshal, eight; John, six; and Frances, five). King probably underestimated his net worth at three hundred dollars. Alabama law allowed free blacks to own property, but less than a quarter of the nonslave heads of households reported owning any property. Many census takers more than likely failed to ask that question, given the general prejudice toward free blacks.[28]

While nonslave blacks did purchase rural land and urban lots, local regulations often controlled the place of residence for free blacks. The Columbus council, for example, decreed that "no person of colour shall be allowed to live separate from the lot where his or her guardian shall live." Setting the fine at ten dollars a day for violating this provision, aldermen displayed their fear of free blacks. They, like other white southerners, remembered that Toussaint Louverture, the leader of the Haitian Revolution during the 1790s, and Denmark Vesey, the chief conspirator in Charleston during the 1820s, had been free. Columbus tax rates also served to discourage "free persons of colour" from living in the city. The council charged free blacks under the age of sixty twenty-five dollars per annum while the annual fee for white males and slaves only amounted to fifty cents. King avoided these fees by living in Girard, and obviously whites did not fear him. While they feared nonslave blacks as a class, they did not see Horace King as a threat.[29]

In spite of their general regulations against free blacks, the Columbus council provided King with revenue that perhaps allowed him to build his house on land he purchased in Girard. In July 1852 King bought two acres of land just east of the property owned by Godwin. Given its location King

One of a pair of floating two-story spiral stairs that flank the entrance to the Alabama capitol, 1934. Horace built the cantilevered trusses that support these stairs, and he may have executed the finished carpentry work on the staircases. *Photograph by W. N. Manning. Library of Congress, Prints and Photographs Division, HABS, ALA, 51-MONG, 1-7.*

appeared to observe the spirit of regulation concerning the residence of free blacks. He paid forty dollars for this tract, considerably more than Godwin spent on his property a decade earlier. In October of the same year, Columbus paid King three hundred dollars to repair three piers. Again state laws were ignored by entering into a contract with King instead of his guardian. In an emergency, as when the aldermen believed the bridge was falling, they ignored the regulations regarding free blacks.[30]

King immediately repaired the Columbus piers and by December of that year began working again in Tallassee, Alabama. He may well have stayed with the Wright family in the vicinity of Old Tallassee, but he worked for two entrepreneurs, Thomas M. Barnett and William M. Marks, who had begun developing a small industrial complex at the falls of the Tallassee in the 1840s. King was there in January and February 1853, and local histories note that Barnett and Marks built a bridge just below the falls in the early 1850s and a second factory in 1854. The exact date of this span is not indicated, but a map dated April 18, 1853, shows a "projected bridge across the river." By April, when the map was published, King's bridge could have been finished rather than just "projected," or King could have been drawing plans for the 1854 factory, which he might have returned to build.[31]

While Horace worked in Tallassee, Jemison sought his advice on and involvement with a much larger project: building the mammoth Alabama State Hospital for the Insane (now Bryce Hospital) in Tuscaloosa. During this period, many states sought to reform the treatment of the insane by building modern asylums. The national leader of these efforts, Dorothea Dix, who frequently corresponded with Jemison, advised these efforts in Alabama. She and other alienists viewed proper physical surroundings as essential for effecting cures; thus, they emphasized the architecture as well as the nature of the treatment.

The Alabama version, designed to house 250 patients, cost more than $280,000 by 1861. The complex consisted of seven connected buildings: a four-story main one flanked on each side by three three-story wings. Males and females occupied opposite sides, with the most afflicted patients in the most distant wings and the less demented closer to the center. This sophisticated facility boasted steam heat, running water, speaking tubes, and a complex ventilation system, all clad in a fashionable Italianate exterior.

While still merely a concept, the hospital attracted Jemison's interest. In addition to his concern for the insane, Jemison was interested in helping Tuscaloosa and in increasing his personal profits. He spearheaded the legislative wrangling that brought the facility to his hometown. After the loss of capital status, the city's population plummeted from 4,500 in 1846 to 1,950 in 1850. He hoped the hospital could help reverse that trend. The legislature approved the project in 1849, but the destruction of the Alabama capitol in the same year diverted moneys to its reconstruction. By late 1852 or early 1853, the architects, contractors, and suppliers for the new hospital were being selected, and Jemison wanted to be involved. He also needed his friend Horace King, who had just completed his work rebuilding the capitol.[32]

Jemison treated King as a colleague and coconspirator. In late 1852 or the first month of 1853, the former slave visited Tuscaloosa, where he and Jemison planned their strategy as partners. On December 9 Jemison wrote Horace: "Since my last I have no further information as to our Insane Hospital at any rate nothing it would be prudent to trust to so very uncertain and often treacherous medium as a letter. I can tell you a good deal, when I see you, more than I would desire to write and will meanwhile get all further information I can. I will look for you certainly by the time you name." Five days later, Jemison wrote again, "I am keeping a look out and gathering up all the information I can that will likely benefit us. I have picked up several items that may be of service to us." He also questioned King about the efficacy and expense of purchasing machinery for fabricating sashes, blinds, and doors at his sawmill. King may have pleaded ignorance or never responded about the costs, because Jemison sent a similar query to Daniel Pratt the next month. Jemison received the contract for the lumber and "for making & laying 5 to 6 Millions of Brick" but not for the carpentry.[33]

In typical Jemison fashion, not satisfied with being the largest supplier, he schemed to capture a share of the other construction contracts for King and himself. He predicted an internal fight among the contractors that would allow them to acquire more work. King apparently provided Jemison with the projected cost of specific jobs. Jemison responded to King on January 22, "I will as far as I can carry out your suggestions, but am of opinion nothing can be done on terms upon which we can make more than wages if that." Just eight days later he added, "I look forward to a falling out amongst partners. Then we may come in for a part of the carpenters work but not likely before. If laborers can be hired in Georgia or East Alabama we better hire there as there are none here to be had. Let me hear from you frequently." About a month later, Jemison still hoped for an opportunity for more work on the hospital. "It is whispered there will likely be a falling out between Miller & Kirk [the contractors]. Miller has proposed to a third person to buy out Kirk, this I have from the party himself. It is also intimated to me that both can be bought out. If this turns out to be so which is very probable and Dr. Drish [Jemison's colleague in the textile mill] is not interested[,] on what terms could we afford to take their contract? If they will sell I am inclined to think we could get a modification of the contract. . . . Set down yr lowest Figures for the work. I am in better spirits about getting it than when I wrote you last."[34]

Many a "falling out" occurred before the conclusion of this project, and some of them involved Jemison, but he never contracted for the carpentry work that would have required importing laborers from the other side of the state. He did lure his chief brickmaker and mason, William B. Robinson, from Columbus, Georgia.[35] Robinson began producing bricks for Jemison from clay on the hospital grounds in March 1853. Even though Jemison never secured the carpentry work, he wanted King in Tuscaloosa. In August he wrote, "I would like [it if] you could find it convenient & to your interest to get out here. Mr. Robinson is also desirous for you to come. I have no doubt we could be mutually serviceable to each other. We are getting along with our Contracts quite as well as we expected." In the same letter he tried to interest Horace in "an additional Dormitory Building proposed to be erected at our University." He enclosed the plans and specifications.

> If you would like to undertake the Carpenters work either by yourself or jointly with me I would be pleased to have you do so. The Job is small & may therefore suit you best to be unconnected with any other. If so have no delicacy or hesitancy in saying so. I will have my own and other Jobs as much as my hands can do while you are engaged at that.
>
> Whether you desire to have any thing to do with the Contract or not I would like you [to] make and send me as nearly as you can an estimate of Carpenters work, Painting, Glazing &c all materials furnished by Contractors &c &c.[36]

Jemison also bragged of the booming economy of Tuscaloosa. "There is a prospect of a number of Jobs in the Building line." And he wrote of the railroad building spirit that could produce employment for Horace. In several of these letters, Jemison mentioned a straw cutter, perhaps a device for preparing straw for bricks, which he and King jointly owned. King sold his interest in this machine to Robinson, and Jemison seemed to regret the sale, as if it signaled less involvement between Jemison and Horace.

King may have acceded to Jemison's wishes and gone to Tuscaloosa, where he built the dormitory, helped with the brick and lumber production for the asylum, or both. The break in their correspondence may indicate King's presence in Tuscaloosa, but King did not build Bryce Hospital. James Madison Hall, which the university trustees began discussing in 1853, was not completed until 1859 and did not involve King as a contractor.[37]

The extant letters illustrate the wide range of Horace's skills. His knowledge transcended bridges to include general contracting, and Jemison depended on King's grasp of the building trades. Jemison's respect and even affection for his friend Horace is obvious. In 1854 the Tuscaloosan recommended King to a potential client in Florence, Alabama, praising his honesty and calling him the "best practicing Bridge Builder in the South." He also noted that he built both Town and Howe truss spans. Introduced in 1840, William Howe's improvements employed much less lumber than Town's and supplemented a series of boxed wooden Xs with vertical iron rods. Their strength, coupled with a reduced requirement for lumber, made them popular for railroad bridges. King's skills did not remain static; as a professional he expanded his expertise as new techniques emerged.[38]

King probably never worked on the Florence bridge. Only two months before being recommended to that town, King apparently arrived in Cahaba, Alabama, to rebuild a "caved in" span. Established as the first Alabama state capital in 1818, Cahaba declined after lavishly entertaining the Marquis de Lafayette in 1825, the same year it lost the capital to Tuscaloosa. Buffeted by that loss, fevers, and floods, the city declined and then slowly rebounded during the late 1840s with the flow of black belt cotton to the junction of the Cahaba and Alabama Rivers.[39]

The owners of the Cahaba bridge experienced a similar decline and looked to King for a rebirth. In a situation similar to Edward B. Young's actions to save the span in Eufaula, all the stockholders in the Cahaba bridge relinquished their stock to Josiah T. Allen, who promised to rebuild the "caved in" structure. On April 28, 1854, the Cahaba *Dallas Gazette* reported:

> The work is now under contract, and the colored man, Horace King, who has gained such a reputation as a bridge builder, is the contractor. He is now daily expected, and when he arrives, the work will commence immediately. Mr. Allen informs us that it can be built in two months.
>
> We hope that Mr. Allen will meet with no unlucky accidents or expensive delays, but will succeed as well as he deserves.[40]

Perhaps King experienced a delay and never reached Cahaba, because Mr. Allen's bridge continued to experience major flaws. In August 1855, slightly more than a year later, Allen closed the reconstructed passage for all but foot traffic in order to erect "arches for the firm support of the Bridge." Allen expected "to finish the work . . . in a short time," while the newspaper claimed it would be stronger than the original, "so that it may last as long as such an expensive piece of work should." A mechanic from Selma repaired the span; he, rather than Horace, may have effected the repairs in 1854, or maybe King rebuilt the trusses and never reworked the piers, which seemed to be the faulty element in 1855.[41] The once-again-repaired bridge stood ready for crossing by October, but a newspaper in Selma, Cahaba's rival to the east in the same county, called the bridge a failure. The Cahaba *Gazette* praised the new span, which "will stand for years to come," and defended the honor of its proprietor. "Nearly every dollar Mr. Allen owns is invested in this bridge, and it was an act of manifest injustice for any public character to attempt to depreciate it."

Perhaps the attack on his business or the stress of trying to maintain a bridge on a capricious, often raging, river finally got to Allen: He died on November 25, 1855, at the age of forty-five.[42]

In the meantime, King had returned to Columbus and in the summer of 1855 joined with James Meeler, a local carpenter, to construct a warehouse for "John Fontaine and others." Fontaine and his father-in-law, Charles Stewart, had operated a cotton warehouse business since 1829. Fontaine also served as mayor of the city and invested in the river trade, a textile mill, and land—extensive acreage in Georgia, Alabama, Mississippi, and Texas. By 1859 he owned "two plantations on the river and sev[era]l hundred negroes & plenty of money." His fever-season home in Summerville, on the hills west of Girard, sheltered his family during the summer, and his brick Greek revival house in town stood on Front Avenue across from his warehouse.[43]

Fontaine served as the lead investor in the Columbus Factory, better known as Clapp's Factory. This textile mill began in the 1830s, the first in the area to tap the waterpower of the Chattahoochee. Columbus merchants financed it, but its name became linked with Julius R. Clapp who served as the superintendent. Slaves provided some of the labor for its gristmill, tannery, and small textile mill before 1865.[44]

Apparently Meeler partnered with King in building a warehouse for Clapp's Factory. Only seventeen receipts and not, unfortunately, an original contract are left to document this joint endeavor. These bills are requests for Fontaine or J. R. Clapp to pay suppliers and charge them against the account of James Meeler and Horace King. Fontaine required both men to sign these documents, presumably to minimize conflicts between the builders. The involvement of Clapp indicates that this warehouse was associated with that factory located three miles north of the city. The project was of significant scale. The brickwork and plastering of two rooms totaled about four thousand dollars, and the nails, angle iron, bolts, spikes, and other iron items cost almost a hundred dollars. They were purchased from July 16 to October 4, which probably represented a rough chronology of the job. Judging from his signatures on the receipts, King was involved in July, early in the process. His role must have been to mortise-and-tenon together the heavy timbered armature, the same work he performed on the Alabama capitol. Then, in the case of Clapp's warehouse, Meeler and his carpenters finished framing in the building.[45]

For some period in 1854 or 1855 King probably worked for the Franklin Bridge Company in Heard County, Georgia, throwing another covered bridge across the Chattahoochee River. The legislature chartered the company on Valentine's Day 1854 and granted it the right to issue twenty thousand dollars worth of stock. The seven incorporators included Charles W. Mabry, who later became King's partner. Mabry and King must have become acquainted when Horace worked on the Franklin span. Born in Greene County, Georgia, Mabry was practicing law in Franklin by 1850. He declared a personal worth of six thousand dollars, that year, but his business activities seemed to reach beyond that figure. Six days after creating the bridge company, the legislature chartered the Franklin and Oxford Rail Road Company, which was to connect Greenville, Georgia, with a point in Alabama on the Oxford line. Only Mabry served as both an incorporator for the railroad and the bridge. The railroad scheme languished while the bridge became a well-knit structure latticed by King. Other evidence places Horace in Franklin in the 1850s. During the 1878 hearing, when King sought compensation for his losses at the hands of Federal troops, he testified to buying a mule in Franklin from a man named Borgus in either 1853 or 1858.[46]

After a decade of freedom during which he worked for others building bridges, warehouses, and factories, King became involved in a project that promised to give him a constant source of money and to enhance his freedom. King's relationship with Mabry created that possibility. By becoming a bona fide partner in a bridge venture, King and his wife and children were able to expand the limits of their freedom, escape a place associated with Horace's servitude, and create their own family home.

CHAPTER TEN

HORACE'S OWN BRIDGE

He built my bridge and had forty shares in it. He was regarded as an honest and truthful man. Stood better than the common free negro.
James D. Moore's Testimony before the Commissioners on Claims, February 1878

It is a covered structure, very well built, 480 feet long, on two main spans.
Maj. Gen. George Stoneman, at Moore's Bridge, to Maj. Gen. William T. Sherman, July 13, 1864

By the mid-1850s King had constructed six large bridges over the Chattahoochee River at five different sites. His next one, his smallest in scale and the only one not at a major town, was the most important to Horace, Frances, and their children. In 1855 or 1856 King first met James D. Moore, who together with King and Charles Mabry of Franklin formed a partnership known as the Arizonia Bridge Company, which built Moore's Bridge across the Chattahoochee on the road from Newnan to Carrollton. These three men brought different assets to their unincorporated enterprise. Moore owned the land. An early settler and farmer in the area, he had married a land-wealthy widow, Caroline Malone, by 1838 and continued expanding his holdings. By the middle of the next decade Moore acquired his riverfront property and began operating an unchartered ferry.[1]

Mabry brought his capital and perhaps his lumber into the group, and King his skills as a builder. This "very well built" structure consisted of two main spans with a total length of 480 feet.[2] Horace presumably treenailed together a Town lattice truss for his own bridge, as he liked to call it. The origin of the company's name remains a mystery. Its spelling only appears in handwritten deeds; thus, it might have been pronounced *Arizona.* On August 6, 1858, Moore deeded all the land, the river bottom, as well as the ferry and bridge rights to the company, which had only the three partners—Moore, Mabry, and King. Alexander I. Robison signed as Horace's trustee.[3] Robison's slaves may have provided some of the labor for the bridge. Certainly the bridge was standing by August 1858: Jemison addressed his May 1858 letter to King in Newnan.

This bridge played a central role in King's life. Horace, in a fashion similar to Godwin, moved from being a mere builder to becoming a bridge owner and a speculator. King said he "took pay in Stock in the bridge"; Moore noted that King owned forty shares of stock, giving King a significant stake—a third interest, or thirty-five hundred dollars of stock in a ten thousand dollar company.[4] Few free blacks controlled such an asset.

King protected that investment by moving his family to Moore's Bridge, where they collected the tolls. They might have come to Carroll County as early as 1858 or maybe after John Godwin's death in 1859, but certainly they were there by 1861 after the war began. The testimony of a Coweta County neighbor, Jeddiah S. Miller, seems to indicate Horace's family resided there before the war. Miller saw King "when ever he was with his family. I met him every week in offices. He worked for me on a factory and mill during the war then I saw him every day." The 1860 census for Russell County does not list Horace's family, though he was listed in 1850. He is also absent from the 1860 Carroll County census. The Kings lived in a small house next to the west end of the bridge. James Moore's home sat on a hill above. Horace's family also farmed twenty-five acres of H. J. Garrison's land on the eastern side of the Chattahoochee River in Coweta County.[5]

Financial motives probably dictated the King family's move. The bridge itself promised a steady income and perhaps this location closer to Atlanta provided more prospects for King's work. The rail line at Newnan, only nine miles away, connected with Atlanta to the north and Montgomery to the south. From Opelika, Alabama, King could go to Girard and Columbus. Moore probably acted as an informal guardian for King. Moore testified that King had a trustee who signed his contracts, as if to say King had not been his responsibility. Robison, however, lived in Columbus; so, King asserted his freedom by expanding his sphere of operation beyond that of most free blacks. The Kings had no legal right to live in Georgia, which like Alabama restricted the movement of nonslave Africans into the state. The survival of free blacks depended on the acceptance and even friendship of their neighbors—white families who knew them and permitted them to retain their freedom. Unknown free blacks seldom received such respect. Viewed in terms of that social more, the Kings' migration to Carroll County represented a highly unusual action.

Freed slaves tended to move away from their former homes to affirm their freedom. At the end of the Civil War, most freedmen left their old plantations. Tens of thousands of former slaves moved to cities and to areas where blacks predominated. They sought to establish a totally different life. King's move certainly differed from theirs: it occurred without force of arms and at an earlier time; and he moved from an urban area where blacks predominated to a rural one where they were a distinct minority. Nonetheless he may have been motivated by the same impulses that caused the freed slaves to leave their homes. Whatever Horace's motive, the move lessened the Kings' ties with the Godwins and with Horace's maternal family, all of whom remained slaves on the Godwin Place. Many nonslave blacks, after achieving some measure of financial independence, bought freedom for their families—but not Horace. His wife and children, the most-often purchased family members, were already free. By 1858 three of Horace's boys—Washington W. (born in 1843), Marshal Ney (born in 1844), John Thomas (born in 1846)—may have been working with him learning their craft, but not Annie Elizabeth (born in 1848), and George (born in 1850) who were still too young.[6]

The photographic portrait of Horace's wife, Frances, depicts a light-skinned, seemingly tall woman with large features, especially her hands. Her face and hair indicate Indian and European ancestors; her intense, dark eyes and erect posture reflect her self-confidence and her resolve. She may have played a crucial role in the decision to live at Moore's Bridge. King's work meant he continued to travel. Frances could collect the tolls, and the move increased her control over the family. In their new home their nuclear family became much more important in shaping the children. Perhaps Frances pushed for the change to separate her children from the slave environment, which must have been felt on their lot next to the Godwin Place. As successful free mulattos, the Kings occupied a precarious status as a true minority. Whites, many of whom certainly resented the Kings' success, treated them the same as black slaves, only with more suspicion. Black slaves, on the other hand, probably viewed these bridge keepers as belonging to a different caste.

King's move or flight to an area between Carroll and Coweta County offers some insight into his aspirations and, perhaps, the pressures felt by his family. He purposely moved them to a region with few blacks. Most nonslave blacks, in the North and the South, lived in urban areas, especially those free blacks who prospered as carpenters, waiters, barbers, and so on. The Kings' migration to a rural, predominantly white area contradicts all the generalizations. In 1850 Carroll County ranked as the eighth whitest county in Georgia, with a black population of 12 percent and only four free blacks. Eighteen free blacks lived in Coweta County, and blacks made up 40 percent of the population. The Kings lived on the border of Carroll and Coweta Counties, and the total number of blacks in both counties accounted for 28 percent of the population. By contrast, a total of ninety-five free blacks lived in Muscogee and Russell Counties, and blacks constituted 51 percent of the population. By 1860 as the number of black Georgians increased statewide, the black population in all categories increased, but the relative racial proportions remained constant.

Obviously King moved because of the financial opportunity presented by Moore's Bridge, but he and Frances certainly could not ignore the racial dynamics. Perhaps the Kings were moving toward passing for white or at least toward separating themselves from the race with which whites associated them. They stretched the limits of being free blacks. The racial atmosphere of their old home as contrasted with their new one may also have influenced their decision. The Moores and their neighbors apparently made them welcome in Carroll County. Their old neighborhood in upper Girard was facing change: a new Columbus bridge discussed as early as 1856 and built in 1858 linked the factories with workers' homes in Alabama. Poor people, who aspired to be planters and were forced to become mill workers, resented free blacks more than most other whites. Perhaps the Kings fled the approach of the mill village.

King kept his house in Girard; he and his family stayed there frequently. Probably between his building Moore's Bridge and his family's moving there, King worked on the Columbus bridge. In April and May 1856 he won contracts to repair the piers and re-side it. In 1857 he also repaired its eastern approach or land bridge. That year marked the beginning of another nationwide depression. The economic downturn had less of an impact on the South in general and little effect on King, as he continued to remain active.

In February 1858 Jemison once again proposed that he and King work on the other Columbus bridge, the one across the Tombigbee. It, like so many other timbered tunnels, had collapsed by 1856. Lowndes county citizens raised nine thousand dollars to replace it in 1858 and gave "the privilege of building" it to a local company, Jemison informed King. They have "written to parties in Georgia" about the construction of a new span. "It has occurred to me they may have also written to you. If so I would like you would make no engagement with or give them any definite information as to probable cost until I can see you. . . . They know nothing on the subject of Bridge building, and if I can get rid of Mr. [Seth] King in the old Bridge . . . I think I may manage so as to get Back some of my losses." Jemison planned a trip to Columbus, Georgia, to see his sister in the near future and expected to consult with Horace at the same time. In case Jemison missed King in Columbus: "I would like . . . yr opinion as to the relative strength of the double & single Truss plans of Bridge which is strongest, and how long a reach may be safely trusted and relied upon under each plan." Jemison never used that data for the Mississippi crossing. He never could "get rid of Mr. King in the old Bridge." Horace King, meanwhile, had his hands full in Georgia.[7]

In either 1857 or 1858 King began working on a bridge near Milledgeville, most likely over the Oconee River span on the Augusta road. The history of that crossing illustrates the ephemeral lives of wooden bridges, even adjacent to capitol cities. Three times—in 1807, 1816, and 1834—the legislature granted the bridging right there. The 1807 act granted Thadaeus Holt two years to construct a span and a ten-year charter. Since the bridge abutted Milledgeville, he paid the city an annual fee to be used for erecting a school or academy. In 1816 James Roussean received a twenty-one-year charter, provided he built his crossing within three years and rebuilt it within two years whenever it became damaged or washed away. By 1834 the wealthy Milledgeville entrepreneur Farish Carter owned an unchartered privilege to a bridge that connected to state lands on the western side of the Oconee. Apparently that span had failed by 1834, and the legislature authorized the sale of that state land, probably to potential manufacturers. In the same session, the legislature granted Samuel Buffington the right to construct a bridge from the town commons of Milledgeville across the Oconee; the act required him and his heirs to pay an annual fee to Milledgeville and to maintain the bridge for thirty years. The Buffingtons failed to comply: Their bridge failed by 1857.

In December 1857 the general assembly enabled the inferior court of Baldwin County to issue bonds for building a bridge; the act allowed the bonds to be given to the town or to be used to buy stock in a company. In return, either the town or the company was to construct a new bridge.[8] Horace King dealt with the resulting municipal or corporate entity and proceeded to assemble the needed materials for a new span. Robison signed his contract, but neither the county nor the city records for this period have survived. A conflict ensued between King and the individuals financing his work. At the same time, Nelson Tift, the founder of Albany, Georgia, finally acted to complete his bridge in southeast section of the state. According to the Horace King legend, Tift convinced the bridge builder to abandon the Milledgeville project and move his timbers by rail to Albany.

A native of Connecticut, Tift worked as a merchant in Key West, Charleston, Augusta, and Hawkinsville, Georgia, before he obtained a charter for the town of Albany in 1838. This site, although south of the fall line, represented the head of steamboat navigation on the Flint River. Rawls, Tift, and Company played the same role as the organizations that created West Point and Florence: They controlled the real estate and obtained permission to build a bridge in 1838. In a fashion similar to other entrepreneurial slaveholders, Tift used some of his skilled slaves to build his grist- and sawmills, essential institutions for settlers and for a capitalist such as Tift. His first sawmill floated in the river and employed the current rather than falling water. Later those enterprises moved to a permanent brick building, also constructed by his slaves. Some of Tift's other interests involved operating a Flint River ferry, cotton buying, river navigation—initially with box boats and then with steam ones—and railroads, the first of which reached Albany in 1857. Not unlike Jemison, Tift held several political offices including a seat in the

Georgia House in the 1840s and early 1850s. By 1858 Tift's public service and his ownership of the *Albany Patriot* had garnered him a statewide reputation. The Milledgeville newspaper noted the end of his tenure as editor in August 1858. Tift also remained cognizant of happenings in the state capital and knew about King's bridge project in Milledgeville, especially since he still needed a bridge. Tift and his associates had obtained a new bridge charter in 1852.[9]

The Albany sources have Tift convincing Horace to move his timbers to Albany by railroad. One of Horace's biographers, James G. Bogle, asserted: "The timbers arrived in Albany on October 10, 1858, and the bridge was completed for crossing on December 20, 1858." That chronology meshes with the accounts from the *Albany Patriot*: By September 23 Tift and Dr. A. J. Robinson of Columbus had signed a contract "for the erection of a first class Bridge. . . . Horace King the celebrated bridge builder, will superintendent the work." Three forty-foot-high "hollow wooden piers, to be subsequently filled with masonry," were to support 350 feet of lattice work. "The trestle bridging will be about 580 feet long, making a total length of 930 feet." The middle pier was almost finished on October 28. By December 2, 1858, the *Patriot* reported the span would be ready for crossing the next week: "The western land-bridge is completed, the lattice-bridge, in two spans, one hundred and fifty feet each . . . , and the eastern land-bridge alone remains to be raised and floored." The editor, who had recently purchased the newspaper from Tift, praised his effort, saying: "The Bridge will be an ornament as well as a benefit to the town and country. Its cost, completed with all its appurtinences [*sic*], including stone Piers, right of way and privileges will be about thirty thousand dollars." Then the writer turned his praise to the real builder: "Horace King, the master builder, deserves much credit for the energetic and faithful manner in which he has prosecuted this work." The *Columbus Enquirer* reported the spanning of the Flint River on December 6, 1858, and proudly noted its "master builder was Horace King of this vicinity."[10]

The bridge might have been completely finished by December 20. Apparently, King and Robison were working on two bridges simultaneously, and when the deal turned sour on one, they simply moved the timbers from one site to the other. Several sources mention that King moved timbers rather than sawn lumber. King and his crew must have already hewed the heart pine and, therefore, saved time by moving it to Albany by rail. Given Tift's sawmill interests, he surely provided the sawn lumber.

The extra expense, more than ten thousand dollars, for this bridge when compared to similar ones resulted from constructing Tift's bridge house, the most distinctive of King's surviving structures. In a fashion befitting Tift the entrepreneur, he had King and his slave carpenters erect a two-story, six-bay-wide brick building that served as the entrance to the span. Rather than stand-

Albany Bridge House, ca. 1930. The most unusual extant King structure is Nelson Tift's Bridge House (1858), built at the same time that King's crews threw the associated bridge across the Flint River. The bridge house's center section, which was originally open, served as the entrance to the bridge. Tift's office and the toll keeper's quarters occupied the rest of the first floor. The upper story was an auditorium known as Tift Hall. In the twentieth century the building evolved from a blacksmith's shop into a garage and auto parts store. *Courtesy of Georgia Division of Archives and History, Office of Secretary of State.*

ing at the end of a street, this impressive structure stood in the row of buildings on the eastern edge of the street bordering the river. Within the building, Tift divided the bottom floor into thirds consisting of his office, the tollhouse for the bridge keeper, and an arched entrance in the center that led to the bridge. To enter the span required a perpendicular turn. Above that entrance, the upper floor—a large open space illuminated by light from sixteen windows and reached by exterior stairs—became Bridge (or Tift) Hall. Its trompe l'oiel interior walls with classical arches provided an appropriate setting for a theater used by traveling troupes and a ballroom for Albany society.[11] Most residents of Dougherty County never saw the inside of Bridge Hall because their interest focused on the bridge. The span lessened the isolation of the residents of the eastern portion of the county, but this connecting link failed to increase agricultural products in that area, at least before the Civil War. Residents on the other side of the river came to resent the tolls, which should have exempted any agricultural products, since the charter specified the same rates as the Columbus and Macon bridges. This antipathy to the tolls and to

Tift combined with racial animosity toward Tift led to the torching of his bridge during Reconstruction (see chap. 13).[12]

Only slightly less intense than the later conflagration in Albany, a controversy raged in Columbus over the recently completed Upper (or Factory) Bridge at Franklin (later Fourteenth) Street at the same time King assembled Tift's trusses. Despite the oft-repeated claim that Horace built every wooden bridge across the Chattahoochee River at Columbus, he actually played a very minor role, if any, in this bridge. And for the sake of his reputation, his absence here was a positive. The construction, ownership, and maintenance of this bridge all sparked debate. The impetus for the span came from an effort to find adequate housing for the city's workers.

By the early 1850s the city's population of nine thousand included four to five hundred mill operatives and their families. Three river-powered textile mills and other industries operated in close proximity to the commercial district and a row of mansions. The city's workers lived in adjacent tenement houses that posed public health problems, especially the section known as Battle Row on Crawford Street. The press also attacked the "enormous price of rents demanded by illiberal owners of dwelling-houses in Columbus."[13]

As early as the January 1852 municipal election, one of several local issues involved moving the mill operatives to Girard. Later that year, the social dislocation produced by the presence of the workers became even more acute when a November freshet washed away the raceway and flumes, thereby closing every factory except Coweta Falls at the head of the dam. "All things considered," reported the *Enquirer*, "this is perhaps the most destructive freshet which has ever occurred in our river."[14] Idle, unemployed operatives filled the city. Frederick Law Olmsted visited the city shortly after the flood. His remarks are the first published description of Columbus as a rough mill town:

> At Columbus, I spent several days. It is the largest manufacturing town, south of Richmond, in the Slave States. . . . The operatives in the cotton-mills are said to be mainly "cracker girls" (poor whites from the country), who earn in good times, by piece-work, from $8 to $12 a month. . . . The labourers in all these are mainly whites, and they are in such a condition that, if temporarily thrown out of employment, great numbers of them are at once reduced to a state of destitution, and are dependent upon credit or charity for their daily food. Public entertainments were being held to be applied to the relief of operatives in mills which had [been] stopped by the effects of the late flood. . . .
>
> I had seen in no place, since I left Washington, so much gambling, intoxication, and cruel treatment of servants in public as in Columbus. This, possibly, was accidental, but I must caution persons, traveling for health or pleasure, to avoid stopping in the town.[15]

By 1856 a plan for segregating the workers in Alabama had been formulated. The legislature devised a scheme whereby the mills financed the construction of another bridge that, after its completion, became city property. Then, in a manner similar to the lower or original Godwin-King bridge, the aldermen would collect tolls, maintain the structure, and regulate its use. Realizing the extent of the controversy generated by this new span, the legislature required a referendum to approve the plan. The factory owners, the wealthy people living uptown, and the *Enquirer*, the voice of the local Whigs, supported the new bridge. The newspaper argued that "the factories need more operatives, and residences both for them and for their present force; that suitable and convenient locations cannot be obtained on this side of the river, but may easily be had on the Alabama side; and that the proposed measure will, by extending the operations and increasing the force of the factories, materially augment the business of the city." On the other hand, downtown residents, landlords scared of losing rents, merchants who feared business flowing across the river, and people—often Democrats—who resented governmental favors for businessmen attacked the idea. The moderate Democratic *Sun* was only lukewarm in its support. According to the *Enquirer*, "'Bridge or no Bridge' was the question . . . and many persons considered it almost as momentous as Hamlet's 'to be or not to be.' A great deal of interest and some excitement was manifested; carriages with flags conveyed voters to the polls; and electioneering was brisk and indefatigable." The bridge faction won "by a majority of 143 votes–Bridge 425, No Bridge 282."[16]

The factory owners began constructing their bridge, without Horace but with Asa Bates, Horace's sometime rival, sometime partner. Bates contracted for a covered bridge, most likely a Town lattice truss, but the industrialists experienced problems raising money. Perhaps after Bates built one rock pier, they changed the plan to an uncovered bridge consisting of eleven trusses, each approximately sixty feet long supported by nine "wooden props" in addition to the one rock pier. The short length of the individual spans allowed these trusses to be a series of short queen posts, much less expensive than a Town truss. Such a bridge was feasible at Franklin (or Fourteenth) Street because it crossed above the rapids and dam; therefore, it required no long spans for river navigation.[17]

Because of the factory owners' inability to pay Bates, he quit, and a series of other contractors, apparently not bridge builders, finished the span. While internal problems persisted within the project, the *Enquirer* waxed eloquent about the prospects of new bridges. "Columbus will soon be emphatically a city of bridges. When the new public bridge . . . at the foot of Bryan [actually Franklin] street, is finished, and the contemplated new Railroad bridge constructed, we will have four bridges across the Chattahoochee . . . spanning the gulf which 'like a narrow sea divides' Georgia and Alabama, and uniting by new avenues the two sister States in commerce, travel and interest. Columbus and Girard will then be almost one city so many streets will be connecting and unbroken highways from one to the other."[18]

Despite its rhetoric about opening up great avenues of commerce, the newspaper stated that the central rationale for the new bridge was so "a great many of the families who so incessantly labor in the Factories, will have an opportunity to live in less pent-up abodes, and at the same time be near enough to their respective posts of employment." The *Enquirer* envisioned private homes sold "at prices [not] too high for families of moderate means, supported by daily labor in the Factories." If this happens "we will no doubt soon see many comfortable dwellings on the other bank of the river, adding to the growth and population of upper Girard, and promoting the health and business of Columbus."[19]

Almost everyone in Columbus, including the mill workers, wanted to move the operatives out of the center of the city. The council, however, balked at accepting the bridge as specified in the original legislation. They refused it because of its financial indebtedness and its impermanent nature. The incorporators offered the span to the city in early November, and when the aldermen debated instead of acting, the factory owners opened their bridge without any toll. The mill men sought to apply pressure by attracting traffic to the upper bridge and costing the city revenue from its lower bridge. The aldermen countered by abolishing the tolls on the lower bridge and appointing a committee of builders—Samuel R. Andrews, J. L. Morton, R. R. Goetchius, John S. Allen, and Asa Bates—to inspect the new bridge. The group included neither John Godwin, who in the last year of his life must have been ill, nor Horace King, who was in Albany. They might not have consulted a free black on such a formal matter; they tended to seek out King when a near disaster demanded his immediate attention.[20]

The committee reported the bridge of ample strength to support the expected loads, but its durability raised significant questions. The nine "wooden props," which they would not even call piers, represented the real hazard. "We are clearly of the opinion," they reported, "that any considerable accumulation of driftwood against these props, in a high and rapid state of the River, would sweep them away and the whole structure would fall." The chief strength of a wooden bridge lay in its trusses, which builders in the Town mode always sheaved with boards and battens. The design of the trusses on this particular bridge meant they could not be covered and protected from the sun and the rain. The committee concluded: "the Bridge cannot last over five years." The city refused to accept the bridge because of its high maintenance cost. Also, questions remained about possible liens against the bridge

Columbus industrial riverfront, ca. 1857. The observer is looking north from the Alabama land bridge of King's 1841 Dillingham Bridge. The sketch clearly shows the railroad bridge but not the Upper Factory (or Franklin Street) Bridge, which was built in 1858. The five antebellum factories—manufacturing textiles, flour, cornmeal, and wood products—line the river. *From John W. Barber,* Our Whole Country . . . *(Cincinnati: H. Howe, 1861), 1:756.*

from builders and suppliers, as well as concerns about whether the initial owners had acquired legal title to the Alabama abutment. The council wanted to avoid another ten thousand dollar bill from the owner of the western bank as had occurred with the lower bridge.[21]

As if to underscore the need for the bridge, in the midst of the November 1858 debate, a disease resembling yellow fever broke out in the crowded workers' tenements known as Battle Row. The word spread that Columbus had yellow fever, but after only a few fatalities the frost arrived and arrested the spread of the infection. The city council breathed a sigh of relief, but they still refused to accept the new span. The issue dominated the January 1859 municipal elections. It engendered so much debate that it transformed traditional party alignments. Rather than Whigs versus Democrats, the People's Ticket, which opposed the city accepting the bridge on the company's terms, fought the New Bridge Ticket, which supported the factory owners and lost the election. When in office, the People's Ticket attempted to affect a settlement but failed. Both bridges simply remained toll free.[22]

King probably had little to do with erecting this poor excuse for a major bridge; thus although his reputation did not suffer from problems with this span, this bridge certainly affected his life. Eventually it brought hundreds of mill workers to live within less than a mile of his upper Girard house, which he still owned and where he stayed when working in the Columbus area. A hard-working weaver earned twelve dollars during her best month; King when working on a contract could net eight times that much. While some operatives might have become friends of the King family, the typical poor white resented a wealthy free black. The factory bridge would have made upper Girard a less comfortable place for Horace and his family to reside. The decision to build this bridge may have motivated their move to Carroll County.

The river vindicated the bridge's inspection committee. In less than five years, the raging Chattahoochee destroyed the upper bridge during the night of February 19, 1862. As the river rose rapidly, the current brought "driftwood, fence rails, etc., and soon large trees and timbers, indicating the destruction of bridges and mills above. During that night the long bridge over the river in the upper part of the city was swept off, happily breaking up and floating off in sections, and being still further separated by the violence of the rapids over which it had to pass before reaching the lower bridge."[23] The destruction of the upper bridge eliminated the competition between the two bridges, and the city council promptly reinstituted tolls on the lower bridge. During the war, King, working for the factory owners, rebuilt the upper bridge just in time to have the Yankees destroy it again. King then rebuilt it again after the war.

King is often credited with building for John Godwin's daughter a carpen-

Spring Villa, near Opelika, Alabama, as drawn in 1934. The home of William Penn C. Yonge and his wife, Mary (John Godwin's daughter), was built in the early 1850s. Local and family historians attribute the construction of this elaborate carpenter Gothic house to Horace King, but no written evidence documents his involvement. *Delineated by J. S. Wiatt. Library of Congress, Prints and Photographs Division, HABS, Survey 16-508.*

ter Gothic style cottage, Spring Villa, located south of Opelika, Alabama. Godwin must have had a hand in constructing this home, however, it lacks any substantial documentation. Godwin appeared to favor his oldest daughter, Mary, who married William Penn C. Yonge of Marianne, Florida. In 1847 and 1848 John gave his daughter six and a half acres in Girard and, in 1851, a slave named Abram. In his 1854 will John wrote, "I will to my Daughter Mary A. Yonge the Sum of five Dollars. She having already more than an Equal Share of my property." Two years later, Godwin allowed Mary and her husband to borrow $4,396.52 from Ann H. Godwin's estate. No other child had similar access to those funds, and Godwin never granted land to his other children.[24] That Godwin built a house for his daughter seems highly probable.

Mary's husband, however, had the financial means to build his own house. He had answered the call of the gold rush in 1849 and apparently returned with some California earnings. He later operated the Chewacla Lime Works near Spring Villa, and Yongesboro, a stop on the railroad between Opelika and Girard, bore his name. The gingerbread-laden facade of the Yonges' house immediately captures the attention of a visitor. The vertical Gothic effect emphasizes the three wall-dormer gables, a taller center one flanked by two

Restoration of the enclosed stairs at Spring Villa, ca. 1934. Family historians, who have never visited Spring Villa, mistakenly equate this tight stairwell with the sweeping circular stairs in the Alabama capitol and make King the builder of both. Another myth has the owner W. P. C. Yonge being murdered by a slave on the thirteenth step, even though he died from natural causes fourteen years after the abolition of slavery. *Photograph by W. N. Manning. Library of Congress, Prints and Photographs Division, HABS, ALA, 41, 2–3.*

smaller versions, topped with finials and decorated with fancy bargeboard, not exactly the typical exterior for King's bridge work. In its scale and interior details, the structure is not particularly grand; only its profusion of sawn work distinguishes it from an average home of the period.[25]

A version of the Horace King legend has him building an elegant curved staircase in this home, similar to one in the Alabama capitol. Twenty-first-century visitors to the site are shown bloodstains on the staircase where a slave murdered his or her master, presumably W. P. C. Yonge, at that very spot. In addition to resembling a typical southern folktale, the facts demythologize the story. The master lived until 1879, fourteen years after the end of slavery. No murder occurred there. The staircase, rather than being a cantilevered graceful staircase, is a small set of enclosed stairs crammed into as tight a spiral as possible. Also, given the intensity of King's bridge building work in the 1850s, he probably had little role in this project.[26]

Toward the end of his life, John Godwin experienced a decline in both his health and his finances. He probably lost his investment in the Florence bridge when it collapsed in 1846. The previous year he had served as a lead investor along with S. M. Ingersoll and seventeen others in starting what became the Mobile and Girard Railroad.[27] Construction progressed very slowly, reaching only fifty-three miles from Girard to Union Springs by 1861. That stock interest surely lost Godwin more money, even though the company paid King to build its bridges. The Russell County deed books reveal the pattern of Godwin's indebtedness. He borrowed $2,744 from W. C. Wright in 1842, and mortgaged fifteen lots for $1,231 in 1848; sixteen lots and five slaves for $2,300 in 1850; eleven slaves for $2,140 in 1851; and two slaves for $950 in 1854. Perhaps he took out these loans to finance his construction activities, or maybe they were serial loans to carry the original obligation from year to year. He also had other smaller debts, such as the note he issued to his slave Henry Murray. Horace might well have provided his old master with financial support at this time.

With the death of his wife, Ann, in July 1854, John gained access to more assets in her estate, and his liens stop appearing in the deed books. He, along with his old friends S. M. Ingersoll and Asa Bates, posted a $10,000 bond for John to serve as executor of Ann's estate. Apparently he and his daughter Mary Ann began borrowing money from it illegally. In July 1856, after John failed to appear as an administrator and provide an inventory, the probate judge removed John as the executor.

John's health must have been failing by 1858 or the Columbus Council would have consulted with him about the new upper bridge. He died on February 26, 1859, about four months after his sixtieth birthday. He did not die penniless, as some renditions of the Horace King legend imply. Godwin still had more than ten acres of land at the Godwin Place within Girard as well as rural land. His slaves were sold for $16,250 in February 1860, while his personal effects brought $654. Horace obviously had a close relationship with the Godwins, and he certainly helped family members, but his assistance has been exaggerated.[28]

A central tenet of the Horace King legend has him aiding a Godwin widow and her orphans. As the Reverend Cherry spins the myth: "His master's widow and children became his [King's] wards at his own option, taking care of the former as long as she lived and cheerfully assisting the latter in their education. Mrs. W. C. Yonge, widow of the late Col. Pen[n] Yonge, noted for intelligence, cultivation and high social qualities, now residing at Yongsboro [*sic*], in Lee County is one of them." Certainly Horace never told Cherry this story. King's shaky sense of chronology, which did cause him to confuse dates, would never have erred enough to have him caring for Ann Wright Godwin as a widow. She died five years before her husband, John, and Mary Ann received ample funds from her mother's estate to fund her own education. Another version of King's graciousness has him allowing one segment of the Godwin family (perhaps Susan Albertha Godwin Allen, who was widowed at some point and had four children) to move into his larger house, while he shifted to a smaller one. Such an exchange could have happened, since King obviously did not need a large house in upper Girard with his family living in Carroll County. The exact details are unimportant. The story conveys the message of King's genuine kindness, which Cherry did not fabricate.[29]

Unquestionably King and his old master had an enduring friendship. Horace acquired the largest portion of John's personal items at the estate sale, and the sons paid for these items. They included:

Sundries	$3.75
Buggy & Harness	41.00
Safe	1.37
Feather Bed	10.22
Book Case	14.00
Table	8.00
Feather Bed	9.75
Mattress	3.25
Bedstead	2.50
Lot Jug Ware	.60
Cutting Knife	11.00
	$105.44[30]

Obviously, Godwin's legacy to King consisted of much more than these few material possessions. King owed Godwin his livelihood and his skills. Godwin's sons as executors assumed the cost of acquiring these items for Horace, and they reaffirmed in the Russell County deed book that King was free and not a possession of the estate or its heirs.[31] By that time, they *owed*, rather than *owned*, Horace. He, not the Godwin family, erected a large monument over the grave of his former master. This marble obelisk memorialized King as much as it did Godwin. The editor of the *Union Springs Gazette* saw the elaborate marker and wrote the following, which other newspapers quickly copied:

ALL HONOR TO HORACE

It is fresh in the mind of the citizens of Columbus, Georgia, and of many of those of this vicinity, that Mr. John Godwin, the great bridge builder of the South, owned a remarkably intelligent and faithful servant, Horace, to whose skill Mr. Godwin was greatly, if not chiefly, indebted for his success in that department of mechanics. A few years ago Mr. Godwin in testimony of his gratitude and appreciation of Horace's services, at considerable trouble and expense, obtained the passage of a bill granting him his freedom. This was when Horace had yet hardly passed the full vigor of his manhood. Last February Mr. Godwin died, Horace, in the meantime and during his servitude, having accumulated a comfortable competency.

Not long since we had occasion to visit the marble yard of Mr. H. McCauley, in Columbus, and while examining the beautiful works of art there collected, one met our eye bearing the following inscription:

JOHN GODWIN
Born October, 17, 1798
Died February 26, 1859
This stone was placed here by
HORACE KING.
In lasting remembrance of the love and
gratitude he felt for his lost friend
and former master.

This was a beautiful marble monument, and cost $275.[32]

King's thirteen years of freedom had provided him with an ample income, and because he probably had more resources than Godwin's children, he bought the marker for John's grave. The inscription expressed the personal feelings of King toward John, his friend. King had no intention of praising slave masters in general or the South's "peculiar institution." The newspaper editors that copied this article, however, probably made that inference. White southerners needed reassurance that slavery had positive results and that free blacks stood with free whites in support of slavery. Such an impression, may have been King's motive in erecting the marker, given the mood of the time.

Horace King's monument to John Godwin, 1859. King respected Godwin and certainly owed his old master his livelihood and his freedom. But the prominence of King's name on the marker and the timing of its erection, coming when the Alabama legislature debated reenslaving freed blacks, may indicate that King's motive transcended a desire to honor Godwin. King's words praising his former master might have been intended to show him as a friend of slavery, thereby allowing him to keep his freedom. John Godwin's marker is still the most impressive one in the Godwin Cemetery, which is located beyond the western edge of the original city of Girard. *Photographs by John Lupold.*

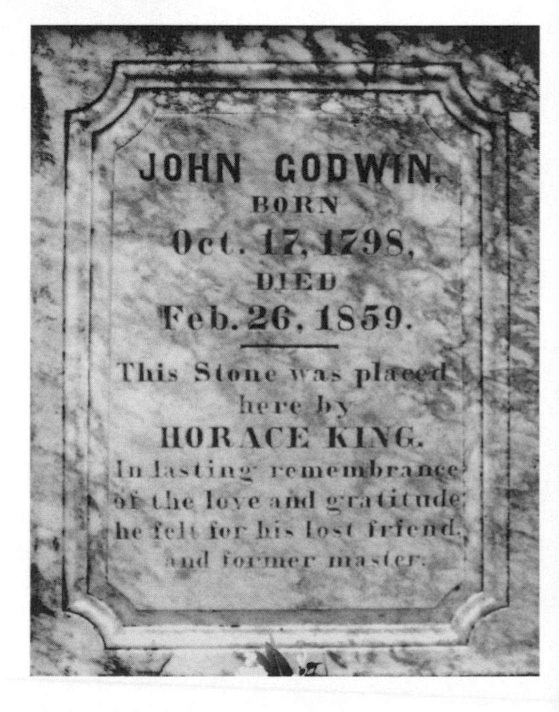

The sense of impending doom that enveloped the antebellum South grew even darker during the 1850s. This act by a former slave represented a ray of light on an increasingly darkening canvas. By 1860, white southerners felt besieged by the rhetoric of abolitionists, the actions of John Brown at Harper's Ferry, the popularity of *Uncle Tom's Cabin* in the North, the success of Free Soilers in Kansas, the growth of the Republican Party, and a myriad of other external threats. The internal threat came from slaves and from their supposed allies, free blacks. Some historians interpret the decline in the number of slaves in urban areas during the 1850s as an attempt to move slaves back to the countryside where they could be better supervised, away from the pernicious influence of nonslave blacks. Many cities sought to increase the regulation of free blacks, especially in reaction to their new, seemingly more arrogant attitudes toward whites.[33]

The prosperity of free blacks, particularly in major southern cities, led them to assert new roles for themselves during the 1850s. Some free blacks displayed their wealth in clothes and carriages, bought property for speculation, formed new churches, exhibited less deference toward whites, and even tried to pass as Caucasians.[34] King participated in this trend in terms of gaining wealth and buying property, but he did not join a church at that time. The First African Baptist Church broke from the First Baptist Church in Columbus in this decade and became an independent congregation, but King probably did not participate. His religious affiliation in this period is unknown. Perhaps the Kings almost seemed to pass as white in Carroll County.

In direct response to the new attitudes of free blacks, some southern newspapers and legislators called for the expulsion or reenslavement of all free blacks. Those attacks intensified after the Dred Scott decision of 1857 asserted that African Americans could never become citizens of the United States, and the Supreme Court stripped blacks of any hope of becoming free citizens. In earlier decades the upper slave states attempted to eliminate free blacks by sending them to colonies in Africa. By 1858 these efforts had failed, and the emphasis throughout the South focused on reenslaving free blacks. A newspaper in Jackson, Mississippi, presented the attitude of many: "a *free* Negro is an anomaly—a violation of the unerring laws of nature—a stigma upon the wise and benevolent system of Southern labor—a contradiction of the Bible. The status of slavery is the only one for which the African is adapted." In October 1858, at the very time King was presumably moving timbers from the capital city to Albany, Representative Moore of Clarke County (Athens) introduced a measure to compel all free persons of color to leave the state by January 1, 1860, "or if they chose to remain, they should pass into a state of servitude, as between master and slave." It failed by three votes. "Mr. Moore instantly gave notice that he should move to reconsider the vote." Such was the atmosphere in which King strived to improve his income and the lives of his wife and children. His life outside his family circle and his construction work was not as pleasant as it seems in the Horace King legend.[35]

King must have shuddered when he read of John Brown's raid in October 1859; its repercussions threatened to destroy Horace's world. Brown attacked the federal arsenal in Harper's Ferry, Virginia, intent on arming the slaves and starting a southernwide insurrection. Five free blacks participated in the raid, and to white southerners, those insurrectionists underscored the revolutionary nature of Africans unshackled from the institution of slavery. The next month, Arkansas, which had very few nonslave blacks, ordered all Negro freemen to leave the state by January 1, 1860, or to choose a master. If they chose neither of those options, they were to be sold into slavery with the proceeds going to support the state school fund. In the wake of Arkansas's action, similar measures sparked debates in almost every lower south state. The legislatures in Missouri and Florida passed similar laws in late 1859 that only awaited the signatures of governors.[36]

Public opinion then shifted. When faced with the reality of breaking up free blacks' families and driving skilled blacks from the state, more rational newspapers and legislators reacted against such injustices. In January 1860 the governors of Missouri and Florida refused to sign their states' acts, and the reenslavement effort in other states lost most of its momentum. The radical fire-eaters railed against these weak-kneed moderates, but moderation prevailed in early 1860. After the election of Abraham Lincoln, however, the moderates failed.[37]

At some point during this debate over expelling free blacks, Jemison acted to protect his friend. Richard Bailey, while searching every nook and cranny in the Alabama archives in his quest to uncover the state's black history, found a copy of an undated bill exempting Horace King and the members of his family from any act that forced the expulsion of free blacks from the state of Alabama. This bill, undoubtedly introduced by Jemison, never became a law, because expulsion never passed. At the same time Horace was having his monument for Godwin carved and erected, the Alabama legislature was debating whether to exclude him from the state.[38]

Given the scale of the monument and the prominence of Horace King's name on it, Godwin's grave marker was intended to do more than merely honor his old master. King sought to make a statement about his support for slavery in order to prevent his reenslavement. Even though King and other free blacks avoided expulsion, Alabama and Georgia's nonslave blacks faced tenuous conditions by late 1860 and early 1861 as the Union began to break apart.[39]

CHAPTER ELEVEN

THE RELUCTANT CONFEDERATE

I stood on the Union Side and always begged and talked for the Union.
Horace King's Testimony before the U.S. Commissioners on Claims, February 1878.

The Alabama convention in Montgomery voted to secede on January 11, 1861. That morning King asked his friend Robert Jemison "if he was going to dissolve this Union?" The moderate from Tuscaloosa replied, "never as long as the blood runs warm in [these] veins." Jemison made a valiant attempt to defeat the secessionists. As the leader of the pro-Union delegates from north Alabama, he rose on several occasions to challenge William Lowndes Yancey, one of the South's and Alabama's leading radicals, but to no avail. The earlier secession of South Carolina had sealed Alabama's fate. Offering a choice between supporting a sister southern state or staying with the Union, the fire-eaters were able to carry the South. Everyone thought the military conflict, if any, would be short and victorious for the South. They never expected to fight a drawn out bloody war. Had the assembled delegates foreseen the carnage of the impending conflict, they would have sided with Jemison.[1]

Once the South seceded, even those leaders who had opposed disunion stood behind the Confederacy. King's associates tended to fit that mold. Jemison, for example, became an aide to Gov. John Gill Shorter and eventually succeeded Yancey as a Confederate senator after the fire-eater's death. King's Albany partner, Nelson Tift, also objected to disunion but contributed to the war effort by building ironclads at New Orleans. King's fellow investors in the Arizonia Bridge Company regretted secession but contributed to the ensuing military effort. Charles W. Mabry voted "no" in the Georgia secession convention but then rose to the rank of major in the Nineteenth Georgia Infantry before resigning to serve in the Georgia Senate. James D. Moore, forty-seven in 1861, did not join the military but later testified: "I never voted for Secession but after the war commenced I did all I could for the South."[2]

King, however, never shifted his allegiances. As he later stated, "I stood on the Union Side and always begged and talked for the Union. . . . I adhered to the Union cause. Secession didn't change my feelings any." He was disappointed that Jemison had signed the secession papers after promising to fight it with every ounce of his blood. King later recalled, "I was told often to hush that the people would Kill me for talking" against the Confederacy. "I had no influence at that time because I was a free colored man and I did not have the right to vote." A recent addendum to the Horace King legend by neo-Confederates portrays him as a black Confederate, implying he supported the southern cause. His testimony and his correspondence with Jemison counter that assumption. "I was a colored man and was as much of a Union man from beginning to the end as I dared to be." While eventually forced to work for the South, he always resented it and rejoiced "when the Union was restored."[3]

King attempted to ignore the war and maintain his normal routine. By the start of the conflict, his wife and children lived in Carroll County, farming and tending Moore's Bridge. Apparently King hoped to continue his usual construction work, such as building bridges for the Mobile and Girard Railroad, in which his old master had invested. King probably threw small spans across creeks on that line and, at some point, raised the piers for its major bridge across the river at Columbus. Early in the war, most likely in 1861, Jeddiah S. Miller, a Coweta County neighbor, engaged Horace to build "a factory & a mill," probably small water-powered grist and textile operations. The war offered new manufacturers the hope of making money.[4]

While the war presented prospects of increased profits, it changed the rhythm of King's work. His bridge building activities became impossible to maintain. After the controversial new upper Columbus bridge washed away on the night of February 19, 1862, the factory owners raised money and hired King to rebuild the workers' crossing. William H. Young, the dynamic entrepreneur behind the Eagle Mill, probably helped this effort achieve a more sophisticated level. Unlike the first makeshift structure at this site, these King-built Town lattices rested on proper piers, but their construction moved slowly. By May 1864 King had finished the bridge. But it remained uncovered because lumber and shingles were needed for wartime projects, and the bridge lacked illumination since no gas pipes were available. The stockholders offered the bridge to the aldermen, hoping to transfer its long-term maintenance to the city. As in 1858 the council refused to accept the bridge, the City Council being reluctant to accept the responsibility of rebuilding the structure when it washed away again. But the councilmen allowed both bridges to require tolls.[5]

Progress on this bridge was slow because of King's involvement in war-related endeavors. Despite his unspoken opposition to the new southern government, as one of the most skilled builders in a booming wartime industrial economy, King could not avoid becoming involved with the Confederacy. In 1861 Columbus already ranked along with Augusta as one of the leading industrial centers of the lower South. The war immediately expanded the city's manufacturing capabilities. The existing industries doubled and tripled their output as textile production for the first time continued around the clock. Young's Eagle Mill, already more sophisticated than most southern mills in terms of its diverse products, wove various fabrics for uniforms, overcoats, tents, and india rubber cloth, as well as spinning thread, twine, and rope. A few blocks to the south, William Brown's Columbus Iron Works shifted from fabricating steam engines and household items to casting cannon, shot, and shells.

Given the wartime demands and the shortages of many mercantile goods, some retailers on Broad Street boarded up their windows and became manufacturers. Storekeepers Manly and Hodges produced tents; the dry goods merchant S. Rothschild tailored uniforms on a large scale; the proprietors of a music store, Brands and Korner, made drums and fifes and also prepared india rubber cloth; Prussian immigrant and tinsmith Louis Haiman shifted to swords and bayonets; cotton factors E. S. Greenwood and William C. Gray initially turned out rope and then invested their capital in a factory for J. P. Murray to fabricate rifles; J. L. Morton and M. Barringer changed from milling lumber and building houses to assembling gun carriages.[6]

In order to guarantee a regular supply of goods, the Confederacy appointed a merchant, Maj. F. W. Dillard, to organize the local shops into quartermaster depots. He established clothing and wagon factories as well the South's largest shoe manufacturer, staffed primarily with black labor. The Columbus Arsenal, created by the Ordnance Department, turned out harnesses, knapsacks, ammunition, and cannon. The demand for labor drew rural inhabitants into the city; working in a factory there provided a relatively safe situation during the war. The wartime boom pushed the city's population from ten thousand to fifteen thousand.

The more the economy of Columbus expanded, the more its leaders feared an attack by Union forces. Civic officials and planters throughout the valley shared this fear and envisioned the invasion route as the Apalachicola and Chattahoochee Rivers. Demands to obstruct the river grew, especially after the Confederates withdrew their troops from the town of Apalachicola in the spring of 1862. In April the Columbus City Council sent a representative to confer with the secretary of war about the inadequate defense of the river. The decision to block the river involved reams of correspondence and a large cast of characters, including the governors of three states, cabinet members, various mayors, army and navy officers, and at least three military engineers. Finally, in the fall of 1862 work began on the obstructions at the Narrows, thirty-six miles above the port of Apalachicola. It involved sinking cribs and pilings across the channel and then attaching large chains to the structures in the river. The chains acted like a trash gate on a dam and caught the debris floating downstream, which formed a barricade to navigation. On November 5 the Columbus council allocated $3,000 for this scheme. Workers in Columbus made the rafts, or cribs, to be used at the Narrows, and skilled laborers from the city accompanied this expedition to the Apalachicola River.[7]

Sixteen years later, King remembered being "engaged in building a bridge at Columbus when the Confederate authorities came and asked me to go down the river to obstruct it. I told them I couldn't do it." Obviously, working as a private contractor appealed more to King than the idea of being impressed into Confederate service, but he had little choice. "The next day they came & said I must drop every thing and go down. I was carried down the Chattahoochee River with a Guard[,] staid there three weeks[,] was under guard the whole time[,] could not leave with out a pass." The Confederates needed King's skills. He apparently traveled all the way to the Narrows, where he directed the labor of slaves who were impressed from planters in the lower valley.[8]

Two states wanted King's skills. As he returned from the Narrows, the Alabama governor John Gill Shorter and Columbus mayor J. F. Bozeman exchanged telegrams, asking where was Horace? Apparently, on the twenty-ninth of November Bozeman notified Shorter, "Horace is driving piles low down on Apalachecola [*sic*] river will be back in a few days." By Monday, December 1, the mayor informed the governor, "Horace Godwin is here." King must have received Shorter's message and told him he was going to visit his family before he started working for the state of Alabama. King left Columbus, but then Bozeman realized he needed him on the Chattahoochee. The mayor telegrammed Shorter on the sixth: "Is Horace engaged if so how long if not tell him I want him on this River." Shorter replied the same day: "Horace gone to New[n]an Georgia—Will return Tuesday [the ninth] and go down Alabama river to consult."[9]

King may have done more than consult on that trip down the Alabama, or he may have returned for an extended stay within a month or so. On March 6, 1863, the Alabama Quartermaster Department authorized the payment of $200 to Horace King for his employment in bolstering the state defenses as well as another $365.15 to King for the "use of I. Higginbottom for hire of Negroes at work on state defenses." The exact location, duration, and nature of King's work remain unknown. Blockade-runners continued to operate on the Alabama River until the very end of the war; so he did not obstruct the river as on the Apalachicola. Perhaps he worked on fortifications at Choctaw Bluff on the Tombigbee. Robert Jemison Jr., who served as an aide to Shorter during this period, might well have recommended his friend for the job. If the $200 represented King's pay alone and no money for his crew, then he might have spent a month or more in the service of Alabama, working impressed slaves to protect the state from their liberators.[10]

After working for Governor Shorter, King probably returned to the factory bridge in Columbus, but he soon became involved in another extended project, this time an offensive rather than defensive venture. At the same time Alabama and Georgia were demanding obstruction of the Apalachicola River, the navy began building crafts on the Chattahoochee designed to steam down the river and attack the Union blockade. This quixotic and expensive construction effort became the city's most spectacular and best-remembered wartime effort.

With the Federal blockade threatening southern port cities and their shipyards, the Confederacy moved some of its naval manufacturing to inland sites. Columbus's industrial experience, especially its iron foundries, as well as its location on the Chattahoochee River made it a natural choice for such a facility. In September 1862 the navy leased (and later confiscated) the Columbus Iron Works. Its new head became Chief Engineer James H. Warner, a former Union officer with experience at the U.S. Naval Yard in Gosport, Virginia. Warner's operation produced at least eighteen steam engines for Confederate vessels, some small brass cannon, and iron fittings for navy ships. Its mission also included rolling iron cladding for gunboats. At the same time, Augustus McLaughlin, another former Union officer, established the Navy Yard to the south of the Iron Works. McLaughlin, who had previously directed the construction of the gunboat *Chattahoochee* at Saffold in Early County, Georgia, produced the two-hundred-foot ironclad *Jackson*, or *Muscogee;* the *Shamrock*, a river steamer for hauling supplies; and the *Viper*, a small torpedo boat. The construction activities associated with these craft drew large quantities of materials and labor into what became a futile effort.

King and his crew eventually supplied $16,946.16 worth of labor and timber for these two operations. His principal efforts ($13,455.15) came within the Iron Works, where he developed a friendship with the chief engineer, Maj. James H. Warner. King's affable personality and his professionalism continued to serve him well, even with military officers. King probably preferred working with Warner rather than returning to the Narrows or going back down the Alabama River for Governor Shorter. This assignment also allowed him to return to Carroll County to visit his wife and children on a regular basis.[11]

From May 1863 until April 1864 King, as a private contractor, constructed a stable and a new rolling mill building for the Iron Works. Designed to melt iron, usually railroad rails, and shape or roll it into iron plate, the mill was to provide cladding for the *Muscogee* and other ships. The first rolling mill burned in October 1862, so its replacement became a high priority for Warner.

The extant receipts show that the size of King's crew varied from month to month, from a high of thirty-eight men to a low of ten. The fact that he collected their pay indicates that he controlled them, a real indication of his independence as a builder. He commanded a crew of carpenters, some of whom might have been slaves. Their daily wages varied, with the lowest paid earning between $1.50 and $3.00 and the highest—including Horace—collecting from $3.00 to $5.00 each day. The rates ascended as the war continued but failed to keep pace with the decline in the value of Confederate scrip.

At the same time he was constructing the new rolling mill, King completed the upper bridge and repaired the lower bridge, presumably with the same crew. Beginning in the spring of 1864 King's name began appearing as a worker on the Iron Works payrolls, which means he no longer controlled his crew. He said the Confederates impressed him, but working for Warner was surely preferable to most alternatives. He served as a pattern maker for the first half of January and as a carpenter from February 16, 1864, until the last extant payroll (June 15, 1864). But based on his testimony and the pace of construction at the Iron Works, he continued in its employ until April 1865. Warner befriended King and kept his sons from being conscripted.

A small number of African Americans served in the Confederate army. Some wealthy slaveholding, free mulattoes around New Orleans, many of whom had fled Haiti, readily joined the armies of the South. Their motives and their racist rhetoric paralleled that of their white neighbors. They feared their slaves, and blacks in general, and fought to prevent their emancipation. The Alabama legislature also allowed the free blacks in Mobile to serve in the army, but most free blacks in Alabama and Georgia had no intention of fighting for the Confederacy. King expressed that sentiment and was extremely worried that his sons might be conscripted. As with so many other aspects of King's story, the chronology of events seems contradictory.[12]

King wrote Jemison on March 23, 1864, that he had traveled to Montgomery in an unsuccessful attempt to visit him when he passed through the city on his way home from Richmond. King wanted information about whether the Confederate Congress had passed a bill to conscript free blacks between the ages of eighteen and forty-five. But King also wanted a private session with Jemison to discuss a scheme, perhaps to remove himself and his family from participation in the war effort. "Myself and children have been at work for the Gover[n]ment for more than a year. I would like to know how we stand if we should not continue at work for the Gover[n]ment. Let me know about what time you will be in Montgomery as I would like [to] Come over and see you on some particular business." Apparently King never talked to Jemison, and he did not avoid participating in the war.[13]

When testifying in 1878, King said his sons were threatened with conscription in 1864 and Warner intervened on their behalf, employing them at the Iron Works. Yet in 1864 Horace wrote Jemison that they had been working there for a year. The actual law to conscript free blacks into the Confederacy, not as soldiers but as wagoners and laborers, did not pass until October 1864. Perhaps King acted to save his sons from local conscription efforts, once again substantiating his status and his opposition to the Confederacy.

While the Kings worked closely with Warner at the Navy Iron Works, Horace also provided supplies and services for McLaughlin's Navy Yard, which constructed the gunboat *Muscogee* and other vessels. Horace earned $3,491 from that facility. By the end of May 1863 King had leased mules and teams to the operation. Since he was working for the Iron Works during the same period, in all likelihood Horace's men or perhaps even his sons handled these teams.[14]

King probably directed the crews that supplied timbers to the Navy Yard. These included trees, logs, oak knees, and treenails. The right-angled oak knees became braces that formed the angle between the bottom and the sides of the boat. King was the logical source for three thousand large treenails; these along with iron spikes tied together the beams of the gunboat. Even though he supervised the preparation of some components, King had no role in actually building the *Muscogee*. But he certainly saw some familiar faces delivering the other timbers and sawn lumber.[15]

A steady stream of wood products poured into the Navy Yard. Alexander I. Robison, King's guardian, supplied forty-two pine logs in July 1863—an indication that Horace was not working his guardian's slaves, at least not full time. King's brother-in-law, Henry Murray, identified on the naval receipts as Henry Godwin, delivered timber and lumber. Apparently using the wagon bought for him by William E. Godwin at John's estate sale in 1859, Henry did Navy Yard draying for 114 days during 1863 and January 1864. Presumably, William and Thomas Godwin, who purchased all their father's slaves from his estate, reaped the profits from Murray's labors, even though Henry, who earlier had loaned money to John Godwin, certainly managed to accumulate some money for himself.[16]

The Navy Yard's sawn lumber came from the small fraternity of lumbermen and builders who had always worked with Godwin and King. Asa Bates sold the facility a double-dray wagon but supplied no boards. Bates continued to receive a salary to maintain the county's bridges, and he and his son worked with Dr. Robison to build a bridge at Woolfolk's on Upatoi Creek in 1862.[17] J. L. Morton, the alderman who inspected King's bridges on a regular basis, and S. M. Ingersoll, Godwin's old friend, supplied lumber, but the largest supplier was William Brooks who operated Variety Works on the riverfront. A longtime colleague of Godwin's, Brooks sold John the land that became the

Godwin Place in 1842. In January 1864 McLaughlin chastised Brooks for selling lumber to civilians rather than fulfilling his navy contract. The Navy Yard commander summoned Brooks to his office and threatened to conscript his workers, who had been exempted from the draft. King later identified Brooks as someone who could testify to Horace's unionist tendency, but by that time Brooks had died. In 1871 Brooks, as a friend of the family, had defended Messina Godwin against the other heirs in a fight over John's estate.

While family sources mention that John's sons continued the contracting tradition, Godwin's former business played a minimal role during the war. Judging by the type of timbers supplied by Henry Murray to the Navy Yard, the Godwins no longer operated a sawmill. No other Godwin supplied any material or labor to either the Navy Yard or the Iron Works. The family also relinquished the Godwin Place in December 18, 1862; it sold at auction to Dr. S. Bullock for $1,200, presumably in inflated Confederate tender. The family appeared to be splintering. Wells had ceased working with his brother, John, by 1850 and in 1860 lived in Butler County, Alabama, where as a farmer he declared ownership of property valued at $30,000.

At least four of Wells's and John's sons served in the military, and one died from each family. While Wells had gone into farming, his son Andrew Jackson maintained the engineering bent of the family, serving as "General Hood's Chief Mechanic in his General Supply Train" before being captured in January 1865. His brother Thomas P. was killed in August 1864 during the defense of Atlanta. Their cousin and John's youngest son, Thomas Metternick, died as a prisoner in 1863. Another of John's sons, John Dill, served at the Narrows protecting the obstructions King had placed in the river. The oldest of John's sons, William E. Godwin, died in 1865. Perhaps King aided some of the Godwins during the war, as the Horace King legend asserts.[18]

Whatever his relationship with the family of his old master, King certainly provided for his own family. Their larder in the summer of 1864 exceeded that of most southerners, white or black, slave or free. Apparently he frequently visited his wife and children at Moore's Bridge. He needed a pass, probably from Major Warner, to travel to Newnan by railroad. King had little need for a pass as a free black before 1861, but as the South faced defeat, issues about the loyalty of free blacks combined with the need to conscript all able-bodied laborers made carrying a pass essential for Horace. He needed to prove both his freedom and his service to the Confederate effort.[19]

When Horace's train arrived at Newnan, his sons probably met him with a wagon. On the way to Moore's Bridge, King apparently called on friends such as Jeddiah S. Miller, who lived between Newnan and Moore's Bridge. King had built Miller's factory and mill earlier in the war. Horace must also have visited with James Moore, who lived at the top of the hill, above the bridge in which Moore and King shared a financial interest. But most of his time was spent with Frances and their children. On a visit in mid-May he paid twelve dollars for a new saddle, probably in Newnan, to replace one taken by "Rebel soldiers." A month later, again in Newnan, he bought a new one-horse wagon with an iron axle for "seventy five dollars in gold money" along with a heavy buggy harness. The family had two mules, a mouse-colored one and a bay, both still active and quick at ages eleven and eight and both still fat. Given the money that he and Henry Murray earned hauling for the Navy Yard, King certainly knew the value of a good mule and a wagon. During his respites in Carroll County, Horace always checked his supplies of corn and flour: seventy-five bushels of shelled corn, twenty-five bushels of unshelled corn, and fifteen hundred pounds of flour by midsummer of 1864. They apparently grew all the corn and most of the wheat; he had bought more grain and had it milled into flour, probably at Miller's mill. After Horace returned to Columbus in late June, he sent his family four hundred pounds of bacon, which cost $120 in gold, and two hundred pounds of brown sugar, valued at $40 in gold.[20]

The volume of these foodstuffs must have been exaggerated. Horace reported these as what the Union troops looted from his house. Even if reduced by half, such supplies exceeded those of most southern families by the summer of 1864. Besides providing for his family, Horace sought to follow the example of shrewd Columbus businessmen who transferred their Confederate paper into physical assets that would survive the collapse of the South. King claimed in 1878 that the money for these purchases came from gold he saved before the war, but he must have redeemed some of that increasingly worthless currency which the Iron Works paid him for food and harnesses. Perhaps King sent provisions to Moore's Bridge, because it seemed safer and more remote, less likely to be a target as the war moved into Georgia.[21]

Rivers and bridges are crucial objectives in all wars. Rivers form natural obstacles that usually give defenders an advantage, whereas bridges allow attacking forces to breach that barrier and continue their offensive. Given those realities of military tactics, King's bridges became of prime concern for both armies as the Union forces tightened the noose on the Confederacy. One of the most important campaigns began in Chattanooga in May 1864. For the next two months, Union general William T. Sherman and Confederate general Joseph T. Johnston danced a "red clay minuet." Johnston conducted a successful delaying action, and Sherman was unable either to destroy or outflank Johnston's army.

By early July, Sherman's forces occupied the western bank of the Chattahoochee, just west of Atlanta and its crucial rail junctions. Given the smaller size of his army, Johnston had fought well and had inflicted a high rate of casualties on the invaders, but Sherman outmaneuvered him along the banks of

Gen. George Stoneman, ca. 1863. His federal troops attacked and burned Moore's Bridge and looted the Kings' house and barn during the raid. *Library of Congress, Prints and Photographs Division, Civil War Photographs, 1861–65, 0989.*

the Chattahoochee. Sherman sent two cavalry forces along the western riverbank, one upstream and another downstream, to ascertain the location of Confederate forces, burn gristmills and textile factories, reconnoiter possible fords and bridges, torch those spans not of use to the Federals, and destroy adjacent railroads.

To the south toward Moore's Bridge, Sherman sent Maj. Gen. George Stoneman, who had roomed with Stonewall Jackson at West Point but never equaled the military brilliance of his classmate. By the summer of 1864 caution and conservatism characterized Stoneman's military tactics. His cavalry forces never moved with the alacrity or daring associated with mounted troopers. As they started south from the area of Marietta, his soldiers burned several gristmills and the New Manchester Manufacturing Company's textile factory on Sweetwater Creek. Following Sherman's orders, they also seized its operatives as prisoners of war; these were mostly women, whom Sherman eventually exiled to Indiana.[22]

Sherman ordered Stoneman to proceed far enough downstream to cross the Chattahoochee River with little resistance. His objective was to destroy the Western and Atlantic Railroad that paralleled the river about nine miles to the east. The Confederates at Campbellton constituted too much of a force for Stoneman to ferry the river there, so he continued his slow pace, moving downstream on the west side. He thought the only bridge existed at Franklin, forty miles down the river. A local resident informed the Federals of the existence of Moore's Bridge, and they captured a southern courier en route to the commander of the small detachment at the bridge. The message told the lieutenant to hold the bridge until reinforcements arrived. While pleased by the discovery of this bridge, Stoneman worried about the threat of reinforcements. If he acted quickly, the Federals could cross the bridge, gallop nine miles to Newnan, wreck the rail line, and then quickly return to the safety of the far bank of the Chattahoochee River.

July 13 began as an ordinary day for Frances and her children. The King's corn crop on the far side of the river was laid by and needed no tending on that hot midsummer morning, and nothing demanded her immediate attention. As the heat rose, the dogs dug deeper into their holes underneath the porch, and only a lone buzzard braved the dull burnt-out sky. Her son John Thomas acted as the toll taker, but no one crossed early that day. Elizabeth Gray, a neighbor, brought her wagon for John Thomas and George to repair. While the boys tinkered, Elizabeth and Frances sat and talked. Elizabeth may have been a little envious of the hams and sugar Horace had sent from Columbus. Their chatting naturally turned to war and the rumors of Yankees moving toward Atlanta. But that seemed far away. They could hear the voices of other boys, the Tennesseans who guarded the bridge, swimming and playing, actually skinny-dipping, hidden from view by the deep river bank. Earlier that morning those men had removed some of the flooring from the bridge, just in case a Federal column attempted to use it.[23]

Shortly before noon, nine horsemen dressed as Confederates galloped down the hill through the Kings' yard to the portal of the bridge. When they aimed their carbines at the swimmers, the Confederates realized they faced Union troopers. Some naked rebels swam for it and made it up the far bank dodging the Yankee bullets, but most never ran; they just raised their hands. The Eleventh Kentucky cavalrymen captured about twenty men, including the rebels' wounded lieutenant and James Moore who came down the hill to investigate the ruckus. The southern pickets had no time to light the pine knots and straw they had stuffed into the latticework of the bridge. The few Confederates on the far side of the bridge fled, and Stoneman's advanced force controlled the bridge. They then turned their attention to the Kings' yard and house.

The Federals took the two fat mules and the new saddle, harness, and wagon, along with the Gray's wagon and harness. They looted the corncrib and the house, carrying off the shelled corn and that still in the ear, brown sugar, flour, bacon, clothes, bedding, cooking utensils, and tableware. As Elizabeth Gray fled to her house, Frances King stormed into the yard and confronted the closest Union officer demanding to know "if there was no masons among them. He said yes he was one, what of it. She said my husband is a Mason, my father was one & I think I ought to be protected. The officer made the men put back two sacks of Shorts [wheat by-products]." Sherman's cavalry commanders heard many pleas in the name of Masonic brotherhood, and this request netted meager results for Frances.[24]

Their loot secure, the officers and men began working on the bridge and locating boats to ferry Stoneman's troops across the river. Within two hours the flooring in the bridge had been replaced, and Federals had requisitioned every watercraft in the neighborhood. But Stoneman hesitated. Rather than boldly striking for Newnan, he ordered his troops to bivouac for the night a little north of the bridge and to employ the old trick of building extra fires to convince rebel scouts that he commanded a larger force. Meanwhile, the citizens of Newnan panicked when they received word of the Federals' presence. The town had become a large hospital center, and many of the walking wounded fled. The post commander tried to muster some of them and the local militia to engage Stoneman but to no avail. Instead of local forces, the Mississippi cavalry of Brig. Gen. Frank Crawford Armstrong came to rescue Moore's Bridge. Moving down the east side of the river, his movement paralleled that of Stoneman's. A few days earlier, Armstrong's forces had tricked Stoneman into believing that a much larger force defended Campbellton, and they affected a similar result on the morning of July 14.

After moving all night, Armstrong's advanced force rode to within half a mile of the bridge before dismounting. Without stopping, they formed into an infantry line and advanced on the bridge. At the same time, a Federal patrol started to cross the bridge; the Confederates opened fire from the wood line and rushed the bridge. All the northern troops scurried to the west side of the river. Armstrong reported to Joe Johnston. "I arrived here at 4:00 A.M. Found the enemy in possession of the bridge. . . . I have a small portion of my brigade. . . . I think I can hold them in check until my troops get up. They are working on the bridge. The abutment was knocked down. They have an excellent position and have made breastworks." Armstrong never needed his other troopers. Having waited too long to strike at Newnan and overestimating his enemy's strength, Stoneman ordered the destruction of the boats and the bridge, and an immediate withdrawal. The historian David Evans described the fate of Horace's bridge: "Braving a hail of bullets," a lieutenant from the First Kentucky Calvary "dashed toward the bridge with a flaming torch and thrust it into the dry pine knots and straw the Rebels had stuffed into" the latticework of the bridge. He "ran the gauntlet again and rejoined his comrades just as smoke began billowing from the cavernous mouth of the span. Flames burst through the roof. Timbers cracked and groaned, and after a few minutes the bridge slid off its piling and plunged into the river." The Federals then moved back from the river, looted more houses, and gathered in their patrols. Late that afternoon Stoneman started moving north; fearing an ambush, he continued his march until two in the morning, still fearing an ambush. By the time Stoneman rejoined Sherman's main force, Maj. Gen. John Schofield had crossed the Chattahoochee bridge near Roswell, flanked the Confederates, and sealed the fate of Joe Johnston and the city of Atlanta.[25]

Back at Moore's Bridge, the Confederates reoccupied their positions, though the span no longer existed. Frances may have gone to a friend's place or to a house owned by H. J. Garrison, their landlord on the Coweta County side of the river. She had no means of communicating with her husband. Horace later remembered those days, "I read an account in the paper that the Union army had burned my bridge [Moore's Bridge] and as my family lived right at it, I got permission to come back to look after them, and when I got home the union troops had been gone about two days and all [my food and other possessions were] gone and my family was in distress." If his chronology is accurate, King must have stayed at Moore's Bridge for a couple of weeks, since he aided some fleeing Union forces in the area later in the month.

On July 27 Sherman, still trying to cut all the rail connections into Atlanta, dispatched two cavalry forces to circle Atlanta in opposite directions, with orders to meet at Lovejoy Station, west of Jonesboro, and to destroy the railroad coming from Macon. Brig. Gen. Edward McCook rode west of Atlanta while Stoneman's troops circled the city from the east. Rather than meeting McCook, as ordered, Stoneman divided his forces, leaving a division to screen his rear while he struck toward Macon with his ultimate objective being Andersonville Prison. Because the Federals divided their forces, Confederate major general Joseph Wheeler engaged the three separate commands on three successive days and routed all of them. On July 28 the southerners scattered the detached portion of Stoneman's troops; on the next day, Wheeler captured Stoneman and seven hundred men near Macon; and on the thirtieth the Confederate cavalry caught McCook's forces at Lovejoy Station, where they had destroyed the railroad and other supplies. Outnumbered by the Confederates, McCook's soldiers fled westward. Wheeler caught them at Newnan and decisively defeated them:. McCook lost 950 men (killed or wounded) as well as his pack trains. The surviving Union troops ran toward the Chattahoochee River.[26]

King recalled the ensuing events: "When McCook's men were defeated in Coweta County and scattered, two of his soldiers came to where I and my family was in Coweta County on this [the eastern] side of the river getting ready to move to Columbus, Ga., and begged for something to eat. I gave it to them, I made no effort to report them. There was a company of rebels at the old bridge place and the country was full of them picking up and hunting Federal Soldiers they could hear of. Said Soldiers said they were afraid of being reported. I told them to take the food and hurry on."[27] Horace then brought his family back to Girard, where in less than a year they would again deal with Yankee troops. Compared to the Godwins, the Kings fared better during the war in one regard—they lost no family members as casualties of war. But Frances King died on October 21, 1864, presumably in Girard. According to tradition, she was buried in the Godwin family cemetery, close to where Horace had erected his marker for John.

King's absence from the Iron Works at that time mattered little to Major Warner since his labor force had been "reassigned" to fight in Atlanta. The workers at the two Columbus facilities formed the Naval Battalion, a militia-type unit organized and drilled to protect the home front. The major remained in Columbus when his skilled-workers-turned-infantry fought in Atlanta. They had returned by August only to be sent to protect Savannah later in the year, much to the distress of Warner and McLaughlin. On November 19 Warner wrote that the mobilization of his workers left "only the negro force with sufficient white men to keep them employed—and guard there [*sic*] works." Even so, they finally launched the *Jackson* in December, but she was still not ready to fight in April 1865.[28]

After all the money the Confederacy poured into river defenses and boat construction, the U.S. military thrust against Columbus came overland, delivered by Gen. James H. Wilson's cavalry, one week after Lee surrendered to Grant—an event unknown to either side but irrelevant in any case since the South always claimed it would fight to the last man. The Union raiders needed to destroy this Deep South bastion to break the Confederates' manufacturing capacity and, therefore, their will to continue the struggle. The twenty-eight-year-old "boy wonder" Maj. Gen. James H. Wilson commanded 13,500 cavalrymen—the war's largest mounted force—when he moved in late March 1865 from Union-controlled north Alabama to strike the cities of central Alabama and Georgia. Armed with Spencer repeating rifles, Wilson's troopers routed Lt. Gen. Nathan Bedford Forrest who mostly commanded home guard forces in front of a well-entrenched Selma, another major industrial producer. Montgomery chose not to fight. West Point and Columbus vowed to defend themselves.

As his main body of troopers drove toward Columbus, Wilson dispatched

Col. Oscar LaGrange's brigade up the Montgomery and West Point Railroad toward Auburn and Opelika, both of which fell without a fight. But at West Point, Robert C. Tyler—a one-legged Confederate general convalescing in a hospital there—organized fewer than 300 men at Fort Tyler, a hilltop redoubt designed to control the King-built wagon bridge and the railroad span, the most strategic features in the city. LaGrange bombarded the fort and sent a detachment to secure the bridges. Then he assaulted what became known as Fort Tyler. He summarized his actions in a report: "The garrison . . . was composed of 265 desperate men, commanded by Brig. Gen. Tyler. . . . 18, including the General . . . were killed, and 28 seriously wounded; 218 . . . prisoners . . . commissary stores, machinery from factories, osnaburgs [a type of cloth] etc. were destroyed. Both bridges were burned. . . . Our loss was 7 killed and 29 wounded, Seven hogshead of sugar, 2000 sacks of corn, 10,000 pounds of bacon and other stores were left in charge of the mayor to provide a hospital fund for both parties."[29]

Not surprisingly, bridges also played an important role in Columbus's defenses and in its brief military skirmish. Given the topography of the Alabama hills west of Columbus, the military commanders decided not to burn all the Chattahoochee River bridges but to defend just the east side of the river. The Confederates destroyed the footbridge north of the city at Clapp's Factory and removed the flooring from the lower Godwin-King bridge at Dillingham Street. They preserved King's new factory bridge and the railroad bridge, four blocks to the north. The defensive line protecting those two spans ran from the upper bank of Holland Creek northward up the hills along Summerville Road, overlooking the river. The defenders abandoned the outer fortifications and concentrated all their efforts on this inner line. Columbus appeared ready to protect its bridges and repulse Wilson's troops. The defenders did not realize that the Federals had routed the legendary cavalry of Nathan Bedford Forrest at Selma by charging for the bridge. And Forrest's well-equipped veterans contrasted sharply with the inexperienced two to three thousand defenders manning the defense at Columbus; very few regular troops augmented the home guard—old men and boys and carpenters, pattern makers, molders, spinners, and weavers.

In the middle of the afternoon on Easter Sunday, men from the First Ohio galloped through Girard south of the line of fortification and reached the lower bridge. As they attempted to jury-rig a passage over the unfloored span, Capt. C. C. McGehee of the Naval Battalion, actually a clerk rather than a military officer, braved bullets from the Spencer rifles and torched the Godwin-King span, which, like the upper bridge, was mined with kerosene-soaked bales of cotton waste. Gunfire from the Confederate positions forced the Federals to withdraw. After making a personal reconnaissance and consulting with his subordinates, Wilson decided to shift his attack to the north, down Summerville Road through the middle of the defensive works, and to strike at night. Though they were out of harm's way, the Kings must have evacuated their Girard house just west of the defensive line that was held in part by his fellow workers. King's family probably fled into Columbus, where residents expected an artillery bombardment. Instead they heard small arms fire followed by Confederate cannon.

At about 8:00 P.M. on Easter Sunday, April 16, Wilson began what old Columbusites proudly called "the Last Land Battle of the War-between-the-States East of the Mississippi." Chaos characterized this night skirmish that lasted hardly an hour. As dismounted Federals attacked, the unblooded Confederates tended to fire over the enemy's heads. The Yankees were confused about where to strike as units became entangled with each other in the brush on Ingersoll Hill. One mounted column rode all the way to the bridge before being beaten back. Then the forces in the redoubts and on the line began to flee, running to the safety of the east bank of the river. As their line broke, Confederates raced Federals on foot and horseback, all running for the bridge. "Horsemen and footmen, artillery wagons and ambulances were crowded and jammed together in the narrow avenue, which was 'dark as Egypt,' or 'Erebus,' for that bridge had no gas fixtures and was never lighted. How it was that many were not crushed to death in the tumultuous transit of the Chattahoochee, seems incomprehensible." By 11:00 P.M., when Wilson reached the east bank, the fighting had ended. He had lost twenty-five men. The Federals captured about one thousand Confederates and killed nine people, seven on the Columbus side, including "young Alexander J. Robison," presumably the son of King's former sponsor.[30]

The next day the destruction began. Wilson ordered Gen. E. F. Winslow to oversee the burning of all the industries and between fifty thousand and one hundred thousand bales of cotton in the local warehouses. The Federals seized and torched the torpedo boat *Viper* and the ironclad *Jackson,* which had consumed so many resources and never became battle ready. The crew of the *Chattahoochee* had scuttled it the previous night. The conflagration consumed King's rolling mill building and all the structures in the Navy Yard and the Iron Works. The Federals carefully chose their targets, which did not include any houses. They allowed the flour and corn mills of James Waldo Woodruff and Randolph J. Mott—the city's best-known Unionist—to survive, while they purposely sought out Seaborn Jones's City Mill. Although Jones had died in 1864, his reputation as an outspoken secessionist drew fire to his gristmill. In a similar fashion, they destroyed the presses of the two Democratic newspapers, the *Times* and the *Sun,* while allowing the Whig, antisecessionist *Enquirer* to survive. As the fires raged and the main body of the troops headed

Fourteenth Street (Franklin Street, Factory, or Upper) Bridge, 1886. The original bridge at this site, built by the factory owners to provide passage for their workers to Alabama, was an uncovered bridge with numerous short spans. It failed in 1862. King then constructed a proper Town lattice truss. This bridge became the focus of both armies during the Battle of Girard–Columbus, after which Federals burned it. In 1867–68 King built the span shown in this 1886 view. *Perspective Map of Columbus, Georgia. H. Wellge, Beck, and Pauli Lith. Co. Library of Congress, Prints and Photographs Division, Map Collection, 1500-2003, 75693191.*

Sarah Jane Jones King, ca. 1865. Horace's first wife, Frances, died in October 1864; he married Sarah (or Sallie) Jane Jones in June 1865. She was much younger than Horace but appeared to share his triracial background. When the family moved to LaGrange she became a full partner in the family business, King Brothers Bridge Company. *Collection of the Columbus Museum, Columbus, Georgia; Museum Purchase.*

toward Macon, a mob composed of whites of all classes, former slaves, and some Union soldiers looted the stores on Broad Street.

In the midst of all of this chaos, King probably crossed his Franklin Street Bridge one last time before the Federals burned it on Monday. When he returned to his Alabama home, he found that northern troopers had taken two mules, which he retrieved from a Union officer. He explained to the officer, "That I was not rebel and had nothing to do with it. But was a Union man. And I would be glad if he would return them to me. He, the officer, went and got the mules and returned them himself. He said he had no idea that I could keep them but he shouldn't take them or let his men take them if he could help it. They were never taken from me again." King did not use, as several accounts later claim, any reference to the Masons to retrieve his property.[31]

A pair of mules represented an extremely valuable asset at the end of the war. Those mules would be as essential in postwar building or rebuilding as they had been for King during the war. The officer assumed that in the anarchy of defeat no black man could maintain ownership of two mules. That King could talk the officer out of the mules and then keep them in the postwar chaos attests to his persuasiveness and even more to his position within the community. With the end of the war, King achieved a new level of freedom, and the wake of destruction left by the Federals opened up unlimited construction possibilities for this established contractor. While white elites mourned the passing of the Confederacy and blacks celebrated their new freedom, few people enjoyed as many opportunities as Horace King. Horace also moved to establish a new family. On June 6, 1865, less than a year after Frances died, the rector of Trinity Episcopal Church married Horace and Sarah Jane Jones at their home. Sarah Jane was a mulatto, probably with the same triracial background as Horace and Frances, apparently much younger than Horace and more feminine in appearance than Frances. With a new wife by his side, Horace was ready to mine the opportunities offered by Reconstruction. While best known as an antebellum free black, his most prosperous period probably came after the war that freed all black men.[32]

CHAPTER TWELVE

ECONOMIC RECONSTRUCTION

As we are satisfied that whoever undertakes the contract, they will of necessity be compelled to have the services of Horace King, and should the matter be delayed Horace may be employed at other points, when at present other parties are endeavoring to get him.
Columbus City Council Minutes, May 12, 1865

The reconstruction of the South occurred at three different levels: the physical rebuilding of the war-torn infrastructure, a political realignment with black participants, and the social reforging of the region's racial relationships. Horace King as a contractor benefited greatly from the economic restoration and was involved, though less actively, in the political changes. Because of his prewar status as a free and successful craftsman, King immediately gained an acceptance denied other freedmen after the war. White elites accepted King, certainly not as an equal, but as a professional and a man of integrity. His status benefited him as a businessman and, at the same time, pushed him perhaps unwillingly toward political involvement. While King's political career garnered him notoriety, his work as a contractor remained his first priority.

The Civil War, precipitated by southern secession and fought mainly on southern soil, devastated the region's economy. Freeing the slaves obviously affected large-scale agriculture. Postwar experimentation with labor arrangements continued for several years before sharecropping became the norm. The region's capacity for economic innovation remained limited because abolition eliminated half the invested capital in Alabama and Georgia. Even those portions of the infrastructure—such as mills and railroads—that escaped destruction were worn out after four years of constant use with little or no maintenance. On every southern street, the sight of so many empty sleeves and pants legs provided grim reminders of the real cost of the conflict.

In stark contrast to most southerners, the editors of the Columbus newspapers forecast a bright tomorrow. Scarcely three months after Wilson's

Raiders burned the town's industries and cotton warehouses, the *Enquirer* quoted Alexander Pope:

> What war can ravish, commerce can bestow
> And he returns a friend who came a foe
>
> We are convinced that we shall yet see this truth strikingly exemplified in the restoration of prosperity to our now suffering country, and that COMMERCE and INDUSTRY will be the great agents in the good work.[1]

Most textbooks associate the idea of the New South with Henry Grady and the *Atlanta Constitution* in the 1880s. Actually, editors in other urban centers advocated similar goals almost two decades earlier. Editors of the *Sun* and the *Enquirer* in Columbus and the *LaGrange Reporter* advocated sectional reconciliation, industrialization, and agricultural diversity immediately after the war. As early as the fall of 1865, the *Sun* reported: "Never has the business of Columbus been more active . . . Never were the prospects of the city more bright and hopeful. Do not listen to croakers. There's a better day coming, and what we daily see—the numerous wagons on the streets, the absence of idle Negroes and the general life and industry in every department of trade are not illusions, but facts—the beginning of a happy and permanent prosperity."

These editors, however, tended to overlook the immediate poverty, ignore the paucity of currency and capital, and overstate the rapidity of the recovery. Nevertheless the rebuilding had begun by the fall of 1865. Almost before the dust from Wilson's horses had settled, the merchants and industrialists hired unemployed workers to clean charred bricks as the first step in rebuilding Columbus's central institutions—the mills and cotton warehouses. The price of cotton reached a record high at the end of the war and while it later declined, it remained at a profitable level at least until 1873, thereby, encouraging farmers and planters to plant more cotton. The rebirth of the city's cotton trade demanded the reconstruction of warehouses and bridges. Wagon bridges brought the white staple to Columbus, and railroad bridges connected it with world markets. The arrival of railroads in the 1850s diminished the significance of the river trade downstream to Apalachicola, but riverboats continued to bring bales of cotton to the city wharf for sale to the local cotton merchants.

The city's mills—nothing more than piles of burned bricks in April 1865—still possessed all the requisite elements for recovery: waterpower from the Chattahoochee, experienced operatives, ample markets, entrepreneurship, and even capital in the case of the shrewd industrialists who still had cotton or had transferred their Confederate paper into gold or other assets. For example, during the war S. M. Ingersoll sold William H. Young the Alabama land adjacent to the upper bridge and most likely the lumber for the houses that filled the Eagle and Phenix village. The 1858 Franklin Street Bridge connecting Georgia and Alabama had increased the interdependency between the residents on each side of the river.[3]

The destroyed factories and warehouses, as well as the crippled infrastructure, offered numerous entrepreneurial opportunities for Horace King. With his prewar reputation, King became the rebuilder of choice for companies and towns, but King wanted more. He wanted to establish a new identity, a new position in this society without the shadow of slavery.

King's skills certainly separated him from the average black man who was forced back to the plantation for survival, but his situation illustrates at a microlevel the differing perceptions of whites and blacks after 1865. White businessmen and aldermen viewed King the same after 1865 as they had before the war, but King's expectations changed. Before the war, King made his reputation as a skilled builder, as someone hired to construct excellent bridges or large buildings, but not really as a contractor or an entrepreneur who floated loans and speculated on building projects. The men who hired him immediately after the war continued to view him in the same limited role. They perceived him only as a builder. King, on the other hand, wanted to become a contractor and moved to establish himself in that position. Few southern blacks in 1865 even hoped for such a status. King's peers in such aspirations were the new white industrialists who emerged in Columbus after 1865.

Many individuals shared King's goal of becoming an entrepreneur in what they hoped would be a postwar boom. Columbus industries rebounded with amazing speed, in part because these new industrialists had experienced some success during the war. But the postwar economy presented more challenges than selling products in a wartime environment. The foundry men who worked alongside King in the Iron Works during the war quickly established their own postwar companies. Before the war only two foundries existed in Columbus. During the late 1860s and early 1870s, as many as eight of these companies competed with each other.[4]

These foundry men were only one group of new entrepreneurs who achieved success as manufacturers during the war and continued their enterprises after 1865. Columbus, which had only nineteen industries in 1860, experienced a boom during the war when the number of factories rapidly expanded. In 1865 the Union troops destroyed most of them, but by 1870, 108 manufacturers operated within the city. Most of these were small businesses producing iron goods, specialty textile items, rope, jute bagging, cottonseed oil, carriages, furniture, cigars, beer, ice, chemicals, patent medicines, perfumes, and soda waters. (One of the soda water producers evolved into Coca-

Harper's Weekly view of Columbus, September 19, 1868. This drawing illustrates the centrality of King in rebuilding the postwar city. He superintended the construction of the covered Dillingham Bridge for John D. Gray in 1865–66. Two of the King-built piers of the Mobile and Girard Railroad bridge are visible to the right of the bridge. He won the contract for that bridge in 1869. During the Civil War King and his sons worked at the Iron Works to the right of the bridge. The smokestack serves the gasworks, and the large steeple and sanctuary of the First Presbyterian Church were projects of Asa Bates. *Collection of the Columbus Museum, Columbus, Georgia; Museum purchase.*

Cola.) The difficulties of competing in the new railroad-linked national market coupled with the impact of the crash of 1873 destroyed many of these small companies. By 1880 nearly half the businesses had failed; only eighty-four factories existed within Columbus.[5]

Although he never listed himself as a factory owner, King should be viewed as one of these eager entrepreneurs, and his difficulty in trying to patent an improved saw blade illustrates some of the problems faced by small producers in the national market after 1865. According to family tradition, Horace invented a saw with replaceable or "'inserted saw teeth' and ran the first sawmill by the new process. After the war he sent North to have one manufactured, but then found that another person had patented a similar invention." This account, recorded in King's obituary, continues, "He was thus deprived of the fruits of his labor. The value of the discovery may be inferred from the fact that a gentleman [previously] offered $10,000 for one-eighth interest in it." The description seems to date his innovation as occurring during the war. If so, King did not produce much lumber then. If he had, his boards would have appeared on the list of items he sold to the Navy Yard along with timbers and oak knees. But his tinkering with an improved saw blade indicates his shift from being just a builder to being a contractor, as he began sawing his own lumber. In 1871 he supplied $300 worth of lumber to Muscogee County. He speculated and took monetary risks to increase his profits, and like many of his peers, he experienced financial reverses. He was one of very few African Americans, if not the only one, among these new entrepreneurs in Columbus. He was certainly the wealthiest black member of this circle, at least during his flush times. At some point between 1865 and his death in 1885, he experienced severe economic losses, perhaps because of speculation and the impact of the 1873 crash.[6]

The primary economic activity—growing cotton—was the same after the war as before, though the volume increased after 1865.Beyond the farmers and planters, the critical men in this sphere were the cotton factors—who bought, stored, and marketed the crop—and the commission merchants—who furnished seeds, fertilizer, and other supplies. These businessmen were not newcomers to Columbus or upstarts in this profession. They were part of the old elite. After Wilson's raid their primary objective was rebuilding their warehouses. One of the established cotton factors, John Fontaine, formed a partnership with W. H. Hughes at the end of the war. Even though Federal troops burned Fontaine's warehouse in 1865, Fontaine and Hughes managed to protect a considerable amount of wealth during the war. In February 1866 their combined assets were estimated at $100,000, and by that date Horace King and James Meeler had reconstructed the Fontaine and Hughes warehouse.[7]

Fontaine had employed Meeler and King to build a warehouse for Clapp's

W. C. Bradley Warehouse, originally Fontaine Warehouse, ca. 1925. Horace King and James Meeler built this warehouse and office in 1865; it became the office of the W. C. Bradley Company in the 1880s. The cotton bales had been removed from the lower floor at the time of this picture because of a flood scare. *Courtesy of W. C. Bradley Co.*

Factory in 1855. Their work obviously pleased John Fontaine since he hired them to rebuild his cotton warehouse after the war.

The undated contract for a $10,000 facility allowed Meeler and King to draw advances for supplies but required both builders to sign jointly for all the funds they received. Meeler and King needed Fontaine to pay their suppliers because the builders lacked the capital to finance the materials required for this massive structure. Fontaine and Hughes acted as their own contractors. King and Meeler served as the builders for this edifice that measured 212 feet from north to south and "about" 151 feet from east to west. The contract indicated that certain features, such as wall heights and the iron-covered doors and windows, would conform to those on the Lowell warehouse, a block to the north. Possibly, King and Meeler built that structure before they erected Fontaine's building.[8]

Like the Lowell and other cotton warehouses at that time, this one formed a hollow square with an open courtyard in the middle. The Front Avenue side opened to the public and contained an office, a "Scale Room," and a "Counting Room" for sampling cotton. King and Meeler created elaborate corbeled brickwork on the facade, a testament to the owners' prosperity. Cotton bales were stored in the basement under the Front Avenue rooms and in the two-story side and rear portions. Viewed from the inner courtyard, the building resembled a frontier fort. Two-story covered wooden porches lined the interior space. Massive wooden posts and beams supported these walkways, which provided access to the low-ceilinged chambers that held the cotton bales. Cotton bales entered through passageways in the front or rear that were connected to the center of the warehouse.[9] The nearly solid outer walls protected the valuable commodity stored inside the facility.

Fontaine wanted Meeler and King to finish his storage facility by October 1865, if not before, and apparently they were timely because by December 14, 1865, the *Enquirer* ran ads for Fontaine and Hughes, Warehouse and General Commission Merchants. Fontaine only enjoyed his new warehouse for about a year; he died in November 1866. Hughes was in bankruptcy by July 1872. Such were the economic vicissitudes of the postwar South. The warehouse still stands and probably represents the most significant of Horace King's extant buildings because of its importance to the economic history of Columbus.[10]

This brick structure served as the home base for the city's most important business enterprise, that of W. C. Bradley. As a young man of nineteen, he left his father's plantations on the Chattahoochee in 1883 and came to Columbus. Within two years he and his brother-in-law S. A. Carter controlled the business at Fontaine's warehouse; they later changed its name to the W. C. Bradley Company.[11]

The company's 1899 ad for fertilizers, cotton factoring, and wholesale groceries listed the Fontaine Warehouse as its address, thirty years after Fontaine's death. By that time, Bradley also owned the adjacent Alabama Warehouse, and he hired architect Firth Lockwood to unify and refurbish the Front Avenue facades of these buildings. Additions eventually filled the space in the original interior courtyard, and more windows and doors pierced the exterior walls. Even with those changes, the interior post and beams as well as the exterior of the building, especially along the north wall, testify to the skills of Horace King and his partner James Meeler.

In 1865 a resurrected City (or Dillingham) Bridge was essential for the success of Fontaine's warehouse and for restoring the city's economy. The city council believed Horace King was the only man who could build that bridge. At sessions on May 8 and 12 the aldermen examined proposals from, not surprisingly, Col. Asa Bates and Major J. H. Warner, formerly of the Naval Iron works. Warner was actually fronting for Horace King, but the council apparently did not realize it. At the May 12 meeting, the bridge committee recommended a "speedy action" because "whoever undertakes" the contract would need King, and he would be employed elsewhere if the matter were delayed. They addressed their remarks not to Horace, but to the potential contractor

City (Columbus, Lower, or Dillingham Street) Bridge, ca. 1900. This bridge, more than any other span, is associated with Horace King. His owner, John Godwin, brought him to Columbus after winning the contract to construct the original bridge, which was financed by state loans to the city. Horace spent forty years rebuilding and repairing the piers, trusses, flooring, siding, and roofing of this crossing. After the Civil War, the Columbus city council asserted that regardless of which contractor secured the right to rebuild the City Bridge, he would have to use Horace King. Horace refused to build the bridge as a contractor because he would be paid as tolls were collected. However, he did work as construction superintendent for John D. Gray who won the contract. The fact that this span survived until 1910 enhanced Horace's reputation. *Mike Haskey, Columbus Ledger-Enquirer.*

who would have to hire Horace. The council granted the contract to Warner for $26,000. Payment would come from tolls from the city's flat (or ferry) during construction and from bridge tolls after its completion, until the building cost was met.

At the same meeting R. L. Mott proposed to rebuild the upper bridge at his own expense and to collect its tolls. At any point in the future, he specified, the city could reimburse him for his construction costs and take possession of the bridge. The council rejected his petition, but apparently Mott intended to reconstruct that toll bridge anyway. Given the convoluted history of that controversial span, the council did not move to stop Mott. He represented one of the most powerful economic and political forces within the city. As the city's most outspoken Unionist, Mott had allowed General Wilson to stay in his impressive riverfront home immediately north of the Franklin Street bridge. As a leader of the local Republicans, he exercised considerable influence with the federal troops and over the course of local Reconstruction. He also controlled a multitude of economic investments, including Palace Mills, a large brick gristmill at the southern end of the river-powered mills. While Mott only employed a few operatives, his general interest lay with the mill owners who wanted the bridge rebuilt for their operatives. The move to build an upper bridge led King to take an action he had probably never contemplated before 1865. In a letter to the council, he wrote, "My instructions to Major Warner were to insert a provision in his proposition which was that we would build the Bridge . . . provided there was no steps taken to build a Bridge Up Town which has been the case." King's motives were simple. He refused to wait for his compensation. If an upper toll bridge competed with the lower one, it would take too long to collect enough income to cover his fees. Therefore, King scuttled Warner's plan and forced the council to ask for bids again.[12]

On May 20, 1865, three bidders submitted proposals to rebuild the old city or lower bridge:

B. H. Coleman, W. R. Brown, and George Goulding for $26,000;

T. Jeff Bates for $25,000; and

John D. Gray for $24,000.

Coleman, Brown, and Goulding simply stated that they would follow Warner's specifications. Coleman had worked with King as the brick contractor in erecting the rolling mill during the war, and Brown owned the Columbus Iron Works, so they represented continuity with Warner. T. Jeff was Asa's son. Gray, who won the bid, had been a major contractor on canals, railroads, and other projects throughout the South before the war. He established an iron foundry in Columbus during the war. His brother William, as a partner in the cotton factoring firm of Greenwood and Gray, financed rifle manufacturing during the war and a small textile plant after 1865.[13]

Gray contracted this bridge; Horace King acted as superintendent or one of the supervisors for its construction. The Reverend Cherry's account, based on King's recollections, and other local sources credit the men with those roles. Under this arrangement, Gray paid King a salary. Given the limited number of skilled bridge builders, Gray needed King and his four sons to work on this span, especially since the council wanted it completed in a hurry. Gray had problems finishing this bridge within the allotted time frame of only sixty days. The council agreed to pay Gray in "good current funds" as the work proceeded, but no funds valued at par were available. Strapped for financial resources, the city had borrowed $20,000 from John Fontaine at 15 percent interest, in part to provide money for the project. Gray's involvement with the bridge might have encouraged Fontaine to extend credit to the city. He had provided capital for Gray's railroad construction in north Florida before the Civil War. Fontaine, of course, also needed the bridge to attract Alabama cotton to his warehouse. In addition to funding problems, Gray experienced difficulties in securing sawn lumber. An old stockpile of lumber, originally intended for the Mobile and Girard Railroad bridge, proved to be unsuitable, and the federal government expropriated another expected source from a local sawmill before Gray could obtain it. On August 10, only a month beyond his deadline, the council insulted Gray by demanding he give a guarantee that he would fulfill his contract. His indignant reply informed them that during thirty years of public works with "many contracts of considerable magnitude," he had never failed to complete any of them.[14]

The entire city was eager for the bridge to be finished. The *Daily Sun* provided a running account of its progress. On August 31, 1865, it reported: "The bridge will be completed in about three weeks, foot passengers can cross next week." By September 5, it was "rapidly approaching completion." By September 12, "Many persons passed over the bridge Sunday. Mounting a ladder from this side [Columbus], many passengers crossed to Alabama minus the ferriage fare. Some person, however, did collect 5¢ each from some negroes and greenhorns." Apparently, the ladder was placed underneath the bridge and connected the bank beneath the bridge with the completed framework. "We think by to-night the bridge will reach the banks of this side, so that passers may dispence [*sic*] with the ladder. . . . It may be a month yet before the structure is completed."[15]

The *Sun* continued to record every step in the process. By September 16 it reported, "The floor is being laid from the Alabama end. The frame work of the sides and roof is nearly finished." On the twenty-first, "Wagons, we learn, can pass over the bridge late this afternoon or to-morrow morning. John Gray will collect tolls until the bridge is paid for by council." Finally on September 24, the *Sun* reported, "The sides, roof and railing will soon be completed. The difficulty in procuring lumber has been a drawback all along. At last we have a bridge across the Chattahoochee." The builders moved slowly to apply the final details. Not until February 12, 1866, did Gray inform the council that the bridge was"painted and finished." He had moved much earlier, as soon as people began crossing the bridge, to name Mr. Hines as the bridge keeper. Apparently, the new bridge rate structure still exempted all produce and crops. Pedestrians passed free, while riders of horses and operators of wheeled vehicles paid tolls. Usage remained high. By late November tolls averaged forty-five dollars per day. At that rate the bridge could have netted $10,000 per annum.[16]

Immediately northwest of the new bridge in Alabama lay Holland Creek. Confederate defenders had burned the span across that stream in 1865, and it had not been replaced. Pedestrians crossed by means of a log, but wheeled vehicles heading northward took a long "winding road" before reaching a suitable bridge over the stream. With the upper railroad bridge out of service, railroad passengers bound for Atlanta or Opelika followed this circuitous route to reach the rail terminal in upper Girard. In January 1866 Russell County moved to remedy this situation by awarding the contract to bridge Holland Creek to "Horace King, one of the best builders in this country."[17]

King joined with three other entrepreneurs to extract more money from travelers using the new Columbus bridge. Horace worked for the elite before 1865; after that he joined them as a fellow speculator. The former slave partnered with S. M. Ingersoll, one of the founders of Girard and a colleague of Godwin and King since 1832; Judge Alfred Iverson, a Democratic U.S. senator from Georgia during the 1850s; and R. L. Mott, the area's leading Republican, who owned land and a mill (perhaps a sawmill) west of Girard. On February 13, 1866, the Alabama legislature granted them the right to build a turnpike from the reconstructed lower bridge to Crawford Road, which headed directly west out of Girard. Tolls on this plank road could not exceed those of the lower Columbus bridge, and they would remain in effect for a limited time, until "a sufficient amount of money is raised to pay for the building of a new bridge across the Chattahoochee river." This toll road was a scheme to raise money for rebuilding the upper (or factory) bridge, the rationale being that travelers on this road would have crossed that bridge if it had existed, so they could help pay for it before it existed. Apparently King never built this turnpike, but it illustrates the machinations behind financing a second bridge in Columbus.[18]

In the same month he joined with these members of the old elite, King mourned the death of his friend Maj. J. H. Warner, who became an accidental casualty of Reconstruction. Since Wilson's raid, Warner had worked with the Union navy to save naval equipment at the Iron Works and Navy Yard. He had struck a deal to raise the *Muscogee* and the *Chattahoochee* for the U.S. Navy and to retain 50 percent of their salvage value, a scheme that may have involved King. Warren also planned to move to New Orleans to pursue his career as a navy engineer. His death ended those plans and any future collaborations with King.[19]

At the beginning of Reconstruction, when Pres. Andrew Johnson was still in control, the actions of white federal troops reflected the racial views of white southerners. In Columbus the Union soldiers forced idle former slaves to destroy the fortifications around the city, and the same units burned the shanties inhabited by blacks on the east commons between what is now Sixth and Tenth Avenues. The federal commander enforced the wishes of local white leaders. Only a few black units remained in the South and one of these arrived in Columbus on February 6, 1866.

Several incidents ensued between these soldiers and white Columbusites. Cooper Lindsay, a young townsman, shot a black soldier and was then captured by a detachment of these troops commanded by a white lieutenant. A mob of locals, numbering more than one thousand, challenged the lieutenant, who released his prisoner. Lindsay fled town but tensions remained high. That evening about dusk, a black soldier fired random rifle shots from his hotel window. One round struck Warner in the leg as he was walking home from the Iron Works. The mob quickly reassembled, but moderate town leaders restrained the group from attacking the black soldiers. Warner's wound became infected and doctors could not moderate it, even after amputating his leg. The naval engineer died after three weeks of suffering. By February 22 the black troops had left Columbus.[20]

During 1866 King worked with "Mr. Bates," either Asa or his son T. Jeff, to rebuild the railroad bridge for the Western Branch Railroad that connected Columbus with the Montgomery and West Point line at Opelika. In the 1850s, when this road reached town from the west and the Southwestern (or Muscogee) line came in from the east (by way of Fort Valley), the city fathers sacrificed the north and east commons for progress. They situated the railroad depots on the four blocks of former green space along the eastern edge of the city and allowed the Western Branch to lay its track along the northern commons. It bridged the river just south of Eighteenth Street. King apparently had no role in building the first crossing at this site, but he and Bates reconstructed that span by September 21, 1866, when, appropriately, a lumber train became the first to test the span.

The *Sun* criticized the "officers of the [rail]road" for taking seventeen months to replace the bridge burned by the Yankees because they procrastinated so long before they let the contract. But once construction started it moved forward with more than deliberate speed. The editor who had criticized the company praised the "strong and substantial structure" that "reflects great credit on the builders, Mr. Bates and Horace King. They have pushed forward the work as rapidly as possible."[21]

During an 1883 interview T. Jeff Bates indicated that the 1866 bridge was built primarily by blacks, under the direction of the Bateses and the Kings. T. Jeff stated that he and his father always used black workers. He praised black mechanics but viewed them as being limited. "[T]hey can do what they have seen done before them, but they cannot start anything new. They have to copy." T. Jeff had apparently forgotten about the skills of King and his sons, who worked with the Bateses and regularly won contracts when the two families competed.[22]

In 1866 their joint crews worked rapidly to finish the railroad span. The tracks ran on top of the Town deck truss, which was, by that date, "fully completed except the planking of the sides and flooring it." King and Bates never covered the sides of this span. Slowly, Columbus moved toward regaining all its prewar facilities. "If the upper bridge was built, and it soon will be, the river crossings would be as they were over a year and a half ago—before Mr. Wilson and his 'critter-company' came to town."[23]

Despite the timetable suggested by the newspaper, another year and a half passed before the manufacturers and the city fathers decided how to fund rebuilding the upper bridge. In the meantime, King did not lack for contracts, and he must have directed several projects simultaneously. In late 1866 and early 1867 he apparently rebuilt Clapp's Factory, located about three miles north of town, just north of the present location of the Oliver Dam. He and James Meeler had built a warehouse there in 1855. The postwar mill management, which kept Clapp as the superintendent, had some new investors, including Gen. R. H. Chilton, Robert E. Lee's adjutant general, who became the company's titular president. Employing Confederate veterans in such positions was typical after the war; the practice provided jobs for this class of gentlemen while legitimating industry by connecting the Lost Cause to the New South.

The 1866 wooden mill that King built for Clapp and General Chilton stood four and a half stories tall including a basement and a clerestory monitor running the entire length of the 120-foot wooden structure. An open belvedere, presumably for a bell, graced the top of a small stair tower on the front facade of the 50-foot-wide building. Because the spinners and weavers needed as much light as possible, fourteen windows pierced every story and the monitor. As a result of the number of openings, the walls did not carry the load of the building. Instead, the armature of mortised hand-hewn posts and beams, designed by Horace, supported the weight and absorbed the vibrations of the twenty-four hundred spindles and seventy looms that began operating in April 1867. According to Clapp family tradition, King received $4,000 for erecting this structure; James Meeler and his crew also probably worked on this building. It represented a typical first-generation southern mill. Horace was not involved in raising the more sophisticated downtown, five-story, brick Mill No. 1 of the Eagle and Phenix.[24]

While the productive Eagle and Phenix quadrupled in size and capacity by the 1880s, the Columbus (or Clapp's) Factory failed during the depression of 1882. The vacant mill became one of the most picturesque scenes in Columbus. Nestled at the bottom of wooded hills among riverfront boulders, Horace's building decayed gracefully and became a favorite backdrop for picnics for two decades until an arsonist torched it in 1910. Turn-of-the-century postcards captured this bucolic view of southern industry, and several of those preserved in the vertical files of Columbus libraries noted the date of the fire and the fact that King built Clapp's Factory, adding to the renown of this talented man of color.[25]

In February 1867 King and Jemison renewed their correspondence. Jemison proposed that King rebuild the Tuscaloosa bridge, which Federal troops had destroyed in 1865: "I wish you to build the Bridge[,] also my Mill. We have sold our R.R. to a company who undertake to complete it to this point in three years. This will include the Bridge across both Tombigbie & Warrior Rivers besides sundry minor bridges & a large amount of trestling. If you should want any of these contracts and I can serve you in any way I need not tell you it will afford me a pleasure to do so." Jemison thought that if his rheumatism permitted, he would be in Montgomery shortly and wanted to see King. They might have met and talked, but Jemison's vision of the future of his mill, the Tuscaloosa bridge, and the railroad bridges was overly optimistic.[26]

One postwar political change in Alabama was the formation of new counties—thirteen new ones between 1865 and 1870. One of them, Lee County, provided more projects for King and his sons. This entity made the dynamic rail center of Opelika the county seat of the new political jurisdiction. The county included the Eagle and Phenix mill village and separated it from the rest of Girard. The boundary ran on the northern side of the street, across from King's upper Girard house. Although King was not a resident of the new county, its commissioners hired him to construct their initial public buildings.

In 1867 and 1868 King erected a small two-story brick courthouse in Opelika. Its rectangular shape with the short side facing the street conformed to an urban lot, and its architecture harkened back to earlier influences of the Godwins' work and public buildings in South Carolina. The front facade of Horace's edifice replicated Wells and John's 1838 Muscogee County facility on a smaller scale. The dominant feature was a central two-story portico. Four square piers supported the second floor of the porch, from which extended four Doric columns topped by a classical pediment. Italianate brackets deco-

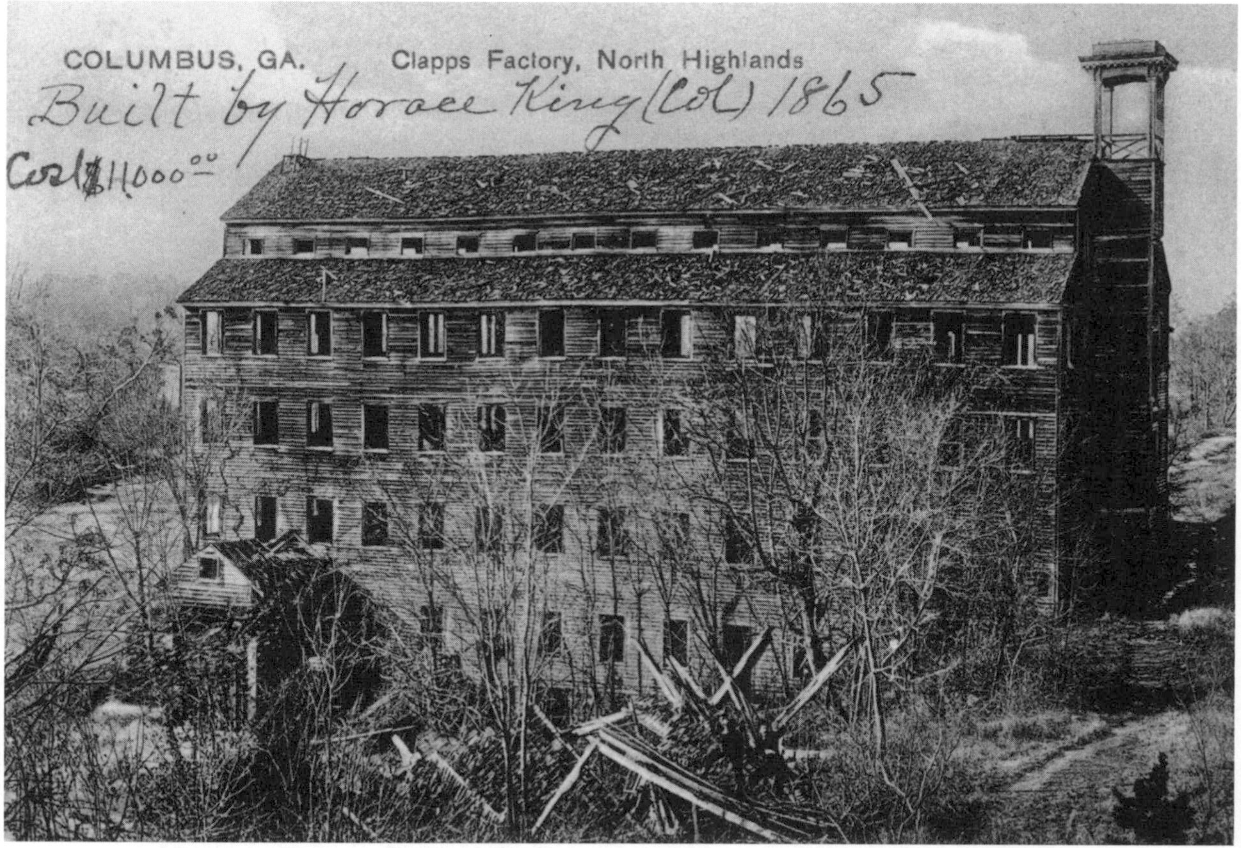

Clapp's Factory, ca. 1900. In late 1866, early 1867, King built the postwar version of Clapp's Factory, which stood three miles north of Columbus above the present location of Oliver Dam. This postcard view dates from about 1900, between the closing of the mill in the 1880s and its burning in 1910. Loretto Lamar Chappell, the longtime and much-beloved Columbus librarian, wrote the inscription on this postcard, which had been sent to her by a Clapp family member. *Courtesy of Columbus State University Archives.*

rated the main block of Horace's building; their style being appropriate to the late 1860s, even though they visually clashed with the classical detailing on the portico.[27]

In late 1867 the industrialists finally hired King to bridge the Chattahoochee River between the Columbus factories and the mill village in Lee County. Mill superintendent W. J. McAlister, who directed the erection of the first Eagle and Phenix mill, served as president of the Franklin Street Bridge Company. The firm, which represented the interest of the factories on the Columbus riverfront, reached an agreement with the city about this span in November 1867. Before and during the war, the city had refused to accept the first two spans built by the mill men at this location; protracted negotiations in 1866 and 1867 attempted to settle all the controversial issues.[28]

The aldermen and the company drew up detailed specifications for the bridge, its funding, and its transfer to the city. Not surprisingly, Horace King won the bid: $17,972 for a bridge to be completed in 100 days, "provided there is no delay from high waters." The actual financing proved to be complicated, illustrating the financial vicissitudes of the period. The company paid King $7,972; its stockholders received the right to control tolls on the bridge for five years or until they received $7,972 with interest at 7 percent. The city agreed to award King $10,000 in bonds for constructing this bridge. Payments were furnished to him as the work progressed.[29]

By 1867 King was a full-fledged contractor. He had no partners and accepted full responsibility for the work. He and his crew provided all the materials and the labor for the job—bricks, stones, timbers, and sawn lumber for rebuilding both abutments and the four piers as well as assembling and installing the five 130-foot truss section. Based on the one-hundred-day contract period, King should have finished the bridge on or about February 10, 1868. In March he told a reporter it would be another six weeks if the weather cooperated. His crew had completed the framework for the trusses coming from the Alabama shore.[30]

On May 5, 1868, the *Sun* reported with a nearly audible sigh of relief, "The upper bridge, Horace King, the builder tells us, will be open for crossing of wagons on to-morrow afternoon. This will be of great convenience to the people of Columbus, and increase the value of the upper city property, and hence benefit all classes." In praising the span, which Horace "thinks . . . the best structure he has yet directed to completion," the editor noted that "[t]he structure appears to be a strong, durable one, that reflects great credit upon the city and the builders—Horace King and his four sons—the best in this country. . . . They have won golden opinions in this section for years—where, if an important and substantial bridge is to be built, they are certain to be employed."[31]

King's building skills needed little if any endorsement; everyone recog-

Lee County, Alabama, Courthouse, ca. 1890. Horace King built a courthouse and a jail for this newly organized county in 1867–68. The courthouse resembles, on a smaller scale, the Muscogee County structure built in the 1830s by King and the Godwins. Unfortunately this courthouse was replaced by a larger one in the 1890s. *Courtesy of the Museum of East Alabama, Opelika, Alabama.*

nized those. He won contracts because of his low bids. If he used only his sons as the skilled laborers, then his cost would have been less than his competitors. In the case of the upper bridge, his competitor was C. B. Harkle and Co., an out-of-town firm. Their bid exceeded King's by $4,000 or 23 percent of his estimate. Beyond the matter of King's reduced rates, the city appeared to take advantage of him by paying with bonds instead of cash. King's compensation became an issue in the city's accepting the finished bridge. The city bonds, originally valued at $10,000, had declined precipitously by the time King was paid for the bridge. Apparently King absorbed the losses associated with the depreciation of the bonds. Even after the bridge was opened to wagons in May, King and his sons continued working on it, replacing the wooden piers with stone ones. In September, Horace received another $3,500 in city bonds, which were likely still of questionable value, and he released the company from its contract. The actual value he received for his work fell considerably below the $17,000 he had bid. On September 28, 1868, the city finally accepted the Franklin Street Bridge, only sixteen years and three bridges after the plan to have the factory owners build the bridge and the city maintain it was first proposed in 1852.[32]

At the same time King finished the piers for the Franklin crossing and negotiated for more compensation, he won the bid for the Mobile and Girard Railroad bridge, the southernmost span in the city. The earlier history of this bridge illustrated the ineptitude of locals trying to run a railroad. The line served to bring cotton to Columbus but not directly. Off-loaded at the Girard depot, the bales moved by wagon across the lower bridge to Columbus warehouses, and their owners only reluctantly agreed to build a bridge to connect the Mobile and Girard line with the Southwestern (or Muscogee) Railroad, a subsidiary of the Central of Georgia, on the eastern side of town. Since such a connection might encourage Alabama planters to bypass Columbus warehouses and ship their cotton directly to Savannah, the city's businessmen blocked approval for the bridge until the late 1850s. Work then proceeded slowly and expensively. Tradition, but no primary document, links King with the prewar construction of the first piers. Between the beginning of 1860 and March 31, 1868, the local company officers spent $49,507.19 on a Chattahoochee River bridge and had only partially finished piers to show for their expenditures. Only a small portion of that expenditure may have been in inflated Confederate dollars. Those figures reflect the company's general financial problems even before the war. By April 1861 the railroad was "much embarrassed with its stock . . . selling at half of its original price." It also "suffered considerable damage" during Wilson's raid.[33]

After 1865 the need for capital, especially to complete the bridge, forced the company to borrow money from the Central of Georgia Railroad, the most powerful economic institution in the state. In 1867 the Central of Georgia agreed to endorse $250,000 of the Mobile and Girard bonds. That transaction came with conditions on how and when this small line would connect to the larger one. In July 1868 William Morrill Wadley, president of the Central of Georgia, became president of the Mobile and Girard. As his first order of business he oversaw the bid process for the new span, not a novel task for Wadley, a former bridge builder himself. A year after Godwin and King had come to Columbus in 1833, the twenty-year-old Wadley migrated from New Hampshire to Savannah and became a blacksmith's apprentice—his only education. He applied those skills to the construction of Fort Pulaski and within six years was superintendent of the project. He then turned to bridge building in Savannah and for the new Central of Georgia Railroad. By 1849 he had become the company's general superintendent; he worked briefly for other rail lines, served as superintendent for Confederate transportation until 1863, and returned to guide the Central of Georgia after the war.[34]

King's innate skills as a builder must have rivaled those of Wadley. His skills as a manager or entrepreneur may have rivaled those of Wadley, but Horace's color kept him from exercising or testing those abilities. At Wadley's second meeting as president of the Mobile and Girard, the company drafted the call for bids for building the bridge. "Proposals will be received until 9 O'Clock A.M. Saturday, August 1st next for building a Lattice Bridge about six

The remains of Horace King's 1867–68 Fourteenth Street Bridge, February 1902. High water washed away the old factory bridge, and the remnants were recorded from both sides of the river. *Above*, the Columbus police guard the entrance to the one surviving section. *Below*, a Phenix City crowd mills around the ruined western approach. *Courtesy of Columbus State University Archives and Gary Doster.*

hundred and fifty (650) [feet] long, divided into six spans—the shortest being about 50 feet and the longest 138 feet . . . in accordance with plans and specifications to be seen at the Superintendent's Office. Also for the necessary Rock Masonry on the two unfinished piers, & timber trestles to carry them up to the requisite height for the superstructure; and trestle approaches at each end of Lattice Bridge; the whole work to be completed by the 15th day of October next."[35]

Ten entities (people or firms) bid on the project: The six local builders were King, Asa Bates, John D. Gray, Barringer and Morton, A. Gammel, and Champayne and England. Barringer and Morton had enjoyed a diversified career since the war. Barringer had gone into the grocery, saloon, and bottling business. By 1870 both partners were being "engaged by railroads," but their bid came in too high and their experience was too limited to win this job. D. W. Champayne of Champayne and England became one of the city's most important builders and architects. In 1873 a credit agent considered him an honest and fine workman. He built the Eagle and Phenix Dam in 1882 and was one of the contractors for the Georgia state capitol from 1884 until 1889. A. Gammel ran a livery stable and an omnibus line in February 1866; he must have planned to act as a contractor and hire a superintendent. Four other men or companies also bid on this job: T. J. and J. L. Grant, O. C. Johnson, Rozel and Nagle, and Daniel Culver. By contrast King had established a reputation as a bridge builder. He submitted a low bid, a practice that, along with his quality work, won him many contracts.[36]

King and Bates submitted the two lowest bids. The six directors decided between those two men. King received four votes, including those of Wadley, the former bridge builder, and the president pro tem R. L. Mott, King's partner in the Girard turnpike. Only two directors voted for Bates. The minutes did not record the actual bid amounts, but King apparently received $14,156.38 for this railroad bridge that was to be constructed by November 1, 1868—an extremely short period. The Bates family tradition has them building this bridge; they and the Kings probably worked on it together. King and his sons alone could never have finished it in three months. Even so, King failed to complete the Mobile and Girard Railroad bridge in a timely manner. On June 19, 1869, the board resolved, "That the penalty incurred by Horace King for failure to complete the Bridge within the time specified in his contract . . . is remitted." King's speculative ventures on this railroad bridge or on the upper (factory) bridge may have produced the financial problems noted in his obituaries.[37]

King may have completed the work for the Mobile and Girard by June 1869. By July he was involved in another Columbus project, rebuilding City Mills, formerly owned by Seaborn Jones. In 1828 this wealthy lawyer, who had already served as a U.S. congressman, moved from Milledgeville to the new-born town on the Chattahoochee. As local and state officials organized the town, Jones purchased the riverfront property immediately north of the corporate limits and built a gristmill, the first local industry to harness the river. Jones owned the facility as part of his industrial holdings, but he never personally ground corn or wheat. Jones, "a man of wealth," leased the mill in the 1850s as he continued to practice law, speculate in land, operate cotton plantations in various locations, engage in politics, and serve another term in Congress. A fire-eating Democrat, Jones, along with his son-in-law Henry L. Benning, advocated secession as early as 1850. Because of Jones's outspoken sectionalism, his was the only gristmill torched by Union troops in April 1865, even though he had died the previous year. By July 1869 several men had purchased this valuable riverfront property from his estate.[38]

According to the *Columbus Enquirer*, a group of merchants and investors planned to sink between $30,000 and $40,000 in a "first class Merchant Mill" at the site: "We understand that Messrs. Duer, Pridgen, Ligon, [W. S.] Stapler, and others, are rebuilding the City Mill. . . . The foundation is already laid, and the frame work about ready to put up. Horace King has the work in hand, and will push it to completion with as much expedition as may be compatible with the excellence that always characterizes his work." Smaller in scale but similar in construction to Clapp's Factory, the four-and-a-half-story City Mills measured forty-five by sixty-five feet with single hewn timbers spanning the smaller width. Mortise-and-tenon joints, still visible in the upper floors, tied the building together. Vertical board and batten siding, resembling the covering on most wooden bridges, protected the structure. The basement consisted of a wheelhouse for turbines; the grinding floor had five runs of stones in 1889. The facility produced primarily cornmeal and grits, as well as limited quantities of flour, shorts, and bran.[39]

Horace's corn mill, as it became known, survived as the plant expanded around it. In 1890 the Richmond City [Indiana] Mill Works built a state-of-the-art, five-story, brick flour mill immediately south of the original plant. The expansion illustrated the nationalization of the construction industry. By that date, even in the South, local builders rarely fabricated a manufacturing plant; national companies like this Indiana one erected turnkey facilities complete with all their equipment. The role of the corn mill changed to an auxiliary one. A corn mill with turbines in the basement had moisture problems. The company remedied that problem in 1908 by replacing timber supports and wooden floor with concrete. In 1914 the turbines under the corn mill began supplying power via a rope drive to a new wooden grain elevator that stood seventy feet to the east. The grain elevator was separated from the rest of the operation because of the greater risk from fire. In 1946 with profits from wartime production, the directors applied corrugated metal siding to the

Left, Horace King's corn mill at City Mills, 1927. This photograph from a *Dixie Miller* article inserted an image of the Corn Mill in the foreground of a general view of City Mills. The article misidentified the Corn Mill as being the 1828 plant, which was burned at the end of the Civil War. *Library of Congress, Prints and Photographs Division, HAER, GA, 108-COLM, 19-3.*

Below, Corn Mill, second floor interior, 1977. The columns and hand-hewn beams installed by King in 1869 are still visible on this floor. The first floor and the exterior sheathing have been changed. *Photograph by David Sharpe. Library of Congress, Prints and Photographs Division, HAER, GA, 108-COLM, 19-1.*

wooden elevator and the corn mill. King's battens disappeared but the original boards were preserved. In that same year, a new generator on the old grinding floor linked to three new turbines in Horace's refurbished wheel pit began producing electricity. The corn mill stopped evolving at that point.[40]

While Horace built larger structures related to the Chattahoochee River, he and his son Washington also repaired and built bridges for Muscogee County. Horace had taught his sons the art of bridge building, and Washington, the oldest, became the most active bridge builder for Muscogee County between late 1868 and June 1872. Even before Horace began representing Russell County in Montgomery, Washington had replaced Horace as the family builder of county-financed bridges. Nearly four years passed before the county recovered enough financially to begin reworking its bridges after the war. Between December 1868 and April 1869 the commissioners paid Washington $2,200 for repairing bridges over Bull Creek. In May and July of the same year he won contracts for two new bridges ($3,500 and $3,000, respectively): at Schumpert's on Upatoi Creek and Lumpkin Road on Bull Creek. The county officials may have had reservations about the younger King; they required a performance bond whose guarantors included Washington's father and two owners of City Mills—W. S. Stapler and a Mr. Pridgen. Washington's work apparently pleased the inspectors, however. In August 1870 he and M. Daucer both bid $3,000 for another Bull Creek bridge. The county selected Washington for the job. In addition to payment for those three bridges, he earned another $3,897 between October 1870 and June 1872 for sixteen different repairs to various spans. Over a period of three years and seven months, Washington grossed $15,597 working for the county.[41]

Washington's share of that revenue was reduced for several reasons. He had to absorb the cost of labor and materials. The commissioners also made some payments, perhaps as much as half ($6,939), in county bonds. Washington certainly did not wait for these to mature, so he must have sold them at greatly discounted prices. Finally, the county paid Horace for some of Washington's bridges and paid Washington for some bridges where Horace was the contractor of record. Presumably, Horace and Washington worked together and shared crews, and the county viewed them as one organization. Even so, Washington's income exceeded that of most young men of color during Reconstruction. Washington spent some of his money buying land in Carroll County and some building a nest egg that would allow him to break with his family and become an independent bridge builder, working out of Atlanta.

Although Washington completed most of Muscogee County's bridgework during this postwar period, Horace finished three jobs in his own name. As a fitting coda for his Muscogee County career, Horace's last bridge contract there came at the expense of Asa Bates. In September 1869 they competed over a span at the Old Morrison Bridge Place on Upatoi Creek. As had often happened, Horace underbid Asa ($3,400 to $3,600). Unlike in Washington's case, however, the county allowed Horace to guarantee the bridge and its maintenance in his own name. By March 1871 King had also constructed for the county a $200 breakwater in Bull Creek. This project harkened back to the breakwater built by Godwin for the city in 1839. Horace's last transaction with Muscogee County came in April 1871, when he supplied $300 worth of sawn lumber.[42]

During the immediate postwar period, King's work was centered primarily in Columbus and in Muscogee County. But he worked simultaneously in other cities, though much of that work is not well documented. Postwar bridge construction in other towns illustrates two major changes that affected the Kings' profession. In towns like West Point and Fort Gaines, public or government entities replaced private investors as prime financiers of bridges, and national bridge construction firms began competing with local builders.

After the war, the city of West Point, Georgia, thirty miles north of Columbus moved to replace its private bridge with a free public span. By late 1868 nearly $15,000 had been pledged for the project through small subscriptions. The city apparently failed to raise all the necessary funds, and James M. McClendon, a wealthy citizen who was serving as mayor, loaned the city some of the needed money. The city hired King to replace his and Godwin's earlier span. By February 1869 "people were passing over" the new bridge. Though the bridge was completed shortly thereafter, the city did not finish paying for this bridge until April 1874.[43] The financial problems of towns and railroads probably cost the Kings a sizable portion of their original contract figures. Such was the price of doing business in the postwar South.

King did not work at Fort Gaines, Georgia, eighty-two miles downstream from Columbus, until the late 1860s or early 1870s, when he assumed the contract for an unfinished bridge there. Despite its location south of the fall line, the town had evolved where high bluffs dominated both sides of the river. Because of those towering banks the bridge piers stood about sixty feet out of the water, and their height created constant problems for builders. The state legislature never incorporated the first span in 1841. Financed by cotton merchants Alfred Prescott and Charles Bemis, and erected by unnamed builders, it stood for thirteen years before succumbing to a freshet in 1854. A new company received a charter but failed to act on it. A ferry served the crossing until after the Civil War.[44]

In 1867 a reorganized stock company contracted with Bonner and Walden of New York City to build a new bridge. Apparently these northern workers failed to respect the Chattahoochee River. In August 1868 a freshet washed

The *Three States* steamboat, loaded with cotton, and the soaring Fort Gaines Bridge in the background, ca. 1910. Horace King did not build the antebellum span at Fort Gaines, where high bluffs made it difficult to maintain a wooden bridge. Three generations of Kings (Horace in 1869, Washington in 1888, and Ernest in 1913) did replace and repair bridges at this location. During that period, the technology of building Town lattice trusses remained constant. *Courtesy of Georgia Division of Archives and History, Office of Secretary of State.*

out the false work and the center pier under partially completed trusses. Thirteen men drowned and several others, including one of the contractors, sustained injuries. The company canceled the contract, and about a year later King accepted the challenge of finishing their work. He built three latticed piers to support the trusses. During the flood of 1875 the pier closest to Alabama failed, bringing down two sections of the bridge. A Eufaula builder, Capt. B. B. McKenzie rebuilt it. The next flood in 1888 brought Washington King to the rescue, and after high water in 1913, Washington's son Earnest resurrected the span again.[45] Despite the competition from northern bridge engineering firms and the introduction of iron and steel bridges, the Kings continued to find work as the master builders of covered wooden bridges.

In 1870 Robert Jemison and Horace King began planning how to secure the contract for rebuilding the Tuscaloosa span over the Black Warrior River. As on other rivers in other towns, this bridge had become a strategic target during the war. In April 1865 Brig. Gen. John Croxton, with a brigade detached from Wilson's Raiders, headed south toward the university town. Their ability to take the city without a protracted fight depended on securing the span between Northport and Tuscaloosa before the Confederates destroyed it. The blue-clad cavalry reached the covered bridge just as the rebels were removing the floorboards from the bridge. Using their six-shot Spencer rifles, the Federals drove the defenders back and easily took Tuscaloosa. After burning much of the university and all of the local industry—including Jemison's sawmill—they withdrew to the north, burning Seth King and Jemison's bridge as they retired.

As Jemison and Horace King began planning to rebuild this span, their correspondence sounded similar to that before the war, but Reconstruction had reversed the roles of King and Jemison. King, not Jemison, served in the legislature, but King still treated Jemison with deference, and Jemison still expressed genuine affection for King. In May 1870 Jemison proposed that Representative King meet him in Tuscaloosa, and they would go together to Columbus, Mississippi, where they hoped to rebuild the bridge. They probably never made that trip; but if they had their conversations would not have focused on politics but on their true passions—sawmilling, bridge building, and money making. Jemison's economic holdings had suffered, while King's financial opportunities had expanded.

King's journey to the Luxapalila in the vicinity of Columbus, Mississippi, would have involved rebuilding Jemison's mill. By December 1870 Jemison's reputation had declined there. "Col. Robert Jamison [*sic*]" and his gristmill were viewed as a credit risk. He was "form[er]ly a rich man but genly owed abt as much as he was wor[th]," reported the Dun agent in Mississippi. "Think he or his wife owns some p[ro]p[ert]y. He is getting old now abt 70 & drink[s] hard. Can't recomd for Cr[edit]."[46] Such an assessment never appeared in the reports from R. G. Dun's credit agents in Tuscaloosa, and judging from Horace's letter to Jemison, he, unlike the Mississippi agent, never saw Jemison as a worthless old drunk.

King and Jemison had trouble remaining in contact after the war. In November 1869 Jemison wrote King in care of a mutual friend, Gen. James Clanton in Montgomery. Presumably, the general should have given the epistle to King when they were in Montgomery for the legislative session but did not deliver it until late March or early April 1870. King responded to Jemison's request by sending him a "bill of lumber" or a list of the boards needed for the Tuscaloosa bridge. Horace also recommended that Jemison fund his trip to Tuscaloosa. Jemison asked about using sweet gum lumber for the bridge, and King assured him "it will make a better bridge than any other lumber you can get." Jemison apparently no longer had access to mature pine timber. Horace then added a postscript: "Please let me hear from You as early as possible as outhers [*sic*] wants my services and I wish to accommodate You before any outher man." Jemison wanted King's services as well. The situation with the Tuscaloosa bridge would prove to be complicated.[47]

By 1870 Seth King had died and Jemison controlled the stock of the old company, but in the postwar period, municipalities moved more aggressively to take control of their bridges. Jemison outlined several plans to King, but all of them involved Jemison's company receiving $30,000 and King rebuilding the bridge. As Jemison explained to King: "The Bridge will be built by some party on some terms or other, and I want you to build it." Jemison queried King about the cost per foot of constructing the main bridge and the land bridge, as well as his willingness to accept city bonds in payment. Horace responded that he could frame and raise the lattice for the main bridge at a rate of $9 per foot and the land bridge for $3 per foot. That figure includes all the cost of labor, but no materials and no work on the piers or abutments.[48]

Like other town councils, the Tuscaloosa aldermen moved slowly toward a decision. A citizen's committee opposed Jemison's plan, but eventually a referendum decided the issue of whether to issue city bonds in order to purchase Jemison's company. The nature of the bridge's potential supporters left Jemison a little chagrined. "If we do [win the election about the bond issue] it will be by the vote of freedmen and other non-taxpayers, those who will be least benefitted by a Bridge." Jemison may not have liked these voters, but their ballots carried the day. Still, the city did not act, and Jemison continued to propose ways to reduce costs, such as combining a wagon and a railroad bridge since Tuscaloosa still lacked a railroad but one was projected across the Black Warrior.

Jemison queried King about the maximum reach of a Town lattice span and suggested that King's span at Columbus, Mississippi, "was smartly over two hundred feet. If so there will be no absolute necessity of a pier in place of the fallen pier. . . . Write me fully your views as to all these matters, for I have great confidence in your judgment."[49] Jemison certainly overestimated the length of the Mississippi span, and his sense of expediency overreached a reasonable estimate of how far Horace could push a span. Jemison estimated the expense of the Tuscaloosa bridge as follows:

My estimate as to cost of rebuilding is as follows to wit.

Lumber 240 M feet @ $2		4800.00
Main Bridge 650 ft. @ $9.00 pr. foot		5850.00
Land Bridge 200 ft. " "		1200.00
Repairs to brick piers & abutments		1000.00
		12850.00
100 M. Shingles	$350.00	
Poles, &c	500.00	
		850.00
		13700.00
To this should be added wooden piers say		1300.00
Making the round total of		$15000.00[50]

In comparison, the 1824 Cheraw bridge (415 feet of latticed trusses; 1,337 feet of land bridge) cost $17,000; the 1832 bridge in Columbus, Georgia (558 feet of latticed trusses; about 250 feet of land bridge) cost $16,000. The cost of Ithiel Town's bridges had declined only slightly in forty-five years.

The Tuscaloosa authorities continued to procrastinate.[51] In January 1871, more than six months after the bond referendum and the subsequent flurry of planning, King wrote Jemison, inquiring about the project:

> It has bin som time Since I have heard from you please let me hear from You, how are you getting and what has becom of the Bridge has it failed Entirely or What are they waiting for, if the Bridge has failed is thair nothing Eals that we can go into to make som money at. I think that I can make som money for You and My Self with Your assistance please let Me hear from you as Soon as You receive this by attending to the above You Will Confor a favor on Your humble Servt.
>
> Horace King[52]

Jemison died before this bridge was built. King had also proposed building a sawmill for Jemison. By 1870 King, the contractor, was either running his own sawmill or was building such operations for clients. Jemison, having lost his sawmill during the war, mentioned his need for a new one. King emphatically told him not to construct one until he arrived in Tuscaloosa, assuring his old friend that a mill he was building in Columbus for only one thousand dollars would begin producing in a few days, and it could serve Jemison's purpose with limited expense. Perhaps King's work with the "inserted saw teeth" came at this point in his career, either in his own mill or in those he erected for other people. Like the Tuscaloosa bridge, Jemison's sawmill never materialized.[53]

Given the warm nature of their friendship, it would have been appropriate for Horace to visit Jemison in Tuscaloosa or Columbus, Mississippi, one more time before Jemison's death on October 16, 1871, but no evidence points to a final reunion. Their relationship had changed very little in three decades, despite the Civil War, the abolition of slavery, and Reconstruction. When King traveled to the Tombigbee at the request of Jemison in the early 1840s, he went to make money for Jemison, John Godwin, and Horace King, the slave. The proposed bridge in 1870–71 offered the same possibilities for two of them. The symbiotic economic relationship between the two rested on Jemison's capital and King's skills. Their warm friendship rested on shared interests and a mutual regard. They enjoyed each other's company.

Ironically, in light of how much Jemison wanted Horace to build the Tuscaloosa bridge, his death allowed King to construct it. The *Montgomery Advertiser* reported on March 3, 1872, that the city of Tuscaloosa awarded the contract for the Black Warrior River bridge to King. With both Seth King and

Jemison dead, the old bridge company had apparently ceased to exist, and the town could act to build its own bridge without having to pay private investors for their bridge rights. Jemison would probably have been pleased that his old friend Horace finally got some money from the project.

Considering all the construction work King received from 1865 until 1872, he should have been rich by the early 1870s. He had reconstructed four major bridges in Columbus, Georgia, and a large warehouse, gristmill, and textile factory, as well as similar structures in other towns. Yet his January 1871 letter to Jemison sounds as if he needed money. Perhaps King bid too low and accepted too many bonds from towns, counties, and corporations. His income may also have suffered because his building career was diverted by his venture into politics.

CHAPTER THIRTEEN

A NOMINAL REPUBLICAN

For members of the General Assembly, Capt. Henry, (Dem.), and Horace King, (colored), a liberal and nominal Republican, have been elected to the Legislature. Well done, Russell!

"Official Vote of Russell County," *Columbus Sun*, November 12, 1870

While the economic and professional concerns of building bridges, warehouses, and factories preoccupied Horace King after 1865, his prominent position in local society also drew him into the maelstrom of Reconstruction, during which the southern states rejoined the Union and the status of newly freed slaves was decided. Many white southerners stereotyped the period as one of misrule and corruption by blacks and carpetbaggers. In reality, native white Alabamians dominated the state, with black officeholders always representing a small minority. Even so, Democrats—then and later—purposely haunted voters with tales of the horrors of black Republican Reconstruction; such tales served the Democrats' political interests. For at least eight decades conservative Democrats remained in power by recounting fanciful stories of the atrocities of Reconstruction rule. The reality of Reconstruction is much more complex than the stereotypes. When examined in detail, Reconstruction defies any attempt to make facile generalizations. Political chaos characterized the period and the Alabama Republican Party was plagued by factionalism.

Reconstruction began mildly enough under the leadership of Pres. Andrew Johnson in 1865 and evolved through several phases. Johnson's Reconstruction plan sought to continue Abraham Lincoln's lenient policies toward the South. Johnson, a native of Tennessee, forced wealthy white southerners who had supported the Confederacy to seek a presidential pardon if they intended to participate in politics, but he never envisioned black suffrage. Moderate white leaders, antisecessionists and unionists, organized the state governments under the direction of Johnson, and they—both the president and the new legislators—did little to help the freedmen. Johnson, in fact, allowed southern legislatures to enact so-called Black Codes—measures that defined freedmen's rights and responsibilities. In general, these codes gave them the right to own property, marry, make contracts, sue and be sued, and testify in court against other blacks. However, the main focus became controlling the freedmen's labor, which was usually restricted to the plantations. The former slaves and many northerners interpreted these laws as attempts to reestablish slavery.

Caring for the needs of the former slaves fell to the Freedmen's Bureau, which was established in March 1865 (for one year) and became a point of contention between Johnson and the Republican-dominated Congress. The bureau's wide-ranging responsibilities included dealing with the misconception of "forty acres and a mule." Many blacks believed General Sherman had implemented this policy in January 1865 with Field Order 15, which appeared to transfer the ownership of the Sea Islands and a thirty-mile strip of land in South Carolina and Georgia from the white owners to the former slaves. Johnson reversed that policy in July 1865, but African Americans, including the Kings, still aspired to land ownership.

By the end of 1865 the fight over Reconstruction began to intensify at the national level. In December, Congress refused to seat the legislators from the South, whom Johnson presented as representatives of reconstructed states. In 1866 Congress overrode Johnson's veto in order to extend the life and expand the power of the Freedmen's Bureau. During the congressional elections in the fall of 1866, northern voters repudiated Johnson and sent an overwhelming Republican majority to Washington.

In March 1867 congressional Reconstruction began with a series of bills passed over Johnson's vetoes. These bills reestablished military rule, disfranchised some former Confederates, and granted the vote to black males. Almost immediately, Horace King became a minor player in restructuring southern society. Part of this process involved registering newly enfranchised blacks. King, along with Thomas Harris, a black "gospel minister," and T. G. K. Quillin, occupation unknown, served as registrars for Russell County. The head of the Freedman's Bureau in Alabama, Wager Swayne, received a deluge of applications from potential registrars. Horace's name appeared in a list of registrars praised by John Keffer, the leader of the Union League in Alabama. This organization, which began in the North, pushed the creation of biracial Republican parties within the South, a development opposed by many former white Unionists who joined the party and other moderate white Republicans.[1]

King and his two fellow registrars appeared to support the aims of the Union League. They registered 2,630 black voters and only 945 whites. Many white Democrats boycotted the process. Swayne paid the registrars twenty-eight cents per voter registered; this duty should have netted King $333.66 if the registrars split the profits equally. Perhaps King acted for monetary rather than political reasons, but he does not appear to have been a reluctant supporter of the Republican Party at this point.[2]

King's first biographer, the Rev. Francis Cherry, writing in 1883, has Horace only reluctantly becoming involved in politics. The brief, laudatory accounts of Horace's life—mostly written by whites with negative views of Reconstruction—paint the bridge builder as being detached from and very dissimilar to other black Republicans. These biographers, who tended to be conservative Democrats, had to explain how their hero could have become a member of the black Republicans. In the middle of their paeans to Horace, his legislative venture represented a discordant note. The most extreme dichotomy between him and his fellow officeholders was drawn by Columbus newspaperman James Jackson in 1895: "In the days of Reconstruction when Ethiopia made a 'Fell Swoop' and took possession of the chief seats in the Alabama temple of legislation, casting dark shadows of 'spike tails' and two and a half story beavers on frescoed walls and ceiling, Horace King . . . was registered as a member from the grand old county of Russell. While regretting the unfortunate state of affairs which then existed, all who knew Horace King will admit that Russell had the brightest and best member ever enrolled on the list of the colored contingent."[3]

Democratic newspapermen never made an objective analysis of Republicans, even thirty years after the event. In truth, King was an honest, conservative man, but he did not distinguish himself very well as a legislator. And contrary to the view presented by Cherry, King did not appear to be a reluctant participant in the Alabama legislature of 1867. Becoming a registrar represented an active political step taken with deliberate forethought.

Furthermore, in August 1867 King tried to establish a black colony in Georgia under the auspices of the Freedmen's Bureau. On August 24 Wager Swayne, the assistant commissioner of that agency in Alabama, wrote his counterpart in Georgia, Caleb C. Sibley, "Mr. Horace King[,] a very respectable and intelligent colored man and a successful master mechanic who now resides in Russell County in this state[,] desires to found a small colony of colored people in Coweta and Carroll Counties Georgia. He proposes to solicit subscriptions among those of his people who are willing to engage in the enterprise, build the necessary buildings[,] etc. by the joint work of the colony and then divide the land among the different heads of families." Swayne asked Sibley to send King a letter of endorsement that could "prevent any active opposition from those hostile to such a settlement." Sibley replied, "Affirming the design of . . . a colony" and gave "assurances" of support to King; "he may rely upon the protection extended to every citizen by the General Government. I desire that he may confer often with this office in the further prosecution of his plan."[4]

King probably neither conferred with Sibley again nor attempted to implement the plan. This scheme may have its origin in Sherman's Field Order

15, which specified that when "three respectable negroes, heads of families" joined together, they could request an area of land within Sherman's proposed reservation. While King's colony had no connection to Sherman's reservation on the coast, its objective sounded similar. King may have been aware of an earlier colony in Dougherty County near Albany that was encouraged by Sibley's predecessor as assistant commissioner, Davis Tillson. Its participants had moved from Wilkes County, Georgia, to southwest Georgia and operated on five hundred rented acres. In March 1866 they petitioned the bureau to send them a teacher. Tillson wanted to encourage this type of venture and even explored the possibility of having northern philanthropists purchase large tracts of land for former slaves. But the Dougherty experiment failed at about the time King proposed his plan.[5]

Caleb Sibley may have had little intention of helping King secure land. He had implemented the wishes of President Johnson in the area of the Sherman reservation and had returned the land to its prewar owners, the white planters. Perhaps the bridge builder thought the Southern Homestead Act of June 1866, which gave preferential treatment to blacks and white unionists, could be used to acquire property, but no vacant land existed in Carroll and Coweta Counties. Would James Moore, with whom King still owned the rights to the bridge, and his other neighbors have accepted a new group of black residents there? Did King consult with them before he began floating this idea? Had King already decided to move back to Carroll County by 1867? If so, why did he represent Russell County in the Alabama legislature? King appears to have been pursuing various options in an attempt to improve his financial and social status. Given his prewar status, he was better able to rise above the whites' concept of the proper social and economic position of blacks.[6]

Because of the Reverend Cherry's negative feelings about Reconstruction, he probably exaggerated King's political reluctance. In discussing King's candidacy in 1868, Cherry wrote: "it was done against his wish and under his most earnest protest." In becoming a candidate, King joined at least twenty-two other registrars who made the same transition. King's entry into politics appeared logical, if King used as models the careers of his prewar business associates, particularly Jemison and Nelson Tift of Albany, Georgia. Those men stood for election because they saw themselves as natural leaders of their communities. If the time had come for blacks to sit in the legislature, then King—as the best-known gentleman of color in Russell County—was the logical candidate. Despite Cherry's interpretation, King may have wanted the prestige associated with holding office. Also, men like Jemison and Tift used their political positions to further their economic careers, though not in a venal or crass manner; their political status simply translated into economic gain. In this new world, why should King not have a chance at the same rewards?[7]

In Horace's first election to the statehouse, a simultaneous referendum on a new constitution overshadowed the selection of legislative delegates. A majority of all registered voters had to approve the new Alabama constitution. White conservatives boycotted the election as a means of defeating the implementation of the radical constitution. The local Columbus newspaper focused on the constitution issue and never bothered to report the names of the candidates in the Russell County house race on the other side of the river. Cherry reported King as being uninvolved, even though Horace acted as "one of the managers of the election at Girard." In that capacity, according to Cherry, King "earnestly urged every voter, on presenting a ticket with his name on it, to 'scratch it off,' which was done at the precinct in every instance but three.—Nevertheless, he was elected over his own protest."

While Cherry's version makes a nice story, it is not true. Gen. George Meade, in an attempt to garner as many votes as possible for the constitution, extended the length of the election from two to five days. Given King's busy schedule as a contractor—he was working on both the upper bridge and the Lee County Courthouse at that time—it is doubtful he could have spent five days at the polls. Cherry's version of the election as supposedly recounted by King illustrates Cherry's bias and the limitations of oral history. If the objective is to learn King's motives, then what King told Cherry twenty years later is not an accurate measure. King may well have relished the prestige associated with being a candidate in 1868 but changed his view by 1883. Certainly Cherry viewed Reconstruction in negative terms by 1883, and King may have done so as well. Thus King may have exaggerated his reluctance to run for Congress.

The actual level of support for King in 1868 is impossible to discern. Rather than reporting the returns for the races, the local newspaper related stories about how little the blacks understood the election and how the white federal soldiers duped the former slaves into giving them money. Although a majority of the registered voters did not approve the new state constitution, the U.S. Congress changed the rules and validated the constitution. By the same act, they confirmed the election of all the winners in February 1868, and that included King.[8]

After King's election, but before he entered the legislature, the Columbus *Sun* included an extremely flattering report about the Kings in an article about the opening of the upper bridge. The new span "reflects great credit on . . . the builders—Horace King and his four sons—the best in this country—freedmen who are irreproachable in morals, general deportment, politics and demeanor, as they are skillful, able and intelligent. They have won golden opinions in this section for years."[9] King's political views were obviously aligned with those of this moderate Democratic newspaper, and its editors wanted to reassure the public that this new Alabama officeholder was no radical Republican.

During that same interim period between Horace's election and his first session in the legislature, Jemison wrote King:"I will be in Montgomery in eight or ten days when I would like to see you." The letter contained mentions of Seth King and Mr. Williams, two of their former associates, both deceased by that date, but Jemison did not discuss a building project. Perhaps his purpose was political, to instruct the new lawmaker on what his role should be in the upcoming legislative session. Jemison mentioned their "mutual friend" Gen. John H. Clanton as a possible point of contact between King and Jemison. The idea of a friendship between King and Clanton raises an interesting question regarding King's possible role in Reconstruction. In May 1867 Clanton, as head of the state Democratic Party, debated Sen. Henry Wilson of Massachusetts; they both sought to convince Alabama blacks to support their respective parties. While the Democrats never seriously pursued the black vote, Clanton's relationship with King may suggest the depths of Horace's conservatism before he entered the legislature. The actions of Klan-like vigilantes in Columbus in March 1868 may also have pushed King in a moderate direction. A group of masked men had broken into the room of George Ashburn, an outspoken radical scalawag, and shot him. The U.S. Army, which still controlled the local government, arrested a dozen men including several prominent whites. As the "Columbus prisoners" they gained national notoriety. White conservatives held considerable power in the area.[10]

Once in office, King never posed a threat to the status quo. Whereas white Democrats may have praised King's career as a legislator simply because he did not introduce or support any radical legislation, in truth, he was an inattentive and ineffective legislator. Of all the facets of his long and productive life, his service in the legislature ranks as his least successful. In July 1868, when King should have presented himself to be sworn in as a legislator, he was busy working on the piers for the upper bridge and preparing his bid for the railroad bridge. He never took the oath of office during his first term.

King's legislative colleagues, recognizing his earlier involvement with the capitol building and hoping to use his expertise as a builder, placed him on the Capitol Committee. Had he joined his peers in Montgomery, he could indeed have provided advice about the structure, but he did not take his seat during the entire first term. Cherry explained: "Being at that time under contract to build a railroad bridge across the Chattahoochee, and not wishing to fail in it, he was not found in his seat during the first session of the term." This "session" extended from July 16 until August 11 and was followed by a called session from September 16 through the month of November. Perhaps King had accepted the nomination because of the prestige but was unwilling to let business interfere with any other obligation. Perhaps feeling embarrassed to join the proceedings late, he never appeared at all.[11]

In November 1868 a committee reported "[t]hat Russell County is without proper representation, it appearing that one King was elected at the February election as a member of the Legislature, but has never presented himself for admission, nor had he ever offered any excuse for his absence during the several sessions of this Legislature." King's name does not appear as a member of the House in the 1868 *Acts of the Legislature*, and his fellow black member from Russell County, J. H. Tyner, claimed to be the only representative of the county. According to Cherry, "Many of his [Horace's] friends, of both parties, urged him so earnestly on the opening of the second session, representing that if he did not serve, an election would be called and a person very objectionable to all good men would be elected to fill his place."[12]

King did present himself at the start of the November 1869 session and only one day late. Even so, he must have been bored with being a legislator, or maybe he was too shy to become involved in the proceedings. He seemed uninterested in the larger issue of Reconstruction and remained focused on local concerns and matters relating to his profession, a trait of many legislators. Cherry explained King's role in the second session and exempted him from his excoriation of radical Republicans. King "was found in his seat, and though quiet, soon exercised an influence which was felt, in the direction of moderation, prudence and conservative wisdom, even in that noted assembly of the strangest elements ever brought together to enact laws for a great commonwealth."

The second session ran from mid-November 1869 until early March 1870. During that period King failed to vote in 22 percent of the roll calls, missing nineteen of eighty-six votes. He consistently voted with the majority on most of these issues. Even when in his seat, King's level of participation fell short of his peers. He only introduced three bills—half the number presented by the average legislator—and none of his bills became law. Of the ninety-seven members, nine introduced no bills; for the remaining eighty-eight, the median number of bills introduced was seven. In general, the level of participation for black legislators equaled that of whites, except that their most active members did not present as many bills.[13]

King's legislative record reveals a man motivated by self-interest (as are many legislators), rather than an elected official who worked for the general good. Unfortunately, only the titles of bills have survived, and the exact content of these measures remains unknown. King introduced his three bills during February 1870, the last full month of the session, and they reflected the concerns of his profession. His bill for the relief of laborers and mechanics appeared to serve the needs of his fellow workers and his Girard constituents.

It was amended on the floor to include "all agricultural laborers and hired hands." Those groups encompassed a larger segment of impoverished Alabamians than did King's original legislation. Even so, the Committee on Local Legislation never reported the bill back to the full house. As amended it probably represented too radical a measure, and King's original intention had been lost.

His bill "to amend an Act to reorganize the Centreville Bridge Company" may well have directly related to his business interests. He might have been representing private investors who were fighting town officials over control of the bridge. In the postwar period communities moved to assume the ownership of bridges, and this bill might represent such a fight. Whatever the purpose of King's legislation, it failed to pass the House, at least in its original form. Rather than amending the charter, on March 1, 1870, the legislature repealed it.

Horace's last piece of legislation is traditionally associated with the conservative Democrats. Introduced on February 19, 1870, King's final bill attempted to "require the court of county commissioners of Russell County to employ persons convicted of crime and sentences to hard labor on the public highways and public works of said county." Establishing a chain gang for Russell County was not an action usually identified with black representatives. Since King's profession concerned maintaining roads, he envisioned this bill as an acceptable method of reaching that end. The House members referred King's bill to the Committee on the Judiciary and it never became law.[14]

Even though he did little for his constituents, King was a quiet, reserved representative, and his peers seemed to respect him. Rather than being reappointed to the Capitol Committee, King was placed on the Federal Relations Committee. This group should have been a pivotal body during Reconstruction since the U.S. Congress actively monitored developments within the southern states. Maybe his position on that committee allowed him to secure the position of census taker for Russell County in July 1870. Once again King seemed to avail himself of an opportunity to make money, even if only a couple of hundred dollars. The mystery about this census is that he is not listed on it. He is not found in the county he represented in the legislature nor in any other county. His sons may have already moved to LaGrange, Georgia, by this date, and the remainder of the family could have joined them, but certainly Horace was still in Girard at this point. Apparently he simply forgot to include himself in his record. He most likely hired assistants to actually collect the data. His signature appears on the pages of the document, but the information is not in his handwriting.

Cherry has King retiring "to private life and what he considered the pursuits of his legitimate sphere" after his first term in the legislature. Cherry then presented an elaborate scenario to explain why King returned to office. Apparently, a white Independent, W. B. Harris, aligned himself with King as a candidate for one of Russell County's two legislative seats. Harris was presumably presenting himself as a moderate alternative to the Democrats or Republicans and sought to identify with and benefit from King's coattails. According to Cherry, Harris took this action "without Horace's knowledge or consent, [and] . . . it was discovered that Mr. Harris's chances for election were not so promising . . . and Horace was urged and prevailed upon to go out and identify himself with the campaign, thereby throwing his personal influence in favor of Mr. Harris." The returns from the election, however, cast doubt on Cherry's interpretation. King polled more votes (1,415) than any of the other six candidates and beat his closest rival by 5 percentage points. He was acceptable to both black and white voters. King must have campaigned in order to receive that many voters. Yet if King urged voters to support Harris, then Horace had no political coattails, because Harris made a very meager showing with only 362 votes.[15]

The Russell County returns thrilled the editors of the Democratic Columbus *Sun.* "For members of the General Assembly, Capt. Henry, (Dem.), and Horace King, (colored), a liberal and nominal Republican, have been elected to the Legislature. Well done, Russell!" The same personal and professional respect that allowed King to be accepted as a free black during the antebellum period must have made him attractive to voters during Reconstruction. His election was no fluke. He actively sought it. However, alleged irregularities at the ballot box almost cost King the election. On November 15, the *Sun* reported that the Girard returns might "be thrown out. If it is, Mr. Hines, Democrat, will be elected to the General Assembly instead of Horace King, colored, moderate and nominal Republican." The results were not thrown out, and Horace returned to the legislature for a second term.[16]

In spite of his strong showing at the polls, King was not motivated to be a more active legislator. Both Horace and his fellow legislator, Democrat B. M. Henry, arrived on the second day of the session, November 22, 1870. Perhaps the train from Girard was running behind schedule. During this session, King returned to the inactivity of his first term. He introduced no bills, and his name only appeared once in the *Journal*'s index—to request a leave of absence. He returned by December 10, but made no mark on the record of this body.

During his last term, 1871–72, however, he introduced five bills. Two separate acts simply allowed Georgians to act as executors of estates in Alabama. His measure to "prohibit the sale of liquors within one mile of the railroad depot at Hurtville" was his first measure to become a law. His effort to incorporate the town of Girard passed on January 19, was reconsidered, and laid on the table the next day. Girard was not incorporated until 1890. King also attempted to amend a section of the state code that related to mechanics' liens. The existing code made written contracts subordinate to mortgages in regard to collecting debts, which placed contractors behind bankers and other creditors. Apparently King wanted contractors' claims to supercede those of bankers, but the Committee on Revision of the Laws never reported Horace's bill back to the floor of the house.[17] After four years of service in Montgomery King had only helped two executors with their estates and created a zone of prohibition in Hurtville—not a very productive legislative career.

His career stands in stark contrast to men like Philip Joiner of Georgia and James T. Rapier of Alabama, who became outspoken advocates for black rights in their respective states. The majority of black officeholders during Reconstruction had been free blacks before 1865 or slaves in "relatively advantaged positions." Joiner, on the other hand, was an illiterate slave from Dougherty County who served in the state legislature. As the leading radical in southwest Georgia, Joiner came into conflict with Nelson Tift. Joiner represented the poorer farmers on the east side of the Flint River who resented the tolls on Tift's King-built bridge. Joiner discovered that Tift's tolls exceeded the provisions of the charter, which prohibited charges for produce. Three weeks after Joiner introduced a bill to revoke the bridge company's charter, Tift's bridge mysteriously burned. Nothing in King's career had prepared him to follow Joiner's lead. He certainly would not attack Nelson Tift or his peers in Alabama; his livelihood depended on their patronage. King, a self-made man, never championed the little man but favored people like himself—either black or white.[18]

Nor could King follow the path of James Rapier, a free mulatto whose white father had sent him to school in Canada. Rapier's education, his work experiences, including a stint as a wartime correspondent for a northern newspaper, and his oratory skills prepared him to assume a leadership role within the Alabama Republican Party. In contrast to King's bill for mechanics, Rapier called for debtor relief for sharecroppers and tenant farmers in the 1867 constitutional convention. After a failed run for secretary of state, Rapier was elected in 1872 to the U.S. Congress, where he remained a consistent supporter of civil rights and equality for African Americas. The benevolence of Rapier's father and the opportunities he received gave him the independence from the white establishment that allowed Rapier to attack the white elites. King's career path wedded him to conservative whites.[19]

Though black voters put King in office, no evidence survives to indicate how his black constituents felt about this mulatto who represented them. The legislation he introduced did not appear to aid freedmen nor did he become a public spokesman for expanding black rights. If he had been an activist in this sphere, white Democratic newspaper editors would have never lauded his skills as a builder. Perhaps after two terms in the legislature his support among blacks was waning, as was their political power in the face of a violent political counteroffensive by white Democrats.

King's political career ended in 1872 when he did not seek reelection. By that date, the Democratic backlash, encouraged by Klan violence and Republican disarray, was gaining momentum. King might have been able to win another contest, but he would have needed to exert more effort in his campaign. Despite his limited legislative success, his white biographers have praised his political conduct, but primarily for what he chose not to do. According to Cherry, King "sustained himself in this hour of trial, in a manner worthy of a wise and prudent man. If he had been actuated by a wide, grasping, personal ambition," he could have "reached a position which, even Fred[rick] Douglass, . . . would have envied. . . . He [King] would never consent to be made a political hack, consequently ever maintained the dignity and commanded the respect of the 'noblest work of God, an honest man.'" African American historian Richard Bailey also engaged in hyperbole in assessing Horace's political career: "That the citizens of his county drafted him showed that his efforts to build bridges to connect the races had been effective. His attempts to solidify race relations in Alabama preceded by nearly one hundred years the efforts of [Martin Luther King Jr]."[20]

Horace King's popularity bridged the two races. His actions in the legislature, however, never entered the realm of improving civil rights for blacks. Perhaps his moderation contributed to a lessening of the extremism of Reconstruction, but he was not a champion of black rights. Even so, the attacks from those behind the KKK masks would ultimately have been aimed at King if he had continued to serve. Their aim was to regain control of politics from the Republicans. No black man, not even a conservative gentleman of color, was to be permitted to hold a position of leadership. Whereas King obviously received praise from conservative whites, his moderate actions in Montgomery may have alienated his black constituents and influenced his decision to leave Russell County and move to LaGrange at the end of his legislative career.

Although King had been an inactive legislator, he may not have abandoned the idea of returning to politics. The LaGrange *Reporter* of October 3, 1878, mentioned him as a possible candidate for Congress. At the age of seventy, Horace may have liked the idea of holding office again. King was probably not so much a reluctant politician as he was an ineffective one, and he almost certainly viewed political service as secondary to his primary concern of making money. But ultimately King's true political motives and ambitions remain a mystery, as do the rationales behind most of his actions.

King's legislative service allowed him to join the Masons. Many freedmen after 1865 attempted to form educational or social institutions—schools and churches. Only a few blacks were able to join fraternal organizations in this early period of freedom, and King was one of this minority. He became a Prince Hall Mason in Montgomery in either 1869 or 1870.

African American Masons belonged to separate Prince Hall lodges, though they used the same rituals, dress, emblems, and medallions as white Masons. Initially organized in Philadelphia in the late eighteenth century, the exclusively black Prince Hall organization spread into the old Northwest by the late 1840s. The Grand Lodge of Ohio created new lodges in the Midwest before 1860, then proselytized in the South after 1865. In 1869 the Grand Lodge of Ohio chartered the Hiram Lodge in Mobile "and seven other subordinate Lodges in this State, which remained under the Ohio jurisdiction until September 24th, 1870." One of the seven other lodges must have been the King Solomon's Lodge, No. 4, in Montgomery that King joined before September 24, 1870. Later misunderstandings about King's Masonic membership spring from the Reverend Cherry who claimed that "Horace was initiated into the sublime mysteries of the ancient order of Free and Accepted Masons, in the State of Ohio, the laws of the craft not permitting it in Alabama at that time." The law in question was not state legislation but a Masonic regulation. King's initiation occurred in Montgomery, not Ohio, under the Masonic jurisdiction of the Grand Lodge of Ohio. King probably participated, in an active way, in the King Solomon's Lodge, No. 4, while he served in the state legislature, but in June 1874 and again in June 1875 his name appears in the lodge minutes as a suspended member because of his nonattendance. By that time, King no longer traveled to Montgomery.[21]

By June 1874 King had moved to LaGrange, Georgia. With this move he returned full time to his career as a builder, where he always excelled.

THE KINGS IN LAGRANGE

You cannot go back on Horace King's judgment as a bridge builder, for he has built more bridges than any man in the State. He said . . . that the old place [for erecting a bridge west of LaGrange] was the cheapest and the best location in the county, for he had examined the river up and down the neighborhood for miles.
"Justice" in *LaGrange Reporter*, September 1, 1881

Before he returned to Montgomery for his final legislative term in the fall of 1871, King had begun to build another Chattahoochee River bridge; this one crossed the river northwest of LaGrange just beyond Cameron's Mill. King's Alabama political career ended in early 1872. That year or perhaps the next, the King family moved permanently to LaGrange, where they quickly became prominent members of the community. The reasons for this change remain undocumented and are subject to conjecture.[1]

The move to LaGrange can be viewed as a continuation of the Kings' original decision to leave the Columbus area in the 1850s when they became part owners and keepers of Moore's Bridge north of LaGrange. At that time, they were probably influenced by three factors: the encroachment of a mill village into their old Girard neighborhood, the promise of a new life in a place unassociated with Horace's servitude, and the lure of making money. The exigencies of the Civil War prompted their return to Russell and Muscogee Counties. Safe positions with the Confederate government protected Horace and his sons from more onerous Confederate conscription. After the war, the possibility of gaining political and economic influence from a legislative career kept Horace in Russell County. Horace's lack of success as a legislator plus the lure of new opportunities probably helped push the Kings northward again.

Before Horace ran for the legislature, he had toyed with the idea of establishing a community or colony of black families. He envisioned it in Coweta County, adjacent to his former Georgia residence, not in Russell County where he lived at the end of the war. Though this scheme never materialized, the general area upstream of Columbus-Girard continued to attract him and his family.

Horace's economic situation may have precipitated the family's move. All three of his obituaries mentioned his financial problems during the last years of his life. The *Atlanta Constitution* noted that he "had accumulated a good property, which, however, he lost a few years ago in contracting." No accessible court files document a major judgment against him in a single case during this period. The jury in an 1873 Muscogee County litigation did rule that King owed an unspecified sum to Robert A. Forsyth, probably a wagoner in Girard, but this case appears to be minor. If he had serious financial reverses, they did not besmirch his reputation. His skills as a builder continued to be lauded in newspapers, especially in LaGrange, where Horace began sharing such laurels with his sons.[2]

Horace's children may have been the driving force behind the move. The sons must have enjoyed their teenage years on the Chattahoochee at their father and Mr. Moore's bridge. According to family tradition, Horace "owned two pole-boats.... known as 'Boxes' numbers '3' and '4', and two of his sons were captains of them." These small man-powered, high-sided vessels hauled cotton bales by floating downstream. The exact dates and location of the Kings' river voyages remains unknown, but they probably navigated their boats between the vicinity of Moore's Bridge and West Point. At some time, Horace, accompanied by two or three of his sons, piloted a bateau down the Chattahoochee River from Atlanta to Troup County. Recollections of such experiences might have seemed more pleasant than those associated with Girard.[3]

Horace's sons, particularly Washington, may have considered making Moore's Bridge their home before they decided in favor of LaGrange. Washington invested his money in Carroll County. On March 12, 1869, Muscogee County paid Washington $600 for repairing the Bull Creek bridges, the second of many such payments. Seven days later Washington purchased fifteen acres in Carroll County for $450; his tract lay close to Moore's Bridge and adjacent to the Chattahoochee River. The next year, during the Christmas season, Horace gave a present to his wife, Sarah Jane, and his children—Washington, Marshal, John, Annie Elizabeth, and George. He transferred his stock in the Arizonia Bridge Company to them. The transaction apparently had little to do with Christmas and more to do with King's financial difficulties or Washington's intentions. Perhaps Horace's economic reverses forced him to protect this asset from his creditors. If so, King copied his old master's actions thirty years earlier when he transferred his ownership of Horace to his wife, Ann, and her uncle. Horace's age—sixty-two—might have prompted him to reassign this stock as a way of excluding it from any litigation over his estate. Washington may also have encouraged his father to transfer the ownership in the bridge if he planned to move there.[4]

By 1869 the King children had reached maturity and could make their own decisions about where they lived. Washington (twenty-six), Marshal Ney (twenty-five), John Thomas (twenty-three), Annie Elizabeth (twenty-one), and George (nineteen) may have simply decided to leave Girard, and Horace followed them north. Even though Horace's name continued to appear in newspaper articles, his wife and sons seemingly formed their own company. Their names, without Horace's, appeared jointly on several legal documents.

The Arizonia Bridge Company partnership of King, Mabry, and Moore had survived the war. Mabry and King cooperated on the Cameron Bridge in Troup County in the early 1870s, and they must have rebuilt Moore's Bridge at an earlier date. The 1869 deed from Horace to his family mentioned shares in a new bridge. Stoneman's troopers had not destroyed the piers; thus, for the cost of lumber, King—or more likely his sons—could have rebuilt the trusses. Since Washington had experience in bridge building by this time, he, with the aid of his brothers, could rebuild Moore's Bridge. Judging by the economic history of Whitesburg, the small community at the top of the hill north of the bridge, a new span may have carried traffic by 1872. Some new development spurred a flurry of commercial activity there in late 1872.[5]

From December 1872 until March 1873 eleven small entrepreneurs launched new businesses in this crossroads community: six general stores, four saloons, and a restaurant. This was not a propitious time to begin new enterprises. The banking house of Jay Cooke collapsed in September 1873, which precipitated a nationwide crash that reverberated all the way to Whitesburg. By March 1874 five of these new local businesses had sold out, one was "embarrassed," another possessed only "very sm[all] m[ea]ns," and another proprietor had "gone to Lochapoka," Alabama. A year later, two more merchants were insolvent, and by the fall another had sold out. By June 1876 only the general store of William McMillan was "doing all right"; all the others had failed.[6]

If the Kings had ever considered returning to Moore's Bridge—just down the hill from Whitesburg—they made a wise choice in moving instead to the larger community of LaGrange. Washington realized a profit on his Carroll County land when he sold it to his partner, James D. Moore, in January 1875 for $600. Four years later the Kings, Moore, and Mabry received some revenue from their bridge when Carroll County purchased it and made it a free crossing. By then the Kings were living in LaGrange.[7]

One of John Thomas King's obituaries placed him in LaGrange by 1870, but his wife's obituary (written before John died) cited the date as 1873. Some of the family's oral tradition dated it as 1872. Washington definitely lived there by 1874 and was well known. While Horace King's sons may have pushed the family toward LaGrange, Horace's old partner, Charles W. Mabry, may have pulled them in the same direction.

Formerly a resident of Franklin and a representative for Heard County in the Georgia legislature during the Civil War, in 1867 Mabry moved his law practice to LaGrange, where he and his wife became active members of the community. On October 7, 1871, Mabry and four other men, who constituted a majority of the stockholders in the Chattahoochee Bridge Company, met in LaGrange. As the contractor and builder of their proposed bridge, Horace King addressed the group, informing them that their crossing would be completed by November 5. This span, located northwest of the city and west of Cameron's Mill, replaced a bridge burned during the war. Benjamin H. Cameron, the company's president who was also a contractor, and Mabry, its secretary, encouraged prewar stockholders to come forward and renew their old stock, which consisted of buying shares of the new issue.[8]

King began this bridge in August 1871 and needed to finish it by November in order to report to the Alabama legislature. Apparently the Kings were ahead of schedule. An editor of the *LaGrange Reporter* traveling to "the hills of Heard County" on October 6, the day before the stockholder's meeting, wrote: "The first object of interest we encountered was the new bridge, recently erected across the Chattahoochee river, by that Prince of bridge builders, Uncle Horace King." Obviously, the local press respected him even though the writer, by calling him"uncle," apparently felt compelled to point out that he was an older black man. In another article in the same issue, the editor omitted the term "uncle" when he called King the prince of bridge builders and praised the county's ordinary for contracting with Horace for the Yellow Jacket Creek Bridge that provided access to Mabry and Cameron's new river span. They appreciated King's knowledge and skills and sought him out for this project. As important members of the local business elite, they may well have encouraged King to move to LaGrange. That Mabry left Heard County (which he represented in the legislature) and moved to LaGrange after 1865, also speaks to the vitality of the postwar city.[9]

Established as the county seat of Troup County in 1827, LaGrange quickly became a center of wealth, slaveholding, and education. While LaGrange lacked the waterpower of Columbus, it benefited by its location on the Atlanta and West Point Railroad. In 1860 Troup ranked as the state's fourth wealthiest county (according to state taxes) and fifth in the number of slaves. Planters tended to live in the town and pursue diversified economic interests. On the eve of the Civil War, LaGrange supported two male academies, a university, and two female colleges. The war hurt the community in both human casualties and economic terms. The abolition of slavery meant a $3.2 million loss of capital in LaGrange alone. Viewed in those terms, this town of formerly wealthy planters, where only six free blacks lived before 1865, does not appear to be a logical residence for a family of skilled black craftsmen. Yet La-

Grange seemed to have very positive race relations, perhaps because of the continuing paternalism of its large planters. For example, immediately after the war Benjamin Cameron, the planter-builder with whom King worked, provided a schoolhouse for blacks.[10]

The difference in racial climate between Columbus-Girard and LaGrange certainly was obvious to the Kings. LaGrange had no incident that compared to the murder in Columbus of the scalawag George Ashburn during Reconstruction. Nor did Troup County have any Klan activities during that period. The attraction of the area could also have been economic. There were more bridges to build in Troup County than Columbus, where the Kings had already resurrected the four major Columbus bridges and many Muscogee County spans. Though smaller, LaGrange seemed more dynamic as it underwent economic expansion. Furthermore, its location provided access to work as far away as Atlanta. King's decision to move there had probably already been made by the time he and his sons built the river bridge for Mabry and Cameron in 1871.[11]

Only a year after he finished his legislative term, King was working on projects in the Troup County area that he had contracted while still serving Russell County. In the early 1870s, according to local tradition, he built the Crowder Bridge at the site that later became the Glass Bridge. In April 1873 he completed one bridge at Alford's Mill on Flat Shoal Creek near West Point and started another one near the Howell place (later known as Cofield's) on Wehadkee Creek. The next month Heard County hired King to rebuild the Franklin bridge, which he had originally constructed for Mabry and the other stockholders in the Franklin Bridge Company in the mid-1850s.[12]

Paralleling other bridging operations during the 1870s, the county assumed control of what had been a private company. In August 1872 the state legislature granted Heard County the right to purchase the bridge privileges and to float bonds for a new span. The commissioners contracted with King for $3,500, and by June 20, 1873, he had arrived in Franklin with his hands and predicted it would take six to eight weeks to replace the bridge.[13]

Those hands included his sons, who played an active role in his construction projects by that date. At age sixty-six Horace probably did not supervise the tying together of the middle span over the river; one or two of his sons—Washington, Marshal, George, or John—most likely insured the proper fit of those final treenails. The Franklin bridge may have been Horace's last major span across the Chattahoochee. While the quantity of his bridgework declined, he continued to contract for other types of structures, particularly businesses and schools. For many African Americans education, which had been denied them during slavery, became a primary objective during Reconstruction. The King family eventually became leaders of black education in

LaGrange, but in the mid-1870s, Horace appeared to specialize in building colleges (actually finishing schools) for white girls.

West Point, the other town in Troup County at that time, which was located on both the river and the railroad, competed with LaGrange as a progressive, urban center and needed its own college. Being already well known there, King was recruited by West Point leaders to build their school. In early 1869 Horace had reconstructed the town's wagon- or footbridge. During the same period, the town cooperated with a group of citizens to fund a female college for Georgians and Alabamians. Initially the officers moved an old building from LaFayette, Alabama, but a storm in the fall of 1873 destroyed that structure. As a result, the city council held a referendum to ascertain whether tax revenues could be used for the hall. Supported by the voters, the aldermen awarded the contract to Horace. He built a typical T-shaped school: a two-story block with classical details on the front, which contained the classrooms, and a rooftop bell covered by a dome. The rear one-story portion included separate rooms for music and art as well as the chapel, an obligatory part of college life.[14]

The Kings had some unfinished business with the city of West Point, and maybe Horace's presence there precipitated action on the matter. On April 6, 1874, the city's secretary and treasurer agreed to make the last payment on the bridge that had been used since February 1869. The town pledged to pay $2,472.60 in American gold coin by January 6, 1875, to W. W. King, M. N. King, J. T. King, A. E. King, S. J. King, and G. H. King. The Kings in return granted "the footbridge across the Chattahoochee River" to the city of West Point.[15]

Horace's children—four sons and a daughter—and their stepmother formed an informal partnership, perhaps called the King Brothers Bridge Company. While that name was frequently mentioned in the family's oral tradition (especially by Theodora Thomas), it appeared only once in the deed records, when its partners dissolved the company in 1888. Horace had transferred his ownership of Moore's Bridge stock to them in 1869. Apparently, he had also transferred his interest in the West Point "footbridge" so that they received the final payment. On the other hand, the sons may have built this bridge without their father's assistance, and local newspapers and histories simply associated it with the more famous builder. By this time, Horace was not constructing major bridges. Maybe King's financial problems meant that any asset he accumulated could be subject to attachment by his creditors, or maybe he was too old to continue building bridges.[16]

The most accessible records for this period—newspapers, especially those from LaGrange—indicate that Horace was erecting buildings rather than bridges, but he might have continued to span rivers in other locales. For ex-

Small Troup County bridge, late nineteenth, early twentieth century. George or John King most likely contracted for this span. *Courtesy of Troup County Archives, LaGrange, Georgia.*

ample, he apparently rebuilt the Tallassee bridge over the Tallapoosa River after the war, and he used his Alabama contacts to gain other work. Montgomery County, Alabama, paid him $2,450 for a small bridge across the Catona Creek in August 1876.[17]

Wherever and whatever the nature of Horace's changing business activities, his wife and children continued to act without him in Troup County. On May 4, 1875, perhaps with the income from West Point, they purchased a LaGrange city lot and a small brick building (19½ by 33 feet) for $635 from James W. Harrison, who lived in an adjacent house. In addition to the sale, Harrison welcomed the Kings as members of the local black elite. A free black before 1865, even though not listed as such in the 1860 census, Harrison plied his barbering trade in LaGrange before the Civil War. His career pattern resembled that of the well-known free black William Johnson, the barber of Natchez, who also turned his capital into real estate. By the 1880s the *La-*

Grange Reporter called Harrison the largest black property owner and the wealthiest black man in the county. Indicative of his wealth and status, he rented the town council meeting room to the LaGrange aldermen in 1881, and in 1886 he served on federal grand juries in Atlanta. His community activities, including membership in the Warren Temple Methodist Church, paralleled those of Horace's children, and Harrison's friendship probably helped the Kings establish themselves in LaGrange and in Troup County.[18]

Harrison's small brick building located on Franklin (now Ridley) Street near the center of town had served as a physician's office in the antebellum period, and the Kings must have used it as an office, perhaps for the King Brothers Bridge Company, even though that name did not appear in the records. Washington King withdrew from the family partnership slightly more than a month after they purchased Harrison's building. On June 20, 1875, the others paid Washington $125 for his share of the lot and building.[19]

Why Washington withdrew from the family enterprise and left LaGrange remains a mystery. Apparently he wanted to expand the scope of his contracting, because he tried to float a loan, probably for a sizable amount. (Horace never requested such a loan; he relied instead on local financiers.) The R. G. Dun Company supplied credit information to people who were in the business of loaning money. On November 16, 1874, a LaGrange-based credit reporter for the company answered a query about W. W. King as follows: "He has no ppty has perhaps $1,000 or more in money. Is considrd hon[est] & wor[thy] of cr[edit] among our people. His habits & capacity are good. He is one of the best workmen in this Co." The inquirer must not have known Washington; he may have been a potential creditor from Atlanta, where Washington moved in 1875. After Washington left LaGrange, he began building bridges on a statewide basis, particularly in the northwest quadrant of Georgia. He did not sever his relations with his family, however; his younger brother, George, partnered with Washington on some of his spans.[20]

As Washington replaced Horace as a large-scale builder in a wider geographical area, Horace became known in Troup County not simply as a contractor but as a respected architect. In December 1877 the Southern Female College laid the cornerstone for its new chapel. Reflecting the civic pride associated with the institution, the *LaGrange Reporter* recorded every detail of the ceremony that involved college officials, appropriate dignitaries, the LaGrange Light Guards, and a contingent of Masons including "the Grand Master of the Grand Lodge, F. A. M. of the State," who lived in Hamilton, a neighboring community. The Masons marched to the site, "preceded and escorted" by the militia unit, and "arranged themselves around the Corner Stone." After the chaplain led a hymn, "the venerable architect, Horace King, placed the [corner]stone in position, and prayer was made by the chaplain." The grand master then received and deposited the artifacts to be encapsulated within the stone.[21]

Laying a cornerstone is the greatest honor that can be bestowed on any Mason. Horace, because of his skills as a craftsman and architect, had transcended the color line. Granting this honor to a black man also speaks to the liberality of LaGrange's race relations.

King's building at the Southern Female College, constructed for its president Ichabod F. Cox, reflected the contemporary Victorian style executed with a Gothic feel. The elaborate brick facade had double porches supported by paired columns, numerous gables, dormers, and finials. Unfortunately, the structure has not survived. The college experienced internal administrative strife and moved to College Park, Georgia, in the mid-1890s. The King-built edifice, damaged by a series of fires, continued to function in a drastically altered state as the Render Apartments before being razed in the 1960s to create a site for the post office.[22]

Ironically, at the same time the community recognized this "venerable architect," Horace was involved in a minor lawsuit involving a rival institution, the LaGrange Female College. It had hired him to replace the roof on its main building. Initially begun by Benjamin H. Cameron (whom Horace had worked for in constructing the Chattahoochee Bridge Company span), this building was a replacement for the more classic, colonnaded Dobbs Hall, which burned in 1860. An architect's drawing, dating from after the war, envisioned an elaborate edifice with three second empire–style towers. King's work never approached the details of this vision. As late as the 1930s only one tower, asymmetrically placed on the right side of the building, had been completed. King's reworking of Dobbs Hall survived until consumed by fire in 1970. The partial fulfillment of the plan stemmed from the college's economic problems during the 1870s.[23]

The building committee hired King, as he testified, "to put on the roof, repair the Doors & Windows and they were to pay me in installments . . . as they collected it." Such provisions appeared to be typical for the period. King ordered the lumber from Burket Atkinson and Company of Coweta County, either a sawmill or a lumberyard. Most likely Atkinson had collaborated with the Kings on other projects. At the same time King ordered the lumber for the college, he included approximately $150 worth of sawn boards for a very short bridge. King "continued to work in good fashion" installing a roof structure on Dobb's Hall that spanned eighty feet. The committee informed him they had no money, but he apparently continued building until they failed to make payments. They owed him close to $1,500 when he stopped working. Atkinson sued King for the unpaid cost of the lumber, standard litigation for contractors in the midst of a depression. King responded that he paid Atkinson for his bridge lumber and his contract with the college called for them to pay for their materials. The chairman of the building committee "said he looked upon Horace King as being a responsible man and knew nothing to the contrary." But he claimed that King was to absorb all of the cost of the lumber. Initially, King failed to issue a defense in the case, and the court awarded Atkinson $1,080.80 to be paid by King. King appealed, and a year later, in June 1877, a jury decided for King and dismissed the earlier judgment. That a higher court ruled in favor of a black man again demonstrates that King was a respected member of the community.[24]

As another indication of the respect King enjoyed in LaGrange, the *Reporter* noted his last political flirtation in October 1878. "Horace King, the well known builder, says he will run for Congress in this district. He is a very worthy colored man, and a man of such good sense, that we shall be surprised if he does not come out of the race, before he fairly gets in." Toying with running as late as October of an election year implies he intended to stand as a Republican. The fact that this Democratic newspaper calmly reported this trial balloon underscores their faith in the conservatism and "good sense" of Horace not to rock the political boat. Whether he ever seriously considered running is not known, but he must still have had a flicker of political ambition at seventy-one.[25]

King's political connections might explain the 1878 hearing convened to evaluate his Civil War claim against the Federal troops that raided Moore's Bridge in 1864 and Girard in 1865. A federal law passed on March 8, 1871, allowed compensation for property confiscated during the Civil War for the use of the Union army or navy, when seized from southern Unionists. A later law (March 11, 1872) established a hearing procedure to assess the validity of such claims. King was probably encouraged by the claims of Columbusites at the end of the war. W. R. Brown, who owned the Columbus Iron Works before the war, asserted his loyalty to the Union and argued his property had been unlawfully seized by the Confederacy, thus the Federals in 1865 had destroyed the property of a Unionist. Brown's action was not exactly parallel to King's assertion, but King understood the possibility of compensation when Congress passed this legislation in 1871. King was one of the few African Americans offered this opportunity. Horace's friend and associate C. W. Mabry acted as one of the witnesses for King's original petition. This case also underscores how King was scrambling to find money: His claim amounting to $2,030 covered seven mules (five in Girard), a wagon, harnesses, foodstuffs, and twenty-seven thousand board feet of lumber.

In what was probably a typical hearing, the clerk of superior court for Coweta County, E. C. Palmer, collected the testimony of the witnesses—

Horace, his daughter, Annie Elizabeth, James Moore (a partner in the bridge), J. S. Miller (a friend and neighbor), and Elizabeth L. Gray (a neighbor visiting the Kings at the time of the raid). The commissioners based in Washington, D.C., reviewed the depositions collected at Moore's Bridge, while clerks reviewed captured Confederate records to ascertain King's role in the war. The commissioners denied King's claim as a Unionist because he had received $249.87 for elm hubs and spokes supplied to the Confederate Quartermaster's Department in Columbus in December 1863 and $105 for harness and leather sold to the Confederate authorities in Cuthbert, Georgia. No one discovered his lucrative relationship with the naval facilities in Columbus. The commissioners also noted that some of Horace's mules were returned and that neither the supplies nor the other items were actually used by the U.S. Army.

Apparently still in need of money, King continued to contract for homes, stores, and ecclesiastical structures. Horace could not afford the luxury of retiring. In the spring of 1879 he built a house for the Honorable R. D. Render on the Woodbury Road, two to three miles east of Greenville, Georgia; it later became the Jesse Porch home. In the summer and fall of the same year he erected two stores on the east square in downtown LaGrange: James Loyd's storefront occupied by T. S. Bradfield's drugstore and the adjacent E. R. Bradfield Building. In a case that resembled the dispute over Dobbs Hall, Henry C. Butler sued E. R. Bradfield for the cost of the lumber used in his building. Bradfield argued that Horace King was liable for that expense ($129.38), but the jury found in favor of Butler and ordered Bradfield to pay Butler. Between February and July 1880 Horace built the Hamilton Methodist Church. A relatively small project, it grossed $1,000 for King, whom the *Reporter* called "the celebrated Negro bridge builder of LaGrange."[26]

While he had omitted his own name in taking the 1870 Russell County census, the 1880 enumerator for Troup County listed him as head of an extended family of mulattoes: Horace, his wife, his two sons, his daughter (Annie E.) and daughter-in-law (Julia), and four grandchildren (John and Julia's children). The marked difference in the age between Horace (who was seventy-two) and his wife (thirty-five) may explain her involvement as one of the shareholders in the joint ventures of Horace's children. It served as an insurance policy for Sarah Jane. In a similar fashion Horace's daughter, Annie Elizabeth, participated in the business as an equal partner with her brothers, an uncommon occurrence in that era. Annie Elizabeth had married a horse trader named Cooper but by 1880 was divorced.[27]

John Thomas had married Julia Sanders in 1871, before they moved to LaGrange. By African American standards of the time, Julia's parents—a railroad mail clerk and a doctor's nurse—were the same middle-class status as the Kings. Horace's name was preserved in the family: John named his only

Horace King surrounded by his sons, ca. 1870. This montage appeared in H. F. Kletzing and W. H. Crogman's *Progress of a Race*, first published in 1898, but the photographs date from the 1870s. In the Kletzing book, this page appears with no other discussion of the Kings; it may be advertising a King construction firm. Either to honor their father and brother or to enhance the company's reputation, the three living brothers included Horace, who died fifteen years earlier, and Marshal, who died even earlier. In this 1969 reprint of the 1902 edition, the publisher reversed the names of Marshal and Washington at the top and John T. and George at the bottom. Courtesy of Albany State University Library.

son after his father. John and Julia eventually had eight more children, all girls. The first three were Carrie, Annie, and Juliett. The fourth girl was named Johnnie; John and Julia must have relinquished all hopes of a boy as a namesake for the father. Their assumption proved correct; Julia had four more girls, Grace, Georgia, Estelle, and Florence.

Missing from the 1880 family circle in LaGrange were two sons: Washington, who had moved to Atlanta, and Marshal Ney, who had died the previous year. The *Reporter* noted his passing: "Marshall [*sic*] King, a well known colored citizen of LaGrange died suddenly this morning. He had got out of bed and stated to kindle a fire, when he fell dead without any premonition." After noting the cause of death as a heart attack, the editor concluded with the usual praise accorded to a King. "He was a son of the well known builder, Horace King, and was held in high esteem by all classes of our community." According to the family's oral history, Marshal had been educated at Oberlin College, perhaps in their day school during the Civil War, but no Oberlin records confirm this assertion. And wartime conditions would have made such an experience unlikely. Marshal worked as a builder with his father and brothers. He purchased a lot on Depot Street in March 1879, then, according to the family Bible, married on September 16, only nine days before his death. Marshal had already earned the respect of the community. Nine years later, when the LaGrange City Council designated a King Street, the aldermen cited the naming as honoring "Horace and Marshal King." (The city later changed the street signs from King Street to Horace King Street, ignoring the son.)[28]

The extended King family, as shown in the 1880 census and praised by the LaGrange establishment, represented a typical American family but an exceptional African American one. As a family of color they achieved remarkable success in a racist society. Despite the general prejudices against them as blacks, the Kings managed to acquire property and respectability. But if the Kings are considered as mulattoes and antebellum freedmen, their accomplishments are slightly less extraordinary.

An extensive study of post-1865 African American property owners in North Carolina noted the importance of prewar status and racial origins in shaping land-owning patterns for blacks in the late nineteenth century. At least half of the 1870 black landowners in North Carolina had been nonslaves, and they tended to have mixed parentage. Furthermore, land ownership became particularly strong for nonagricultural, urban blacks. The Kings also fit the pattern of their peers in North Carolina in that blacks there represented 12.5 percent of all urban contractors within the state.[29]

Even though Horace's name appeared on the 1880 census as head of household, by this date the sons most likely made many of the family decisions. Five years later, one of Horace's obituaries observed that he "saved nothing for his

Annie Elizabeth King, the only daughter of Horace and Frances King, ca. 1870. Her brief marriage ended in divorce, and she lived with her brothers in LaGrange, where she participated as a full partner in the King Brothers Bridge Company. *Collection of Thomas L. French Jr.*

old age [and] was dependent upon his sons." His dependency evidently came from his reverses as a contractor. The sons' successes, however, rested squarely on the foundation and piers erected by the father. Their skills, their reputations, and presumably, their personalities as businessmen, came directly from their father. They apparently did "inherit" his "acquired" characteristics, and those traits set them apart from other blacks. In terms of acquiring property—the primary goal of most Americans—the sons surpassed their father.

George, John, and Marshal had made financial investments in their new hometown by 1879. Neither Horace nor his wife ever purchased any property in Troup County. Apparently the only real estate Horace ever owned was the lot in Girard. Even at Moore's Bridge, he lived on Moore's land. John Thomas bought a piece of property on Greenville Street in February 1875 and another on Depot Street in November 1876. The entire family may have been living on that Greenville Street property in 1880. By the 1880s both George and John (and presumably the entire family, including Horace) lived at the Bean Place, twenty-five acres purchased by George for $1,099 in 1886. The deed identified the site as "where George and John now live." This land was later incorporated into Dixie Mills. The substantial two-story, foursquare house at 508 Greenville Street, built by John, came later. Its battered Craftsman-style columns resting on brick piers testify to its twentieth-century origins.[30]

By 1881 the sons had eclipsed Horace as the bridge builders in the family, but the community still respected King's knowledge. Horace's span over the Chattahoochee near Cameron's Mill only lasted about a decade before a flood washed it away. The rebuilding produced a debate within Troup County over the appropriate location for the new crossing. Rather than forming a county boundary—the case for many rivers—the Chattahoochee flowed through the center of the county and, therefore, taxpayers living in different sections of the west bank of the river demanded the bridge be adjacent to them. People in the northwest corner of the county wanted the span rebuilt at Cameron's Mill, where the private company had maintained the earlier bridge. Other residents on the west central edge of the county wanted a new crossing at Lewis's Ferry. When the county commissioners voted for Lewis's Ferry, irate citizens attacked the county officials. King believed rebuilding at the old site was cheaper. Under the pseudonym "Justice," a proponent for rebuilding at Cameron's Mill wrote a newspaper piece entitled "A Bridge of Sighs," where he outlined the insanity and expense of bridging at another location. He invoked Horace as his authority. "You cannot go back on Horace King's judgment as a bridge builder, for he has built more bridges than any man in the State."[31]

"Justice" and Horace lost the argument, but George King received the contract for the new bridge. The process by which the son obtained this job fol-

John Thomas King and Julia Sanders King, ca. 1871. Julia's father was a railroad mail clerk and her mother a nurse. These photographs probably date from the time of John and Julia's wedding, shortly before they moved to LaGrange. Julia and John had one son and eight daughters. *Courtesy of Troup County Archives, LaGrange, Georgia, and Collection of Thomas L. French Jr.*

lowed the same pattern as his father: He worked cheap. The Kings did quality work and usually at a lower cost than their competitors. In August 1881 the county received two bids for the new Chattahoochee bridge. "Mr. George B. Forbes bid $8,800 and George King $8,600." Then the newspaper learned "that the contract has been awarded George King at $6,000, he having reduced his bid to that amount." Apparently, George finished the bridge by July 1882 and received an extra $500 as a bonus.[32]

In the fall of 1883 an elderly Horace King traveled to Girard and met the Reverend Cherry. In a drugstore there, King reminisced about his productive life. By that time, Horace's son George had eclipsed him as the major bridge builder in the area. The next fall George King constructed Mooty Bridge across the Chattahoochee in Troup County. In early 1885 George also won the contract to cover the West Point bridge that he and his brothers had erected in the late 1860s or early 1870s.[33]

Horace's health was failing rapidly, and he probably never inspected these last bridges. Realizing his end was near, Horace's family convinced him to join the Warren Temple Methodist Church. Other Kings, especially John, were already active in that congregation. Having presumably made peace with his maker, Horace died on May 28, 1885—about three months shy of his seventy-eighth birthday. The *LaGrange Reporter* trusted "that the grace of God bridged for him 'the narrow stream of death' and that he now rests from all earthly cares and labors in the peaceful land beyond the flood." The obituaries appearing in the Columbus, LaGrange, and Atlanta newspapers indicate a respect afforded few African Americans in the 1880s.

The *Columbus Enquirer-Sun* remembered him as Horace Godwin and asserted that "no colored man in the south was long or more generally known than" King, who "drew all the plans and . . . made the contracts" for bridges built by his former master. "Mr. Godwin had the utmost confidence in [him and] would send him with the hands to any part of the south to make contracts and build bridges." The editors exaggerated somewhat in crediting King with inventing the lattice bridge in about 1850. But viewing the world from Columbus, Georgia, that probably seemed to be valid: Horace had built most of those bridges. After telling of his caring for Godwin in his old age and erecting his tombstone, the editor ended with an assessment of his financial condition. "Yet he died a poor man, but had the respect and confidence of those that knew him."[34]

The briefer account in the *Atlanta Constitution* with a LaGrange dateline praised the accomplishments of "Uncle Horace King." "He and his sons built all the foot bridges now spanning the Chattahoochee river, and many in adjoining states." Misidentified as "always . . . a freeman," King, "who possessed considerable intelligence, [had] . . . accumulated a good property, which, however, he lost a few years ago in contracting." It closed with praise of his two sons in LaGrange, "who are staunch, good citizens, and own considerable property." The *LaGrange Reporter* made Horace "the benefactor of many communities in facilitating their social and commercial intercourse by means of many bridges of which he was the architect." Being a weekly rather than a daily, the LaGrange account appeared after the others and corrected them, noting that King did not invent the lattice bridge but did "originate what is known as 'inserted saw teeth.'" The *Reporter* did not mention Godwin but accurately dated King's emancipation as 1846 and mentioned his Masonic membership as well as his service in the Alabama legislature. It reported his financial reverses—"saving nothing for his old age"—and that his sons "cheerfully supported their feeble parent." His hometown obituary closed with this description: "His funeral was largely attended by his colored fellow citizens and his death is sincerely regretted by whites and blacks."[35]

The King family tradition memorialized the honors paid to Horace as follows: "When he died, the town merchants, to a man, turned out around the square, hats over hearts, paying tribute as King's funeral procession passed." Ironically, King's name lives on partly because of erecting his master's tombstone while his own family buried Horace in an unmarked grave next to his son Marshal. Obviously, King was respected in his lifetime, and his passing evoked tributes to this amazing gentleman of color who, despite the prejudices of the time, had erected his own enduring monument—one that proclaimed his accomplishments as a craftsman and as a man. Because of his exceptional reputation, Horace King remains today the best-known African American resident of the lower Chattahoochee Valley.[36]

THE KING LEGACY

George King is always "a whole team."... He is the son of the famous Horace King, the veteran bridge builder, and is one of the thriftiest and most industrious colored men in Georgia.
LaGrange Reporter, March 26, 1888

John King . . . won an enviable reputation as a bridge builder. . . . Among his own race he was a constructive leader and wielded a wholesome influence.
LaGrange Reporter, November 12, 1926

The story of Horace King does not end with his death in 1885. His continuing legacy occupies a central niche in the African American tableaux within the local history mural of the lower Chattahoochee Valley. The successful careers of Horace's children and grandchildren as builders, businessmen, and educators in LaGrange and throughout the state undoubtedly enhanced the father's reputation and caused his name to endure. Their biographies illustrate how well Horace transmitted his technical skills and values to his family. The story of this prominent family, primarily set in LaGrange, also illuminates the activities of the black elite during the late nineteenth and early twentieth centuries.

Horace undoubtedly taught his sons how to build. The oldest, Washington W., participated in the family's bridge building business with his brothers, sister, and stepmother when they moved to LaGrange, but he left soon afterward and moved to Atlanta. Washington did not sever his relationship with his family, however; he and George, the youngest son, continued to work together. In May 1883 they won the contract from Bartow County to build the "wood work or superstructure" for a span over the Etowah River at Howard's Shoals.[1]

Another short Bartow County span plays an important role in local folklore and illustrates the power of the Horace King legend. In September 1886 the county awarded a contract for "a Lattice bridge across Euharlee Creek at Lowry's Mill" for $1,300 to John H. Burke and Washington W. King. The bridge survives today and is significant because of its longevity and its association with Horace. A Web site identifies "Horace King, the freeman [*sic*] builder" as the designer of the 1886 bridge. "However," according to this source, "[Horace] King was too ill to be involved in building the structure. Credit for building the bridge goes to son Washington King and another man." Horace was not "too ill" to participate in this project; he had been dead for a year when this span was built. Despite this, others have insisted that Horace King built it. This beloved bridge is too precious to have been constructed by the lesser-known son. The other builder, John H. Burke, was probably a local white contractor, and he is never mentioned.

The Bartow County spans represent a mere sampling of the many bridges Washington erected throughout the state. Two other bridges he built have survived into the twenty-first century. In the year of his father's death, Washington constructed the 228-foot Watson Mill Bridge on the line between Oglethorpe and Madison Counties. Today, completely refurbished, it stands as the centerpiece of Watson Mill State Park. In 1891 Washington threw what became known as Effie's Bridge across the north fork of the Oconee River at Athens. The Clarke County ordinary paid him $2,470 for erecting this 162-foot span. In 1965 this bridge was moved from Athens to Stone Mountain by cutting it into three sections. Shortened to 151 feet, it is supported by three steel I-beams and has been re-covered and reroofed.[2]

Washington married Georgia Swift, and they had two children: Annadell, who, according to family tradition, taught Latin at Atlanta University, and Ernest, who continued the family tradition of bridge building. He was the third generation of Kings to work on the Fort Gaines bridge: Horace in 1868, Washington in 1888, and Ernest in 1913. The final reconstruction lasted until July 1926, when the bridge fell, after being eclipsed the year before by a new concrete span.

While Washington worked all over Georgia, the next oldest son, John Thomas, and his brother George concentrated on bridges and general contracting in LaGrange and the Troup County area. They stepped directly into their father's shoes as builders. A significant difference between Horace and his sons was the degree to which John and George participated in community activities. Although Horace served as a state legislator after the war, he never participated in local politics. That difference may have been the result of the greater opportunities afforded blacks after the war and the more active role they assumed in creating their own social, civic, and educational organizations.

John, more than any of Horace's other children, played an active role in the religious and educational life of the community. His growing family, which eventually included eight daughters and a son, probably motivated his involvement. His most significant activities revolved around the Warren Temple Methodist Church, where he quickly became its leading layman.

The Warren Temple congregation was the result of efforts by John Caldwell, a former southern Methodist minister who became one of the state's most successful "religious scalawags." After the war, Caldwell and a handful of other white Georgia preachers sought to create a biracial denomination (though with segregated churches) in Georgia as part of the (northern) Methodist Episcopal Church, which sent religious and educational missionaries to work among the former slaves. Caldwell had organized the Warren Temple church by February 1866, but he enjoyed his greatest success in August of that year. During an eleven-day camp meeting in LaGrange he attracted a crowd of five to six thousand freedmen who were exhorted by both white and black preachers. Although Caldwell was an active Republican—a political as well as a religious scalawag—his message to blacks was one of moderation. Perhaps Caldwell's conservative actions tempered racial conflict in LaGrange, lessened the white backlash, and allowed the town's black elite to enjoy more freedom of action. After Caldwell moved to Delaware in 1868, the Warren Temple congregation functioned under the auspices of the Freedman's Aid Society of the Methodist Episcopal Church, which continued to send white ministers to the church. By 1873, when John T. King joined this congregation, it represented a conservative alternative for people of color, compared to the more radical African Methodist Episcopal (AME) Church.[3]

The minister and stewards of the Warren Temple Church quickly recognized the leadership talents of John King and gave him responsible positions within the congregation. John, in turn, relished being a leader. On January 17, 1874, King became superintendent of the Sunday school, which oversaw two hundred people in eight Bible classes and fourteen spelling classes. King also assumed the position of secretary for the church, as well as for the circuit and district organizations. Eventually he acted as a lay representative to statewide meetings. By 1880 John served as chairman of the church's board of trustees, a position he held for more than forty years. In terms of responsibility he ranked second only to the white minister.[4]

As demonstrated by its spelling classes, the church played a leading role in promoting black education. Warren Temple, with financial support from the Methodist Freedman's Aid Society and the local community, launched an effort to build a black school in 1875. The Rev. Robert Kent, presiding elder of the Whitesville Circuit, made an impassioned appeal for education as a means of preventing black sharecroppers from being cheated by white landowners. The Rev. J. H. Owens, John King, and fellow black businessmen such as Alex Swanson and J. Sanford Pitman contributed money and helped convince the town council to provide city bond taxes for the school. The

Washington King's Watson Mill Bridge, 1885. Washington's span as it appeared ca. 1970 (*top*) and today as the restored centerpiece of the Watson Mill State Park near Comer, Georgia. *Courtesy of O. Lee Moon, Watson Mill Bridge State Park, and Thomas L. French Jr.; photograph by Edward L. French.*

LaGrange Reporter's editor, John T. Waterman, also raised funds for this endeavor. John superintended the construction of the school on Hill Street in 1877. Though it operated under several different names, including the LaGrange Seminary and the John King School, it remained under the supervision of the church—and John King. Family tradition stressed John's commitment to the aphorism "ignorance breeds poverty." The land purchased by John and George King in the 1880s lay near their church and school.[5]

Churches and schools were always the basic institutions created by freedmen. In LaGrange, the black elite, apparently with the support of the white establishment, created more atypical organizations. About 1878, after creating a school, the local black citizens also organized both their own militia unit and a black fair. The militia unit served as a social outlet for the black elite and as a community organization. Far from being seen as a threat by the white community, its suppers were supported by whites to raise funds to purchase uniforms, and it cooperated with the white LaGrange Light Guards as firefighters and as police on festival days. Service in this organization earned members an exemption from street taxes.

In the fall of 1878 an African American fair, initially organized by whites, began functioning as an annual event staged the week after a statewide agricultural fair in Macon. During this event blacks hosted gala masquerade balls and parades. Whites were not allowed to compete with blacks in the horse races because whites could purchase more expensive, faster horses. They could, however, take part in the ox races.[6] By the fall of 1884, this event was called the "Colored People's Fair," and it operated "under the management of George King, lessee of the grounds of the Western Georgia Fair Association," the same site used for white fairs. The Kings lived on property adjacent to the fairgrounds, and George presumably managed this event for several years. The readers of the *LaGrange Reporter* in the mid-1880s also learned of George's skills in raising rabbits and mules. As the Kings were becoming more involved with community affairs, their bridge-building profession underwent a major change.[7]

By 1886 bridge builders began the transition from the wooden age in which Horace flourished to the era of iron and steel. As if the riverside communities had waited for the death of the great builder of timber spans, the city of West Point bought an iron and steel bridge in 1886. Viewed within the context of the lower Chattahoochee Valley, the West Point span ranked second only to Columbus in the volume of traffic carried by its bridge, and the river seemed to attack West Point with more ferocity than it did Columbus. High water in 1882 moved the railroad bridge several inches downstream, but the wooden bridge held. Four years later, a particularly vicious flood struck the community leaving thousands homeless, the railroad bridge damaged, and the King-

Know No Failure, 1898. The philosophy of Booker T. Washington was alive and well at the LaGrange Academy. John King, a founder and builder of this school is sitting on the left end of the second row. This school, sponsored by the Warren Temple Church, was known for a while as the John King School; it was renamed Hill Street School when it became part of the public school system. *Courtesy of Troup County Archives, LaGrange, Georgia.*

built footbridge destroyed. Troup County and West Point shared the expense of replacing the span. Their officeholders carefully studied their options and decided on an iron bridge.

By the 1880s metal spans—usually a combination of iron and steel members—became popular, even in the South. In Columbus, Mississippi, where Jemison and Horace had planned to build a wooden crossing after 1865, the city combined an iron railroad and a wagon/pedestrian bridge into one structure. In Tuscaloosa, Alabama, where Horace built a wooden bridge in 1872, the King Iron Bridge Company of Cleveland, Ohio, erected three sections of tubular arched metal bridging for the city in 1882. The King Company was one of the largest highway bridge works in the nation during the 1880s. In 1886 a competing firm, the Chicago Iron and Bridge Company, provided a modern span for the city of West Point for eight thousand dollars—only five hundred dollars more than George and John King had bid to rebuild it as a wooden bridge. The use of an iron bridge more than justified the extra money. The postwar wooden bridge lasted twenty years at West Point. The Iron Bridge, as West Pointers labeled it, survived the ravages of the Chattahoochee for twice as long, until the mighty "Pershing Flood" of 1919.[8]

The advent of mass-produced iron bridges meant fewer opportunities for building wooden bridges. George, unlike the peripatetic Washington, worked in the Troup County area on smaller spans. John, the LaGrange family patriarch, pursued a variety of enterprises as a contractor and owner of several businesses. In 1885 John entered into a partnership with Dallis and Edmundson, white merchants (cotton warehouse owners and builders), who provided part of the capital for a ten thousand dollar variety works that produced "doors, sash[es], blinds, chairs, mantels, cheap furniture, and many other things." John King moved his "machinery" from "his steam ginnery" into this new facility operated by "King & Co."[9]

Exactly who the members of King and Company were remains a mystery. The company may have included all the family members in LaGrange or maybe just John. The *LaGrange Reporter* identified the company as the builder of a bridge across Flat Shoals Creek at Alford's Mill near West Point on June 5, 1884. In 1888 the family dissolved the King Brothers Bridge Company when it sold the office that had been purchased in 1875. John also operated a lumberyard on Railroad Street near his Greenville Street home. His son, Horace H. King, remembered riding on the green wagon with gold trim that John used to deliver lumber and building supplies in all parts of LaGrange. John was also a successful contractor, although on a smaller scale than George. He was an important LaGrange businessman for more than forty years.[10]

From 1886 until 1890 George and John profited from numerous building contracts. They built a new addition for the Troup County courthouse that housed the offices and the vaults of the clerk of court and the ordinary. George constructed the new Thornton Building on the west square in LaGrange, while on the east square John added a story to E. R. Bradfield's store, which Horace had built in 1879. The ladies of the Presbyterian Church, having netted $115 from their ice cream business, engaged John to build a twenty-seven-foot steeple for their church.[11]

In 1887 George continued in his father's footsteps when LaGrange Female College hired him to build a new boarding wing. George also erected a warehouse for J. G. Truitt; three years later, the same client hired him to construct a two-story building for speculation. The King brothers, whether as partners or separate contractors, became the builders of choice for many LaGrange businessmen and for the Troup County commissioners. In 1888 John constructed a two-room house at the county poor farm and a bridge over Long Cane Creek on Sulphur Springs Road. During the same period George won the county contract for spans over Blue John Creek at Rachel's Ford, over Long Cane Creek at L. T. C. Lovelace's, and over Whitewater Creek at Henderson's Mill. These structures could have been replacements for old spans or new crossings superceding fords. At Rachel's Ford on the Hamilton road, for example, George's thirty-foot bridge replaced a dangerous ford over Blue John Creek, only two miles from LaGrange; it also included four hundred feet of rock approach-ways. In July 1888 the *LaGrange Reporter* noted that George King had constructed most of the wooden truss work on every bridge in the county.[12]

By 1888 George and John Thomas were more than just skilled craftsmen or foremen supervising crews of carpenters. They were businessmen who participated fully in the economic life of the city. When LaGrange rallied to build its first textile mill, the Kings played an active role. The LaGrange textile experience paralleled that of the Carolina Piedmont where, during the 1880s, entire communities staged drives to raise capital for steam-powered mills. Such civic efforts became an oft-told tale of southern industrialization. The actions of George King, however, made the LaGrange story unique.

A LIBERAL COLORED MAN

> Among the subscribers to the Cotton Factory is George King, who comes up nobly to the rescue with $2,000. Whether building bridges, fighting fires, or helping public enterprises with his well earned cash, George King is always "a whole team." It will be remembered that, when a railroad was agitated, he was one of the first to take a liberal amount of the stock. He is the son of the famous Horace King, the veteran bridge builder, and is one of the thriftiest and most industrious colored men in Georgia.[13]

Following the example of other enterprising businessmen, the Kings profited from their investment in the nascent mill. George and John were hired

LaGrange Mills, the city's first textile factory, ca. 1900. George King invested two thousand dollars in this mill; then he and John constructed the building in 1888. *Courtesy of Troup County Archives, LaGrange, Georgia.*

to build the new factory, and in May 1888 they began laying its foundations. George continued to reap benefits from the industrialization of the city when, seven years later, he sold twenty-two acres of land to Dixie Mills for four thousand dollars. George or perhaps the family had originally purchased the valuable land as an investment because of its proximity to the center of the town and its rail frontage.[14]

By 1890 George was receiving accolades that resembled those accorded to Horace. The *LaGrange Reporter* called George the "Champion Bridge Builder." His 110-foot bridge over Wehadkee Creek at Cofield's, fourteen miles from LaGrange, replaced a span built by his father in 1873. A freshet in April 1886 destroyed it, but George replaced it four years later. It survived into the 1960s, when Callaway Gardens moved the structure to the front of the Sibley Center where it spanned a small pond. In the 1980s the bridge was moved to a field away from public access, where it still stands behind a chain-link fence. All the references to this bridge in the Callaway Gardens literature attribute its construction to Horace rather than to George. LaGrange historian Forrest Clark Johnson discovered its true history. The Red Oak Bridge Creek in Meriwether County, Georgia, which local histories claim Horace constructed in the 1840s, may suffer from a similar confusion between father and sons.[15]

After 1888 the LaGrange press included fewer mentions of building projects in general, thus, the business activities of the Kings became less visible to the public and, therefore, to the historian. In 1890 George raised a building for J. G. Truitt and in 1892 laid the granite foundation for the LaGrange jailhouse, which now houses an art museum. In the same year "Prof. Alwyn Smith" engaged John T. King to construct a "stylish and well arranged" home. The brothers certainly acquired other jobs that were not cited in local newspapers. John most likely built other houses, while George repaired old or fabricated new trusses for county bridges.[16]

John and George also worked on Glass's Bridge, the last major wooden bridge across the Chattahoochee River. The early history of this bridge site, about ten miles north of West Point, remains hazy. Horace may have built a bridge there (at Crowder's, later Bentley's Mill) about 1873, but that span apparently never received any notice in the *LaGrange Reporter.* This crossing presumably failed long before 1896. According to the oral history of the Glass family, who lived adjacent to the bridge, before 1896 "Button" Hairston, a former slave, operated a ferry there, which was owned by William Hairston, who was white. In 1896 Nunn and Hudson Construction Company of Gainesville, Georgia, rebuilt the bridge for Troup County.

Critics immediately denounced spending tax money on this "bridge of folly," asserting that it only served eight county residents. They also raised questions about its shoddy construction: "In reality, before this bridge was thirty days old, George King had been employed to make it safe and keep it from washing away and it is reported had received over two thousand dollars for doing this work which should have been done by the contractors." Nunn and Hudson must have underbid the Kings for the original job. Opposition to the bridge escalated beyond mere verbal assaults. On the night of September 18, 1896, incendiaries burned the new span. Despite widespread opposition the county commissioners decided to rebuild the bridge immediately.[17]

According to Glass family oral tradition, the original contractors returned to rebuild the bridge. Jim Nunn apparently lived with the family and supervised the construction, but "two Negro carpenters from LaGrange, John and Horace King, had charge of the labor gangs." They must have been the brothers John and George. This account seems to indicate erroneously they had also built the original span that burned.[18]

The second 1890s installment of this span stood the test of time. Its six truss sections resting on five rock piers and two rock abutments carried traffic until 1956. As other major wooden spans failed or were replaced, this structure became the longest covered bridge in Georgia, and writers often identified it as the oldest, a distinction actually shared by shorter crossings such as Red Oak Creek or Washington's bridges. The bridge also continued to be incorrectly associated with Horace, although local sources accurately linked it with his sons. In the mid-1950s the Georgia Department of Transportation constructed a concrete replacement and tried to demolish the older wooden bridge for safety reasons. The wooden bridge proved to be so sturdy that the engineers resorted to burning it—an ironic end for this bridge with a phoenixlike history.

Glass's Bridge may have been the last major bridge built by George, but he continued to erect smaller ones. He died in 1899. According to family tradition he was in camp working on a bridge when he died. His brother John's career as a contractor and businessman continued for another twenty years. In 1897 John served as architect and builder of the new Loyd Building on the southeast corner of the courthouse square in LaGrange. The nine-bay-wide building consisted of three modern glass-front stores on the ground floor, two occupied by the mercantile establishment of J. L. Bradfield and one by the LaGrange Hardware Company. Beneath the decorative parapet and behind the nine windows on the second story was "the handsome lodge of the Masons. They will spare nothing to make their hall attractive and pretty." John may have been selected to erect this building because he was a Mason.[19]

In 1909 John T. King's name appeared in R. G. Dun and Company's credit listing as a contractor with an estimated pecuniary strength of between one thousand and two thousand dollars, and a general credit rating of fair. The company gave John the highest possible ranking for someone with fewer than two thousand dollars in assets. Despite this modest assessment of his financial strength, when he died fifteen years later the newspaper called him "one of the best known negro men in the city." By the early 1920s he apparently no longer operated a lumberyard; his letterhead identified him as a contractor and builder, with bridging as a specialty. Even though his specific business activities shifted over the years, an involvement with hydraulics remained a constant. In 1887 he advertised himself as an agent for a new force pump; in 1921 he supplied water systems—gasoline pumps and rams—for country homes. Apparently, John King specialized in moving water as well as in bridging it.[20]

Another constant in John's life remained his church, for which he served as a contractor in a protracted project to replace its wooden home with a brick edifice. The process began in 1899, and both in reality and metaphorically it illustrates the perseverance necessary to create black institutions of that day. Although the church hired architects from New Jersey (Benjamin Price and Max Charles Price of Atlantic Highlands), the foundation was laid by volunteers from the congregation using stones farmers hauled to town. King quit the project at one point because he opposed mortgaging the property but returned as superintendent after the masons and bricklayers walked off the job. After rallying the workers, King as a "worshipful master together with the members of Mount Gilliar Lodge #212" laid the cornerstone in 1903. Work continued until 1905 when fire consumed the walls. Undaunted, they rebuilt the walls. Pews were finally installed in 1908 and the congregation began

Glass Bridge, Troup County, ca. 1950. This span was built in 1896 and immediately burned by arsonists who viewed it as a "bridge of folly." George and John King rebuilt it. The span survived until 1956; by that date it had become the longest wooden bridge in use in the state. When the Department of Transportation replaced it with a modern bridge, they had to burn the Kings' wooden structure because it was so stout. *Courtesy of Troup County Archives, LaGrange, Georgia.*

John T. King (*middle*) and two unidentified workers at King's lumberyard and sash mill, ca. 1920. *Courtesy of Troup County Archives, LaGrange, Georgia.*

meeting in the church even though the roof leaked and most of the windows, lacking panes, were covered with boards.

The cornerstone year of 1903 was an active one for John Thomas and for his church. John finished Unity Methodist Church, a small wooden building in the shape of a Maltese cross, which served the families of mill operatives living in the Unity Mill village. Construction proceeded more rapidly on this church than on Warren Temple since the mill owner, Cason Callaway, did not have to beg and borrow every dollar needed for construction.[21] That same year the Warren Temple Church relinquished control over its school when it rented its building to the new city public school system. That ended John King's twenty-five-year involvement with that institution, but the Kings still

The framework of an unidentified bridge under construction. This photograph, marred by a child's crayon markings, is the only one that documents the King family practicing their craft. According to family tradition, the man standing on the shore is Horace, but that individual might well be one of his sons. The span resembles one of many constructed by the Kings at Fort Gaines, Georgia. *Courtesy of Troup County Archives, LaGrange, Georgia.*

Unity Methodist Church, LaGrange, Georgia, 2003. John King, who built churches for white as well as black congregations, constructed this Methodist church in the Callaway-owned Unity Mill village in 1903; it now serves as a Baptist church for a Hispanic congregation. *Photograph by John Lupold.*

remained active in education. One of John's daughters, Carrie F. Thomas, served as the last principal of the school. Her husband, Augustus A. Thomas, a North Carolina–born printer, supervised it when it became a public school. Their daughter, Theodora, who served as the family's primary oral historian, worked as an administrator and instructor in several Georgia communities before enjoying a long teaching career in LaGrange. Another of John's sons-in-law headed a black school at Kowaliga, twenty-five miles north of Montgomery. Another relative worked as a teacher at the West Virginia Collegiate Institute. In 1921 I. Garland Penn, one of the officials of the Methodist Board of Education for Negroes, wrote John that he regretted not being able to hire his daughter, Mrs. Benson, at the school at Meridian, Mississippi, since the president of Alcorn College had praised her work there.[22]

John's service to his church and its educational endeavors won him recognition at the state and national levels. The same Methodist initiative that created Warren Temple in LaGrange also established Clark University in Atlanta. John served on Clark's board of trustees from 1893 until 1918, when he became ill. His son, Horace Henry, held the position from 1919 until 1922.

John returned to the board in 1924 and remained there until his death in 1926. His involvement at Clark may indicate that he also worked as a contractor in Atlanta.

John's skill as an architect-builder and his activities within the Methodist Church garnered him an extremely significant job in terms of African American history—the construction of the Negro Building at the Atlanta Exposition in 1895. I. Garland Penn, an associate of King's in matters relating to Clark University, served as chief of the Negro Department for the exposition. In the *Bulletin of Atlanta University*, Penn explained "that two colored men, Messrs. King and Smith, have the contract for the building. King is from Georgia, and Smith from North Carolina. . . . The contract was let at $9,651." The *Atlanta Constitution* listed the Georgia contractor as John F. King of LaGrange, but the reference is surely to John Thomas King; LaGrange had no other contractors of color named King. The King family tradition also maintained that John built it, an assumption reinforced by his long-term relationship with Penn.[23]

In 1895, only three decades after Sherman's march to the sea, Atlanta's leaders staged their exposition to trumpet the progress made by the premier city of the New South. Holding it only two years after the Chicago's World's Fair underscored the boldness of Atlanta's boosters. Along with the separate structures showcasing electricity, women's activities, fine arts, the state of Georgia, and the U.S. government stood what the *New York Observer* called "the most unique" venue. The Negro Building showcased the great progress made by blacks only three decades removed from slavery. Designed by Bradford I. Gilbert, its most prominent feature consisted of a multistory glass facade framed by rectangular pavilions at both ends and another set of pavilions flanking a pedimented entrance. In the middle of the edifice stood the largest pavilion with a open-air observation deck. Both the central door and the pavilions contained large arch-headed windows. In contrast to the more classical forms of the Fine Arts and Electricity Buildings, the Negro Building, with its glass wall and banners atop all the hip-roofed pavilions, had the feeling of a less formal fair building. The *Observer* noted, "Every timber . . . was laid by Negro mechanics. The chief contractor is a most intelligent and trustworthy businessman of means and reputation." The exhibits also were "the product of the industry of colored workers" including "models of thirty or more patents" recorded by African Americans. The New York press hoped the building and exhibits "may be made to mean a great deal in the solution of the race problem." In hindsight, their prescient words seemed to set the stage for the appearance of Booker T. Washington.[24]

In addition to the Negro Building and its exhibits, the exposition organizers needed a black leader to proclaim black progress and stress the peaceful nature of race relations. They invited Booker T. Washington, the conserva-

Negro Building, Atlanta Exposition, 1895. John T. King served as one of the two contractors for this facility designed to showcase the achievements of African Americans in the South since 1865. King's involvement in this project indicates both the contacts he had in Atlanta and the respect he enjoyed as a contractor. *Courtesy of the Atlanta History Center.*

tive head of the Tuskegee Institute. His address became the most memorable aspect of the fair and his best-known speech. In the main exposition hall, with a sheet dividing black and white spectators, Washington argued for equal economic opportunity for blacks, who would, in turn, accept segregation in matters that were strictly social. He pleaded for the right of blacks to work in textile mills alongside whites, while conceding that they had no desire to sit next to whites in opera houses. The whites cheered his social concessions and ignored his primary concern, economic equality. Washington used the phrase "separate but equal" to indicate the social status blacks would accept. The following year the U.S. Supreme Court used the same phrase in its *Plessy v. Ferguson* decision, which declared segregation constitutional.

Washington's philosophy regarding education and economic progress for blacks rested on the premise that becoming a skilled worker or craftsman offered the best avenue for black advancement and acceptance by whites. While no surviving evidence suggests that Washington ever used the King family as an example of this premise, he must have known of Horace, Washington, John, and George. They illustrated the validity of Washington's ideas in that both Horace and his children were treated with respect and even at times honor.[25]

But the Kings were far from typical. Not only did their skills exceed that of most other black craftsmen but also their role as bridge builders separated them from craftsmen of any race. Ultimately Booker T. Washington simply argued for the icon of the self-made man—that all Americans, even blacks, could pull themselves up by their boot straps. Horace succeeded because of his skills, his hard work, his demeanor, and the essential help of John Godwin and Robert Jemison. The latter Kings succeeded because they followed Horace's model.

Horace's sons enjoyed the same type of respect afforded their father. As an illustration of John T. King's standing in the community, his 1926 obituary occupied almost a third of a page in the *LaGrange Graphic*. More than twenty years later the *LaGrange Daily News* ran a piece entitled "John King, Bridge Builder, a Man Who Left behind Monuments to His Genius." This 1949 piece, which repeated much of the praise in his obituary, seems to have had no specific reason for appearing at that particular time. No bridge or black institution was being built. Perhaps by discussing the Kings, the editor hoped to improve race relations in LaGrange. The Kings might have served as an example of how whites could respect blacks, but their lives did not have much meaning as a model for the average African American in Georgia or Alabama—either then or earlier.[26]

CHAPTER SIXTEEN

THE LEGEND OF HORACE KING

Even here and now today we need their bridges,
even now a way to get from shore to shore,
they built their covered crossings,
though some have fared not well,
but the bridge from King to Godwin never fell.
Allan Levi, "The Bridge from King to Godwin," *Rivertown* CD, 1996

One hundred twenty years after Horace met the Rev. F. L. Cherry at Smith's drugstore in Browneville, Alabama, the Horace King legend is alive and growing. King was an extraordinary builder, and his actions as slave and free black were certainly unusual, but why does he remain in historical terms the best-known builder and one of the most recognized African Americans in the lower Chattahoochee Valley? Synergistically, the building stones of the legend have combined to make him larger than life. And while this transformation of King from man to legend defies precise explanation, some light may be shed on that process by illuminating the components of the myth as they have evolved over the last century. These elements include the quality of King's work; the survival of some of his spans and buildings; the lore of covered bridges in general; King's own conservative actions, especially erecting Godwin's marker; the boosterism of various towns; the growth of black history; the emergence of the neo-Confederate movement; and the seductiveness of the bridge metaphor as applied to race relations.

The Horace King legend begins with one essential fact: King was an excellent bridge builder, engineer, and architect. As early as 1840—twenty-five years before the end of slavery—a newspaper credited Horace, the slave, as cobuilder (along with John Godwin) of the Florence, Georgia, span. By that date King was serving as the engineer, while Godwin handled the business affairs. As a freedman King became the bridge builder of choice in the Chattahoochee Valley and beyond. Not only did he consistently bid two hundred dollars less than Asa Bates and other competitors but also his craftsmanship set him apart from others. Horace, his sons, and his grandson Ernest remained the princes of wooden bridges so long as bridges remained covered into the 1920s.

Several of Horace's major bridges outlived him by as many as thirty years, while some smaller ones survived for more than a century. History either forgets those crossings that failed within a decade—Eufaula and Florence, Alabama; and Columbus, Mississippi—or credits him with building the succeeding spans. Local pundits ignore the fact that his sons either built or rebuilt many of the minor spans in Georgia and Alabama attributed to Horace. The work of his sons, particularly in LaGrange, coupled with the prominence of the King family within that city, kept the father's name alive. Horace became a cultural icon in several communities—LaGrange, Phenix City, and Columbus—and his reputation became a major point of historical pride in those cities. His oldest son, Washington, also perpetuated the King name for generations; several of his major bridges remain tourist attractions into the twenty-first century.[1]

The romance of covered bridges also contributed to the recognition of the Kings as bridge builder. Covered bridge aficionados—a significant subset of antiquarians—have continued to praise the architectural and building skills of the Kings because, to repeat the obvious, many of their spans survived. As a result, even into the twentieth century, many generations of backroads wanderers experienced the cooling shade, the musty, historic smells, and the admirable craftsmanship of a King timbered tunnel. Those travelers marveled at how a small crew of men armed with presteel technology could span such distances with pieces of wood held together only by pegs. Their wonderment has served to memorialize Horace and his sons.

Beyond the Kings' tangible accomplishments, the events and symbols of Horace's life have captured the imagination of writers, and their fascination with him has increased since his death. King became a legend even in his own lifetime, partly by erecting a monument to his former owner, John Godwin. While Horace undoubtedly loved and respected Godwin, he may have honored his master on the eve of southern secession, as racism reached a peak, in order to assert his loyalty to the South's most important institution—slavery—and, thereby, to escape reenslavement. Whatever his motives, southern newspapers, at that time and later, passed the story around, quoting the text of Godwin's marker and enshrining King as a noble man of color.[2]

Godwin's marker and King's status as one of the original residents of Girard led the Rev. Francis Cherry to interview Horace in the 1880s and include a lengthy biography of the bridge builder in his local history of Opelika, Alabama. Other local newspapermen followed his example. A Columbus journalist and historian, W. C. Woodall, who frequently spun local myths, published his version of King's life. King's story, however, seemed to become a point of particular pride for Girard–Phenix City and LaGrange. In the 1930s writers from those cities working for the Works Progress Administration

Johnson Mill Bridge, after 1900. Local history sources identify Horace as the creator of this bridge, but no evidence links him to the span. Harris County paid W. W. King $2,290 for building an unspecified bridge in 1906; it might have been this one. *Courtesy of Georgia Division of Archives and History, Office of Secretary of State.*

Theodora Thomas, 1979. She carried on the family tradition of education, serving as a teacher in several cities including LaGrange, where she was also a school administrator. She became the most important source for the King family oral tradition. This picture was taken at the dedication of the Horace King marker in Phenix City. *Photograph by Thomas L. French Jr.*

composed sketches about King and his family. Such works kept the family in the public consciousness. Later, King's marker to Godwin appeared in the "Ripley's Believe It or Not" newspaper column. Even today, the Godwin monument serves to perpetuate his memory.[3]

Some accounts written outside of the Chattahoochee Valley treated King as an anomaly—a slave who built large bridges—and did not even bother to include his last name. On a 1958 Albany, Georgia, historical plaque describing the history of Nelson Tift's bridge house, Horace, not Horace King, was cited as the builder. In a similar fashion, Robert S. Starobin in his *Industrial Slavery in the Old South* identified him only as Horace in the text, though the surname King appeared in parentheses in the index.[4]

But most misinformation about Horace has erred on the side of exaggerating his accomplishments. One newspaperman credited Horace with inventing the king truss, an arrangement of posts or timbers that probably evolved at the dawn of human bridge construction. Many communities have assigned King the honor of building their covered bridges, even though no primary evidence supports the assertions. LaGrange and Phenix City, on the other hand, had valid reasons to claim the Kings as their own.[5]

In the 1970s researchers in both LaGrange and Phenix City found new materials relating to King. In 1978 an effort spearheaded by Susie Fowler, working for the local historical society and the Troup County (Georgia) Archives and using information provided by Theodora Thomas, identified and marked the graves of Horace and his son Marshal for the first time. Continuing investigations in LaGrange newspapers at the Troup County Archives finally pinpointed the date of King's death. The King family Bible listed King's death as occurring in 1887 rather than 1885. The discovery of his 1885 obituary in the LaGrange newspaper finally allowed researchers to unearth others published in Columbus and Atlanta.[6]

Harold S. Coulter's *"A People Courageous,"* published in 1976, devoted the equivalent of several chapters to John Godwin and Horace King, with more emphasis on the latter than the former. At approximately the same time, William H. Green, a professor of English at Chattahoochee Valley Community College in Phenix City, began investigating King. Ultimately he wrote a novel, not yet published, that focused on Horace. His research produced valuable information about King, and Green, an excellent raconteur, has increased the community's knowledge of King. In 1979 increased interest in the bridge builder led Phenix City and the Historic Chattahoochee Commission to erect a historic marker to Horace at the corner of Dillingham and Broad Streets, one block west of the Lower or Dillingham Street Bridge, which Horace constructed three times, and one block south of the Holland Creek bridge, a ravine he spanned not long after the Civil War.[7]

By the 1970s and 1980s two seemingly oppositional movements championed Horace King—the advocates of Black History Month and the neo-Confederates. As more institutions and newspapers began to observe Black History Month, King became a popular figure for recognition. His story fit the objectives of early popular black history, identifying African Americans who made discoveries, inventions, or other contributions to society. In 1972, for example, an African American newspaper, the *Washington Star-News*, ran a series entitled "They Had a Dream"; one piece featured Horace King in "Slave a Frontier Bridge Builder." It included the story of King "deeding four acres of land in Phenix City to Godwin's daughter, whose family had come on hard times." Unfortunately, no such deed ever existed.[8]

King's actions appealed to some writers because they cast slavery in a positive light. Godwin only freed one of his slaves and the basis of his action was certainly economic, but these qualifications do not diminish the telling or retelling of his story as an act of altruism. For example a 1996 South Carolina brochure (*Pepper Bird Pathways*) that listed events of interest to tourists also included a brief article on King entitled "Slavery or Friendship?" The piece emphasizes King's South Carolina origins and exaggerates how much he did for the Godwin family. Columbus, Phenix City, LaGrange, and even the Opelika newspapers have long published articles about King. But in recent years, during Black History Month, Horace has become a popular subject outside of his immediate homeland, perhaps because of his moderation and the readily available materials about him.[9]

Today, the neo-Confederate movement glorifies the antebellum South as

Albany Bridge House and Flint River overlook, September 2003. Volunteers, coordinated by Links, an organization of African American women, are building a river overlook to honor Horace King. The thirty-two-foot overlook will be patterned after one of King's bridges at the site of the span he built for Nelson Tift in 1857. *Photograph by Elliott Minor, AP/Wide World Photos.*

well as the Confederacy and, in general, attempts to minimize the connection between racism and secession. Specifically, neo-Confederates seek to identify African Americans associated with the southern war effort, especially when such service could be interpreted as voluntary. Their Web sites and publications tout Horace King as a Confederate even though his testimony in the 1870s contradicted their assumptions. The following quotation is an example of his exaggerated southern loyalty: "Former slave, Horace King, accumulated great wealth as a contractor to the Confederate Navy. He was also an expert engineer and became known as the 'Bridge builder of the Confederacy.' One of his bridges was burned in a Yankee raid. His home was pillaged by Union troops as his wife pleaded for mercy." More accurately, King accumulated a great deal of worthless Confederate paper when he was forced to work for the navy. The Union troops raided his outbuildings, not his home, and his wife pleaded for the return of the mules and grain, not, per se, for mercy. Nevertheless, the endeavors of neo-Confederates have increased King's visibility.[10]

African Americans in the South do not generally view their history in the same terms as neo-Confederates or the journalists who select the honorees for Black History Month. Most blacks would not select Horace King to honor as their representative, especially if they had read the Reverend Cherry's or similar accounts that openly suggest King as a proper conservative role model for blacks. Their heroes are more likely to be men like the radical Republicans Philip Joiner and James T. Rapier who demanded sweeping social changes during Reconstruction.

The research on King became more intensive and more critical in the 1980s with the work undertaken by the father-son team of Thomas L. French Jr., land surveyor–landscape architect, and Edward L. "Larry" French, landscape architect. While teaching mechanical drawing at Pacelli High School, Tom guided his students in constructing a scale model of a covered bridge. That experience led French and his son to write *Covered Bridges of Georgia.* Bill Green, in pursuit of Horace King, needed to understand covered bridges, and he turned to Tom for assistance. Their collaboration converted Tom into an avid King researcher, and he began assembling materials from courthouses, archives, and local historians. His and Larry's biography of Horace that appeared in *Alabama Heritage* in 1989 became the standard account of King's life. Karl-Heinz Reilmann, a member of the Alabama Society of Professional Engineers, discovered more significant information in the process of successfully nominating Horace King to the Alabama Engineering Hall of Fame in 1988. The data collected by the Frenches, Green, and Reilmann, together with the King family oral histories, formed the basis of Tom C. Lenard's film *Horace: The Bridge Builder King.* Lenard's own endeavors also added to the storehouse of knowledge about King.[11]

By the late 1980s King's name was becoming associated with modern bridges. In LaGrange, the city upgraded the roads between the center of town and the malls on the eastern edge of the city. In the process, they constructed a bridge that extended the existing King Street (from his family's old neighborhood) across the railroad tracks. That span logically became the Horace King Bridge, and its 1988 dedication became an occasion for erecting a historic marker to the bridge builder, sponsored by the local historical society and archives.

In Columbus the following year, the Sons of the Confederacy, having learned of King's wartime efforts, asked the city council to rename the Fourteenth (Franklin) Street Bridge after Horace—a nice gesture except that the 1922 concrete span had already been condemned, placed on the list of the most dangerous spans in the nation and scheduled to be replaced with a new one at Thirteenth Street. During the proceedings one of the councilmen noted the problems with the bridge and suggested the new bridge be designated to honor King.[12]

With the turn of the new century a Horace King renaissance seemed to be sweeping the Chattahoochee Valley. The Corps of Engineers christened a picnic shelter on West Point Lake the Horace King picnic shelter. The owners of a building at the corner of Broad and Dillingham in Phenix City floated the idea of using that space as a small museum devoted to King. Suggestions for using the old City Mills in Columbus as a home for various historical activities have also circulated and include a King–black history museum in the old corn mill. A plan being implemented in Valley, Alabama, involves building a new covered bridge with treated lumber and steel bolts for treenails over Moore's Creek on the Chattahoochee Valley Railroad Trail. The new covered bridge, which will contain material about King's life, is being promoted as a tourist attraction and as a significant part of the city's black history program. Travelers between Columbus and Phenix City are reminded of King as they cross the new span at Thirteenth Street, dubbed the Horace King Friendship Bridge.[13]

Over the years, the bridge as a metaphorical device has inspired some soaring rhetoric. For example, Bill Green's words at the dedication of the Phenix City marker: "Laborer and legislator, his life was an astonishing symbolic bridge—a bridge not only between states, but between men. Like one of his stately Town lattice bridges, Horace King's life soars above the murky waters of historical limitations, of human bondage and racial prejudice. He did not change the currents of social history, but he did transcend them and stands as a reminder of our common humanity, the potential of [the] human spirit, the power of human respect."[14]

The bridge metaphor as a representation of racial harmony has always been popular. Local singer-songwriter Allan Levi captured that spirit in "The Bridge from King to Godwin." His line "but the bridge from King to Godwin never fell" is an accurate description of the personal relationship between two men. King and Godwin rose above slavery because of their pursuit of money. They did not lessen racial prejudice in the Chattahoochee Valley. Other than as an exceptional example of benevolence on the part of one slave to his master, King never created harmony between the races. Even as a legislator, he did not try to improve the condition of his fellow blacks. But King would probably be pleased that his life and his bridges have become symbolic of healing the scars between the races.

Stripped of the myths and metaphors, Horace King, the man, still stands as an extraordinary black architect, builder, and craftsman, who as a slave had the extreme good fortune to have John Godwin as a master and Robert Jemison as a friend. Horace used his skills to achieve the American dream of securing some wealth and security for his family in an age when few black men enjoyed such opportunities. As a slave and as a freedman, King stretched the economic and social limits that restrained businessmen of color in the Deep

Contemporary Horace King Sites in Columbus and Phenix City

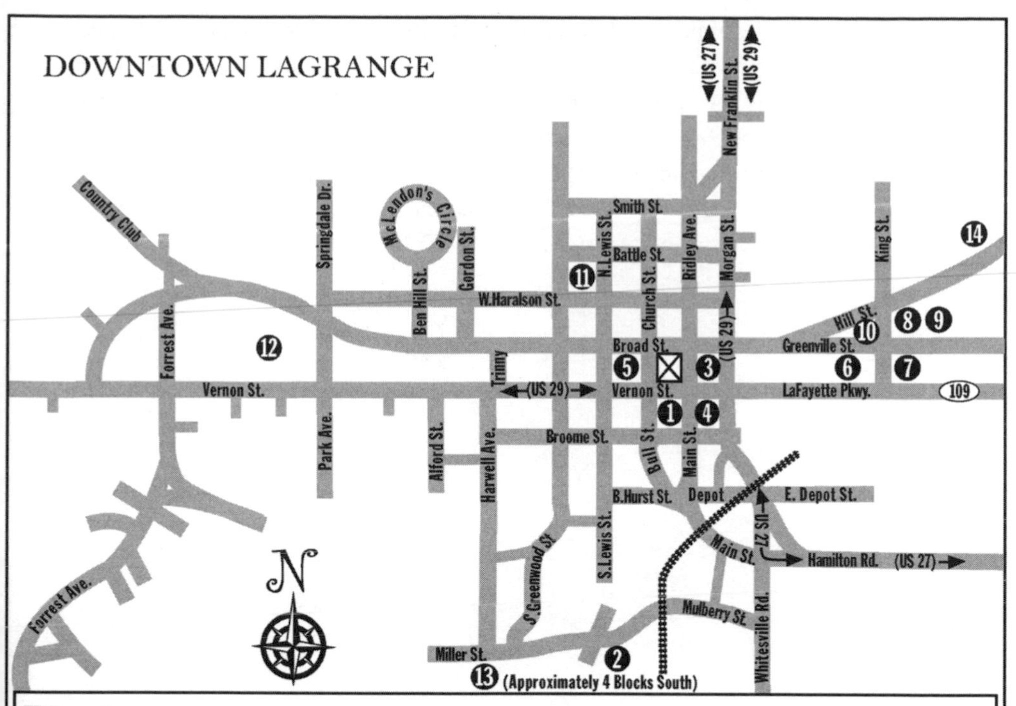

☒ Town Square

1. Troup County Archives, 136 Main Street. Houses King information and photos.
2. Graves of Horace King and Marshal Ney King, 103 Miller Street, just west of Stonewall Cemetery.
3. East LaFayette Square buildings, all built by Horace, John, or George. On south end of block, 101 LaFayette Parkway was constructed by George King.
4. Chattahoochee Valley Art Museum, 112 LaFayette Parkway. George King built the foundation for the county jail here in 1892.
5. Mansour's, 26 W. LaFayette Square, built by George King. Facade on north side of Broad Street side is still visible.
6. Thyme Away Bed & Breakfast, 508 Greenville Street. According to his descendants, Horace King lived here.
7. King Bridge, Horace King Street. Dedicated in 1989 in memory of Horace King.
8. King Street. Named by LaGrange City Council to honor Horace and Marshal King in 1888.
9. John King home, 607 Greenville Street.
10. Washington King Home, 516 Hill Street.
11. Plant House, 203 N. Lewis Street. Built by John King as parsonage for the First Presbyterian Church.
12. Smith Hall, LaGrange College. George King added the south and west wings and the Victorian-style porch.
13. Primera Iglesia Hispana de LaGrange, 802 Truitt Street. Built by John King for Unity Methodist Church.
14. Graves of John King and family, Eastview Cemetery.

Sites identified by the staff of the Troup County Archives

Contemporary King Family Sites in LaGrange, Georgia

South. Just as he understood the ultimate limits of a bridge span, King seemed to grasp just how far he could push himself into areas usually reserved for whites. But the society's racial constraints prevented him from bringing any blacks other than his family into this privileged zone.

The seductiveness of Horace King's story is powerful. No doubt he will continue to be one of the best-known historic figures in the Chattahoochee Valley in the twenty-first century.

NOTES

Introduction

1. The following information about construction of the Holland Creek Bridge and the weather that day is based on news in the *Columbus Enquirer-Sun*, October 7, 1883. The physical description of Horace came from the Reverend Francis L. Cherry's biography of him; Cherry's is based on his photograph. The dam for the Eagle and Phenix Mill—the economic and political fulcrum of the city—was constructed in 1882. W. C. Bradley and his partner, S. A. Carter, bought the old Fontaine warehouse in 1885; it had decorative parapets by 1886. The changes in the Holland Creek Bridge are clearly visible in a comparison of the 1872 and 1886 perspective maps of Columbus, Georgia.
2. Cherry, "History of Opelika." King's biography is found in *Alabama Historical Quarterly* 15, no. 2, 193–97. In 1996 the Genealogy Society of East Alabama reprinted the *Quarterly* version as a separate book that retained the same pagination. Its introduction included a sketch of Cherry's life.
3. The obituary for Horace's son John Thomas King contained quoted paragraphs without any notation of their source; *LaGrange Graphic*, November 26, 1926. The same paragraphs about Horace and the family appeared in Lane, Bridges, and Jones, "The King Family," 368–71. Even though this sketch was included in the WPA (Works Progress Administration) slave narratives, it was not based on an interview with a living slave but on family recollections. A draft of this family history and another account of Horace King, probably based on WPA work in Columbus, can be found among the drafts of WPA sketches in Box 20, Folder 15, Alva C. Smith Collection, Columbus State University (CSU) Archives, Columbus, Ga., Smith was involved with the local WPA effort.
4. Minutes of Columbus City Council, May 12, 1865.
5. Horace King Interrogatories, Near Whitesburg, Carroll County, Georgia, February 28, 1878, Records of the U.S. House of Representatives Case File of the Southern Claims Commission, Commissioners of Claims, no. 19661, National Archives, Record Group 233. Photocopy available at Troup County (Georgia) Archives; hereafter cited as King Interrogatories.
6. Based on the correspondence in the Robert Jemison Jr. Collection (William Stanley Hoole Special Collection Library, University of Alabama), Robert S. Starobin mentions King, but refers to him only as Horace in the text. Someone added the name King in parentheses in the index but kept the entry under Horace. Starobin,

Industrial Slavery, 30, 107–8, 171–72, 173, 317; Davis and Hogan, *The Barber of Natchez*; Johnson and Roark, *Black Masters.*

7. Johnson and Roark, *Black Masters*, 23–25.

CHAPTER ONE. Pee Dee Origins

1. Cherry, "History of Opelika." Cherry (193, 197) mentions the Catawba roots of Horace's mother. Her age is based on the later inventory of John Godwin's estate. In the 1880 census Horace listed his mother's birthplace as South Carolina.
2. The following summary of the history of the Catawbas is based on Merrell, *The Indians' New World*, 92–225.
3. Cherry, "History of Opelika," 194, 197.
4. King must have identified Edmund King as his father to Cherry, but the Reverend (or more likely King) did not identify his occupation or mention when he died.
5. Davis and Hogan, *The Barber of Natchez*, 16–19; Johnson and Roark, *Black Masters*, 5–7.
6. Quoted in McFeely, *Frederick Douglass*, 13.
7. King Interrogatories; the date 1830 comes from Cherry, "History of Opelika," 193; Rudisill, *Doctors of Darlington County*, 44–45.
8. Several researchers have pursued this connection. The most documented came from Kenneth Vance Smith. His genealogical material on the Godwin family was originally posted on a Web site (www.flash.net/~kensmith/godwin.htm) that no longer exists. Hereafter cited as Ken Smith's Web site. Paper copies of the Web site information in possession of the authors.
9. The 1810 census for Chesterfield District does not confirm this idea. It only shows one male older than forty-five years in Miles's household, Miles himself. The ages of the five other males appear to conform to Miles's children. In the early 1820s a King plantation existed in the southeastern portion of Chesterfield District, but its owner remains unknown because of the lack of deed records. See the map for Chesterfield District in R. Mills, *Atlas of the State of South Carolina.*
10. Three separate genealogists working on the Miles King family—Mrs. Virginia P. Lidwin, Mrs. R. T. Murfee, and Mrs. Madean Sims Rush—all placed Edward, son of Miles, in Alabama, either in the 1830s or the 1840s. See Miles King folder, Darlington Historical Society, Darlington, S.C. In 1810 an Edward King of Chesterfield District received a passport from the Georgia governor's office to travel through the Creek Nation, which still occupied the western fourth of the state. This Edward may have been the son of Miles. Also receiving a passport at the same time was William Wright of Marlboro District, who might have been the uncle of Ann Wright, the wife of John Godwin. Potter, *Passports of Southeastern Pioneers, 1770–1823*; citation from Ken Smith's Web site.
11. McFeely, *Frederick Douglass*, 28.
12. King in his interrogatories mentions Dunlap as his purchaser; the fact that his mother and siblings were part of the Godwins' holding in 1837 suggests that all three of them passed from King to Dunlap to Godwin in 1830.
13. *Cheraw Intelligencer and Southern Register*, June 4, 1823.

14. Inhabitants of Cheraw Petition (0010 003 ND00 05117 00), no date, South Carolina Department of Archives and History, Columbia.
15. Sneedsboro was also spelled Sneedsborough and Sneydesborough.
16. The Coastal Plain is underlain with sedimentary rocks through which rivers can carve a level or gently sloping course, whereas the rocks of the Piedmont are metamorphoric and produce rivers characterized by a series of falls, especially at the fall line. The Pee Dee River does not mark the boundary between the two states but dissects the state line at a perpendicular, so Sneedsboro lies north of Cheraw.
17. Superintendent of Public Works, *Plan and Progress of Internal Improvement in South Carolina, 1820*, 121–22.
18. An indenture, dated April 3, 1808, between William Johnson and the sons of the deceased John Godwin passed the title for these four lots to Godwin's sons. Anson County Deed Book N and O, 13.
19. Medley, *History of Anson County*, 73–77.
20. Neither Mary nor John Godwin was listed in the census for Anson County, N.C., or Chesterfield District, S.C., in 1810 or 1820. The name John "Goodwin" is listed in both Chesterfield and Anson for 1820; the most likely candidate to be Horace's master was the young gentleman (between sixteen and twenty-six years old) in Chesterfield. He was living with a woman over forty-five (i.e., old enough to be his mother); no other females were listed in the household. Two young males (under ten years of age) may have been his nephews. This John "Goodwin" owned no slaves. The John "Goodwin" in Anson County, an even more improbable fit, owned only one slave.

    On September 20, 1823, Wells and John Godwin were involved in litigation with Robert G. Daniel relating to a sheriff's sale of lot no. 148 in Chesterfield District, a typical action for builders. Citation from Ken Smith's Web site.
21. Edgar, *South Carolina: A History*, 162.
22. The team boats required fifteen days to move three hundred bales of cotton from Society Hill, south of Cheraw, to Georgetown. Ads for the steamboat *Maid of Orleans* and the team boat *Yadkin* appeared in the *Cheraw Intelligencer and Southern Register* in June 1823. Rogers, *History of Georgetown County, South Carolina*, 227.
23. In November 1823 a committee of Cheraw citizens attended the legislative session to request money to improve the Great Pee Dee. At the same time, the "Grand Jury of this District, has presented the obstructions in this river, as a public grievance, and recommend an appropriation of $20,000 . . . for improving navigation." *Cheraw Intelligencer and Southern Register*, November 14, 1823.
24. Balloon framing is usually dated from the 1830s and the development of Chicago. The scarcity of trees on the prairie made the use of heavy-timber frames impractical, so sawn lumber shipped in by rail was used for the mass-produced houses. Condit, *American Building Art*, 22–24.
25. From Ken Smith's Web site.
26. Will of Sarah Wright, recorded March 31, 1817, Marlborough District Will Book A, 99. Inventory of her estate, May 13, 1816, Case I, Box 12. In 1837 John Godwin transferred the ownership of five slaves to his wife and her uncle, William Carey Wright. Russell County Deed Book B, 65.
27. Ann was born in 1808 and John in 1798. Pervis, John, Petition (0010 003 ND00

0111800), no date, South Carolina Department of Archives and History, Columbia. William Wells might have been a relative through one of Godwin's maternal lines.

28. Lyon published the *South Carolina Spectator*, the *Cheraw Intelligencer*, and the *Pee Dee Gazette* during this period. John's joining the church at this time is a supposition. The provenance of the Bible, which became the family Bible, is described in Mrs. Virgil C. Curtis Sr. and Mrs. Clyde Brown, compilers, "The John Godwin Bible," *Tap Roots—The Genealogical Society of East Alabama Quarterly* 2, no. 3 (January 1965). The Bible was in the possession of Mrs. Clyde Brown of Phenix City, Alabama. Information about Mason Risley Lyon from Ken Smith's Web site.
29. Linder, *Medicine in Marlboro County*, 145.
30. Marlboro County Deed Book N, 107.
31. Despite all their careful planning, the Moffetts did not enjoy their house for very long. By 1837 James had died; Maria had married Joel Winfield; and they had moved to Alabama, the next destination for many residents leaving the Carolina Pee Dee region. Marlboro County Deed Book O, 268–70.
32. William R. Godfrey Collection, South Caroliniana Library, University of South Carolina, Columbia, S.C.

CHAPTER TWO. Ithiel Town's Bridge

1. Other slaves, subcontractors, and suppliers are enumerated in the accounts kept by the Cheraw Bridge Company, but not these later bridge builders. William R. Godfrey Collection, South Caroliniana Library, University of South Carolina, Columbia, S.C.
2. Gies, *Bridges and Men*, 86–87.
3. Brown, *Bridges*, 44–45.
4. L. Edwards, "Evolution of Early American Bridges," 13; Hopkins, *A Span of Bridges*, 80.
5. Examples of English suspension bridges include Thomas Telford's spectacular 579-foot span for the Menai Strait Bridge between Wales and the island of Anglesey and Isambard Kingdom Brunel's design for the Clifton Bridge at Bristol. Finley quoted in Kranakis, *Constructing a Bridge*, 36–37. Kranakis compares the practical Finley with the more theoretical French engineer-scientist Claude-Louis-Marie-Henri Navier. This book also illustrates the pragmatic American approach to bridge design.
6. Hindle, "Introduction: The Span of the Wooden Age," 3–12.
7. Danko, "Evolution of the Simple Truss Bridge," 9–30.
8. The Swiss covered some of their bridges to protect the timbers from climatic extremes as early as 1333, but this practice did not transfer to colonial America. Palmer's bridge stood until 1875, when it was destroyed by fire. Kranakis, *Constructing a Bridge*, 75–77; Pease, "Timothy Palmer," 97–111; Gies, *Bridges and Men*, 105–6.
9. Pease, "Timothy Palmer," 98.
10. Ibid.; Danko, "Evolution of the Simple Truss Bridge," 6–7, 60–65. This Burr bridge measured 1,355 feet in ten spans and was the second bridge at the site, a flood having demolished Wade Hampton's bridge, which dated from 1790.

11. Newton, *Town and Davis Architects*, 15–75, esp. 42–47; *Dictionary of American Biography*, s.v. "Ithiel Town."
12. Danko, "Evolution of the Simple Truss Bridge," 126–48; Kranakis, *Constructing a Bridge*, 77, 79.
13. Town's own 1821 description of his patented bridge argued that it was cheap; when covered from the weather it would last seven to eight times longer than uncovered ones; its pieces could be sawn at "common mills"; it required "no iron work"; and it had "less motion than is common in bridges." Newton, *Town and Davis Architects*, 44–45.
14. Cheraw bridge folders, William R. Godfrey Collection, South Caroliniana Library, University of South Carolina, Columbia, S.C.; *Cheraw Intelligencer and Southern Register*, June 18, 1824.
15. The ferry owner, Samuel Gillespie, supported the building of the bridge. "Petition to . . . Senate, and House . . . South Carolina," no date (ND 4847-01), South Carolina Department of Archives and History, Columbia.
16. *Statutes at Large of South Carolina*, 1791, 337–39; 1792, 344–45; 1798, 390–91; *Digest of the Laws of the State of Georgia*, 419–21, 468–69.
17. *Acts of the General Assembly of the State of Georgia*, 1834, 1:48–50. Town later served as governor of Georgia. The complete *Acts* can be found at www.galileo.usg.edu.
18. The following assumptions about the construction of the bridge are based on the records of Cheraw Bridge Company, ca. November 1823–July 1825, Godfrey Collection, South Caroliniana Library, which consist of various accounts belonging to Hearsey, Coming, and King. These documents (approximately one hundred items), chiefly bills, receipts and account sheets, enumerate the expenses incurred by these men and later reimbursed by the company; cite wages paid laborers; list quantities of sawn lumber, hewed timbers, and shingles; and document other agreements. William R. Godfrey, who in a later reorganization became a major stockholder in the company, preserved these records.
19. Hearsey charged the company for his expenses in coming to Cheraw, so presumably he lived somewhere else. His role as a banker cited in "Bridge Company Issues Stock for First Bridge across River," *Cheraw Chronicle* Bicentennial Edition, [no date], 4E, clipping in Cheraw Town Hall. Other large bridge companies also organized banks; the sale of the Augusta bridge in 1819 included a bank. Announcement of the sale, January 20, 1819, Henry Schultz Collection, South Caroliniana Library, University of South Carolina.
20. Russell was paid $1,000 for his services. He appeared to keep the accounts of who was paid what, but the individual vouchers requesting payment for specific men were signed by William. William, who received $429, may have been in day-to-day control of the men, and Russell may have supervised the overall construction. On March 23 (probably 1824), room and board for Town and his horse cost $3.00; a more extended stay ending on June 10 cost $13.75 just for his board.
21. The records indicate fourteen men were brought from Georgetown and four—Hathaway, Cook, Pickering, and Winslow—from Charleston. Three of the latter were paid $177.70 for their work from November 1823 until May 1824—the highest rate paid any workers. The company also absorbed their travel expenses.
22. The price was approximately $10 per thousand board feet. Today similar timbers would cost $1,000 per thousand board feet, or a hundred times more.

23. The shingles cost $869. The joints between the siding of random width boards were covered by 928 two-and-a-half-inch battens.
24. *Cheraw Intelligencer and Southern Register*, June 18, 1824. The modern example for such camber is an unloaded flatbed truck that bows upward until it is heavily loaded.
25. Undated petition of Seth King et al. to General Assembly [ND-5881-01], South Carolina Department of Archives and History, Columbia. George T. Hearsey remained active in the Cheraw area; he signed a petition in December 1827 requesting the legislature to fund improvements to the Pee Dee. Petitions to the General Assembly [1827 00088], South Carolina Department of Archives and History, Columbia.

    An undated "List of Stockholders in the old Cheraw Bridge Co." is probably from after the Civil War when the bridge was being rebuilt. It shows T. E. B. Peques as the largest stockholder after Jane King and W. Godfrey with twenty-five shares. A letter from J. W. King to Mr. Searcey, May 9, 1861, identified Jane O. King as Seth's niece, Letterbook 9, Robert Jemison Jr. Collection, William Stanley Hoole Special Collections Library, University of Alabama, Tuscaloosa; hereafter cited as Jemison Collection, UA.
26. South Carolina, vol. 9; Chesterfield; and Alabama, vol. 24, 102, R. G. Dun and Co. Collection, Baker Library, Harvard Business School. In 1861 a credit reporter for R. G. Dun identified Seth as a bridge architect and part-time resident of Tuscaloosa.
27. In a letter from Robert Jemison Jr. to Horace King, February 5, 1858, Jemison mentions his litigation with Seth King with no indication that Horace would know him as a relative of his old master. After the Civil War, Jemison tried to locate Seth King and wrote others asking about his whereabouts. At about the same time he corresponded with Horace about other matters but never asked about Seth. Letterbook 8, and Jemison to S. H. Williams, Fall River, Mass., May 18, 1868, Letterbook 12, Jemison Collection, UA. Apparently Jemison also instigated a Dun inquiry into Seth King's holdings in Cheraw after 1865.
28. See clippings, *Cheraw Chronicle*, [no specific date] 1939, and the *Cheraw Chronicle* Bicentennial Edition, [no specific date] 1976, both in Cheraw Town Hall.

CHAPTER THREE. The Missing Years and Other Mysteries

1. Cherry, "History of Opelika," 193; Russell County Deed Book B, 85 [May 5, 1837].
2. Johnson's master freed him as an extension of the emancipation of Johnson's mother. Davis and Hogan, *The Barber of Natchez*, 14–19.
3. Johnson and Roark, *Black Masters*, 13.
4. *LaGrange Reporter*, November 26, 1926; Lane, Bridges, and Jones, "The King Family." This work is obviously based on conversations with descendants of John Thomas King. The sketch is a secondary account and should not be confused with a WPA slave narrative.
5. Horace King Collection, Columbus Museum.
6. Gerber, *Black Ohio and the Color Line*, 3–24; Love, "Registration of Free Blacks in Ohio," 38–74.
7. William H. Green, a former professor of English at Chattahoochee Valley Community College who for years pursued King as a subject, made this statement con-

cerning the Fugitive Slave Law in Tom C. Lenard's film *Horace: "The Bridge Building King,"* Auburn University, 1996.

8. Grimshaw, *Official History of Freemasonry,* 189–213, 288–90; *Formation and Proceedings of the M. W. Grand Lodge of the most Ancient and Honorable Fraternity of Free and Accepted Masons for the State of Alabama* (Mobile: Thompson and Powers, 1871), 19. Available at the Prince Hall Grand Lodge in Birmingham. This record was discovered by Carol Holland.
9. Gerber, *Black Ohio,* 19; Minutes of Columbus City Council, October 15, 1835.
10. Fletcher, *A History of Oberlin College,* 1:356, 528–36, 698–700. Perhaps a later family member attended Oberlin and that action became associated with Horace and his son.
11. Marlboro County Deed Book M, 172.

CHAPTER FOUR. Boom Town on the Chattahoochee

1. Louisville was also planned by the state, but it was not near a navigable river. "Town and City Planning," in Heath, *Constructive Liberalism,* 151–56.
2. The state maintained control of the four-block-wide commons on the south, west, and north sides of Columbus until after the Civil War.
3. Hall, *Travels in North America,* quoted in Lane, *Rambler in Georgia,* 82–84; Arfwedson, *United States and Canada,* quoted in Lane, *Rambler in Georgia,* 104–5.
4. King and Godwin worked together on the Columbus, Mississippi, bridge in 1842. Their partnership is mentioned in Lane and Bridges, "A Sketch of the King Family," [WPA typescripts], Alva C. Smith Collection, CSU Archives. In 1862 Asa Bates and another contractor, William Champion, testified in support of the actions of Messina Godwin in regard to John Godwin's estate.

   Bates served as a town commissioner and became one of captains for the volunteer fire company. In 1831 a large house he was constructing for Charles Stewart burned and Bates had to assume half of the loss. As a result of that experience he helped organize the fire company. Later in the 1830s he served as sheriff for Muscogee County. Martin, *Columbus, Georgia,* 1:21, 27, 29.
5. Telfair, *History of Columbus, Georgia,* 47–49.
6. T. Jeff Bates, Asa's son, testified that his father always used black labor. T. Jeff Bates interview in *Report of the Committee of the Senate,* 4:491; Sweat, "The Free Negro in Ante-Bellum Georgia," 145–53.
7. The two corporations related to bridges at Irwinton, Alabama, and Florence, Georgia; both spans involved Godwin and King. The church was the German Lutheran Congregation at Ebenezer Creek, the successor to the Salzburgers, a group of pietistic German Lutherans who played an important role in Georgia's colonial history. These statistics came from the Georgia Legislative Documents in Galileo, University System of Georgia, Web site, which has full text entries for every Georgia legislative act.

   During the 1830s only a place known as Lopahaw received similar beneficence from the legislature as Macon and Columbus. In 1836 the legislature granted $800 for a bridge at Lopahaw on Coffee's Road, which extended from south central Georgia into Florida. The act had no provisions for repaying that sum, a very unusual charter. Perhaps this span related to moving troops during the second Seminole war. *Acts of the General Assembly of the State of Georgia*, 1836, 1:75.

8. By 1840 Gazaway Lamar, a wealthy Savannahian, owned the Augusta bridge, and in that year the town purchased it from him and began operating a municipal bridge. Perhaps, they modeled their actions after Macon and Columbus. *Acts of the General Assembly of the State of Georgia*, 1840, 1:163.
9. *Acts of the General Assembly of the State of Georgia*, 1824, 1:90–91; 1828, 1:39–41; 1833, 1:281; Heath, *Constructive Liberalism*, 154–55. The Macon bridge had washed away by late 1832 and the legislature allowed a ten-year grace period before Macon had to resume payments.
10. *Acts of the General Assembly of the State of Georgia*, 1831, 1:232–34, 236.
11. Ibid.
12. The model might have been a set of plans. "To Bridge Builders," *Columbus Enquirer*, January 21, 1832.
13. *Columbus Enquirer*, March 10, 1832
14. Miller et al., *World of Daniel Pratt*, 12–17, 47–49, 79.

CHAPTER FIVE. To Throw a Bridge across the Chattahoochee

1. Since the town of Columbus built this bridge the extant city council minutes provide more information about this bridge than any other Godwin and King built across the Chattahoochee—another reason for examining it in some detail. Unfortunately, a gap in the minutes from 1828 until October 1832 includes the period when the bridge contract was consummated and construction began.
2. Using the 270,735 board feet consumed by the 415-foot Cheraw bridge as a guideline, a commensurate amount for the 560-foot Columbus bridge would be 360,000, but the Columbus one had much less land bridging, so the 300,000 board feet is only an estimate.
3. No ads for steam sawmills appeared in local newspapers in the spring of 1832, as they would in later decades. Steam engines were not plentiful in Georgia in the 1830s; in 1838 the state had only twenty-three stationary engines and eleven of those milled rice on the coast—a much more valuable crop than lumber, which could not justify such a capital investment. Pursell, *Early Stationary Steam Engines*, 74–75; Wik, *Steam Power*, 18–19.

   Mills in Maine and Georgia operated either one vertical blade or perhaps a gang of vertical saws. The rotary did not come into general use until about 1860. In 1840, 677 sawmills operated in Georgia with an average annual production valued at $68 per year or only 6,800 board feet per year per mill. Defebaugh, *History of the Lumber Industry*, 1:489; 2:23, 53.
4. J. T. Williams, in 1955, recounting the construction of the Glass Bridge in Troup County during the 1890s, mentioned this process. It had probably changed very little in the previous half century. Tom Swint, "Glass Bridge Is Falling Down and with It, a Georgia Era," *Atlanta Journal*, March 14, 1955, reprinted in *Valley Daily Times-News*, March 18, 1955.

5. Horace King's obituary in the *Columbus Daily Enquirer-Sun,* May 30, 1885, noted he was known as Horace Godwin "in the early history of Columbus."
6. In 1830 John and Wells lived in the same household, which reported six slaves. Wells married in 1831, and by 1840 his household included six slaves, whereas John owned eighteen. In 1832 John and Wells together supplied between six and twenty-four slaves for the bridge. Ken Smith's Web site.
7. Based on an interpolation of the figure of $15 per month for such labor in 1845. Robert Jemison Jr. to John Godwin, July 4, 1845, Letterbook 1, Jemison Collection, UA.
8. Cypress might have been desirable for poles but was not as plentiful on the Chattahoochee at the fall line as on other southern rivers. Large cypress swamps occupied the riverbanks immediately below the fall line on many southern rivers, but such swamps do not exist on the Chattahoochee where a late uplift in geologic terms elevated the banks on the Chattahoochee as far downstream as the present location of Fort Gaines.
9. Martin, *Columbus, Georgia,* 1:32; *Columbus Enquirer,* July 7, 1832.
10. The stones measured about eighteen inches in depth, two feet in height, and at least two feet in breadth. Iron bars clamped the top course of rocks together. The details about the construction of this bridge are based on the July 17, 1841, contract to rebuild it after the Harrison Freshet. The 1832 contract has not survived; since the 1841 contract states the bridge will be rebuilt on the same plan, the two spans should have been similar in appearance and construction details. The primary differences between the two involved the piers at both ends. The 1832 span as originally constructed lacked an eastern abutment, and its westernmost pier was a wooden post support that was replaced by a stone pier in 1841. Muscogee County Deed Book B, 218–19; Minutes of Columbus City Council, April 1, 1841.

    According to the terms of the contract, the eastern abutment originally lacked piles and did not have a strong foundation. Near the end of the construction project, the commissioners made separate arrangements to protect and preserve the eastern abutment. Minutes of Columbus City Council, July 9, July 27, and August 17, 1833.
11. Coulter, *"A People Courageous,"* 107.
12. Smaller bridges had one chord at the top and another at the bottom. In large bridges, such as the Columbus bridge, four chords were used. The primary cord at the very top and very bottom consisted of boards 28 feet long, 3⅛ inches deep, and 12 inches wide; the secondary or interior string pieces (placed immediately below or above the primary ones) measured only 10 inches in width.
13. The modern example for such camber is an unloaded flatbed truck that bows upward until it is heavily loaded. As a point of comparison, the arches in the three spans of the 1824 Cheraw bridge did "not make more than 3 or 4° of a circle." *Cheraw Intelligencer and Southern Register,* June 18, 1824. When King rebuilt this bridge after it was destroyed during the Civil War, the Columbus *Daily Sun* reported: "On this side [of the bridge] is an interval of some fifteen feet which will be filled up by the beams from the floor by to-night. Foot passengers, with the assistance of a ladder, have been passing over the bridge for a week" (September 16, 1865). This suggests

that although the side trusses might have been assembled on the bank, some of the beams connecting the two trusses were installed after the side lattices rested on the false work.

14. From 1828 until 1835 commissioners and an intendant governed the town; it became a city in 1836 after which a council with aldermen and a mayor ruled. Minutes of Columbus City Council, September 30, 1833.
15. *Columbus Enquirer*, August 4, 1832.
16. Martin, *Columbus, Georgia*, 1:23, 33–34.
17. *Columbus Enquirer*, June 22, 1832. According to this treaty, every Creek Indian male head of household received 320 acres of land and every chief, 640 acres. The treaty seemed to hold out hope that the Indians would become U.S. citizens, but the same document also provided for the Creeks' removal from Alabama.
18. Martin, *Columbus, Georgia*, 1:41, 46–47; Minutes of Columbus Council, February 1, 1834.
19. The intendant (or mayor) disagreed with the decision and protested the action of the majority of the commission by calling a public meeting, but the deal held. *Columbus Enquirer*, May 31, June 7, 1834. The Minutes of Columbus Council, May 14, 1834, indicate annual payments together with 8 percent per annum to Daniel McDougald, James C. Watson, Burton Hepburn, and Robert Collins. On March 10, 1838, the council authorized the payment of the unsettled balance to these gentlemen.
20. *Columbus Enquirer*, December 27, 1834.
21. Power, *Impressions of America*, quoted in Lane, *Rambler in Georgia*, 114; Arfwedson, *United States and Canada*, quoted in Lane, *Rambler in Georgia*, 106; Cherry, "History of Opelika," 200.
22. Arfwedson, *United States and Canada*, quoted in Lane, *Rambler in Georgia*, 105; Power, *Impressions of America*, in Lane, *Rambler in Georgia*, 115.
23. Girard supposedly loaned several million dollars to the U.S. government during the War of 1812 when the young nation could not procure foreign loans. Coulter, *"A People Courageous,"* 8.
24. Minutes of Columbus Council, January 17, February 14, June 24, October 1, 1835; November 25, 1836; January 13, February 4, 1837.
25. The list of people and the amount charged for each is enumerated in Galer, *Columbus Georgia*, 161–63; Minutes of Columbus City Council, January 21, 1839; January 9, 1837.
26. The Columbus City Council, also concerned about rain coming through the windows, ordered covers for the windows; Minutes of Columbus City Council, January 19, 21, 1839. Minutes of Columbus City Council, October 12, 1835, June 8, 1839. In 1846 the council applied all the bridge tolls to the city's state debt, which must have consisted of the monies loaned to build the bridge in 1832; Martin, *Columbus, Georgia*, 2:9.
27. Ticknor, *Poems*, 124–26; Power, *Impressions of America*, in Lane, *Rambler in Georgia*, 115.
28. Family tradition has Marshal King, Horace's son, being named after Benjamin Marshall. Actually his name was Marshal Ney after Napoleon's general.
29. Arfwedson, *United States and Canada*, quoted in Lane, *Rambler in Georgia*, 105; Telfair, *History of Columbus, Georgia*, 45–46.
30. Green, *Politics of Indian Removal*, 174–82.

31. Young, *Redskins, Ruffleshirts, and Rednecks*, 102–3, 106, 108–10.
32. Raids on white settlements, such as Roanoke, Georgia, where Indians burned houses but did not shoot at people, and the fact that the central leader of the insurrection, Jim Henry, was captured but never tried seems to indicate complicity on the part of some whites. Coss, "On the Trail of Jim Henry," 55–61. King did mention a renegade Indian in his discussion with Cherry; perhaps that incident made King a hero. Chattahoochee Valley Historical Society, *Valley Historical Scrapbook*, 10–11.
33. "J" [Thomas J. Jackson], "Recollections—No. 27, Reminiscences of Early Columbus Continued," *Daily Enquirer-Sun*, August 11, 1895; Martin, *Columbus, Georgia*, 1:63.
34. Green, *Politics of Indian Removal*, 185.
35. Telfair, *History of Columbus, Georgia*, 43–44; Worsley, *Columbus on the Chattahoochee*, 131–34, 376–77.
36. Court of Ordinary Sitting for County Purposes, 1838–1857, November 9, 1840, p. 31; available in CSU Archives.
37. *Columbus Enquirer*, March 17, 1841.
38. Excerpts from the *Macon Messenger* and *Charleston Courier* reprinted in the *Columbus Enquirer*, March 23, 1841; *Columbus Enquirer*, March 17, 1841.
39. Minutes of Columbus City Council, March 11, 1841.
40. Minutes of Columbus City Council, March 27, 1841.
41. *Columbus Enquirer*, April 21, 1841. The ad was dated January 27.
42. Minutes of Columbus City Council, February 10, 1838. On May 17, 1834, the council purchased bridge insurance from the Augusta Insurance and Banking Company, and from an unspecified company on January 17, 1836; it purchased $10,000 insurance at 1.5 percent on September 29, 1848. Robert Jemison complained of having to pay 2 percent on the Columbus, Mississippi, bridge in 1844–45. Robert Jemison to Jno. O. Cummins, December 17, 1844, Letterbook 1, Jemison Collection, UA. The exact role of the merchants and planters is unclear. They might simply have provided four separate insurance policies, or they might actually have posted money for Godwin, but certainly not without receiving some remuneration.
43. Minutes of Columbus City Council, April 8, 1841. In a letter to the Montgomery *Advertiser*, September 16, 1948, J. M. Glenn, a relative of the Godwin family, asserted that reconstruction of the bridge was the reason for freeing Horace.
44. The council authorized final payment on June 25, 1842.
45. *Acts of the General Assembly of the State of Georgia*, 1842, 1:37.
46. In the cotton years ending August 31, 1844, and August 31, 1845, Columbus received 115,000 and 85,000 bales of cotton, respectively. If only 5 percent of that cotton crossed the bridge it represented 10,000 bales in a two-year period.
47. Worsley, *Columbus on the Chattahoochee*, 197–98; *Macmillan Encyclopedia of Architects*, *s.v.* "Button, Stephen Decatur." Button married into the Hoxey family and could have been related by marriage to Dr. Thomas Hoxey who owned the Lion House, whose design has been credited to Button. He utilized the Egyptian style, which appears in some of the details on this house. After designing the Alabama capitol, Button returned to Philadelphia where he enjoyed a distinguished career.

    Minutes of Columbus City Council, August 19, 1845, noted: "Sec 1st, Commencing at (Pier No. 1) the east end of the Bridge . . . has swa[g]ged over so as to be 4

Inches out of perpendicular and the floor is 1¾ Inches lower on the south side. The Center of this section is 1½ out of perpendicular and the whole frame work of this section at this point has swaged down the River 5 Inches. The floor also at this point is 2 Inches lower on the south side. Sec 2 Commencing at Pier No. 2 . . . the lattice work swaged 4 Inches from perpendicular and the floor level at the centre of this section . . . the lattice work 3¾ Inches out of perpendicular and the floor 2 Inches lower of the South side. The whole frame work . . . swaged down the River 4 Inches. Sect 3 Commencing at Pier No. 3 . . . the lattice work perpendicular and the floor level . . . but at the center of the section the frame work has swaged 2 Inches down the River."

48. Minutes of Columbus City Council, December 27, 1845, February 27, 1846, January 10, 1853.
49. Galer, *Columbus, Georgia*, 170–71.
50. Minutes of Columbus City Council, October 10, 24, 1848. Godwin's 1841 span cost $26.94 per foot; this one would have cost $8.30 per foot. He probably did not propose to build it for that price. In addition to public hands, the committee reported the donation of rocks by John Howard from his factory's raceway, so Godwin probably did not work for $1,300. The council did order that the work be completed for that price. A surviving brick abutment still exists under the present Dillingham Street Bridge; it might date from this repair or from the postwar bridge.
51. The problem must have developed in a short time since the council met every week. The March freshet in 1853 destroyed the factory dam above the bridge, but no flood was noted during the fall. Minutes of Columbus Council, October 19, 1852; Martin, *Columbus, Georgia*, 1:67.
52. Minutes of Columbus City Council, October 11, 1853, February 20, 1854.
53. As early as 1849 the council had been concerned about lighting the bridge. Gas was the logical fuel, since the gas company, organized in 1852, placed its plant at the eastern end of the bridge. The company, in which the city invested $10,000, charged $180 per year to light the bridge and $20 per annum for each street lamp. Minutes of Columbus City Council, September 29, 1849, August 9, 16, 1853; Martin, *Columbus Georgia*, 2:57–58. The city's assets included stock in two railroads and the gas company, the wharf, the bridge, the powder magazine, and seven mules valued at $175 each. Martin, *Columbus Georgia*, 2:87.
54. Minutes of Columbus City Council, April 21, 1856; May 19, June 24, 1856; June 10, 1857.
55. *Columbus Enquirer*, February 20, 1862; *Columbus Enquirer*, December 14, 1863.
56. *Columbus Enquirer*, April 24, 1858.
57. "J" [Thomas J. Jackson], "Recollections—No. 27, Reminiscences of Early Columbus Continued," *Daily Enquirer-Sun*, August 11, 1895. The river at this location has remained a subject of local folklore. Supposedly Civil War cannon were dumped into the Chattahoochee slightly north of the bridge and later the bodies of soldiers and slot machines, which came from juke joints in Phenix City.

CHAPTER SIX. Family Ties

1. Martin, *Columbus, Georgia*. The deed records do not show Godwin purchasing the piece of land occupied by his original house, which given its proximity to Fort In-

gersoll must have belonged to S. M. Ingersoll. The entrepreneurs were John Banks, Daniel McDougald, A. B. Davis, and John Fontaine.

2. Willoughby, *Fair to Middlin'*, 96, 103, 160. Two of the tracts (the north half of section 5, township 17 north, range 30 east and the north half of section 8, township 17 north, range 30 east) lay immediately west of the town, probably adjacent to land Godwin later owned within Girard. The other property (the west half of section 26, township 19 north, range 26 east) was south of Opelika and northwest of Spring Villa, the later, rather ornate house of John Godwin's daughter Mary Ann Godwin Yonge. This Indian land might have become part of her property. See Government Land Office Records, AL5070.184 and 185 and AL2390.131.
3. Martin, *Columbus Georgia*, 1:79. McDougald and his associates paid Marshall $35,000; Columbus paid them $10,000 for the acre of land where the bridge rested. The 1836 sale occurred in November; the purchase by the Godwins and Dodge came earlier in the year.
4. The Godwins owned some of these lots with Dodge and others without him. Early landowners often failed to record their deeds and that could have been the case with some of the Godwin property. However, six lots owned with Dodge are recorded in Russell County Deed Book C, 193–94. The property north of Holland Creek could have been to the south of the nine business lots. Holland Creek flowed into the Chattahoochee River about a block north of the Columbus (Dillingham Street) bridge, and the warehouse apparently stood just to the north of the creek.
5. The corps concluded that although the future site of Columbus (1828) had superabundant waterpower for such a facility, its remote location made the price of shipping goods there too high. The arsenal went to Pittsburgh.
6. Russell County Deed Book H, 571–72.
7. *Howard v. Ingersoll*, Supreme Court of the United States, 54 U.S. 381; 14 L. Ed. 189; 1851 U.S. Lexis 866; 13 HOW 381, May 27, Decided; December 1851 Term.
8. John Godwin's friend and economic rival Asa Bates was a stockholder and presumably the builder of a large warehouse in Columbus. Its investors included Daniel McDougald, Alfred Iverson, Hampton S. Smith, Stewart and Fontaine, and James S. Calhoun, the town's most prominent entrepreneurs and merchants. *Acts of the General Assembly of the State of Georgia* 1835, 1:133. "History of First United Methodist Church" by Mrs. Lucile J. Ward incorporated into Coulter, *"A People Courageous,"* 329–30.
9. John and Ann had purchased the tract in 1839 for $1,600, and Wells paid them $800 for their half in 1842. Russell County Deed Books C, 312, D, 467. The boom had passed and John paid only $400, or about $6 an acre, for the settlers' lots. Russell County Deed Book D, 381.
10. In contemporary terms, his property lay to the south of Fourteenth Street from the intersection of Thirteenth and Fourteenth Streets westward to the present site of the Godwin Cemetery.
11. Their life spans were as follows: William E. (1824–65), Mary Ann (1826–unknown), Napoleon (1828–60), Messina (1830–98), John Dill (1834–85), Sarah Ashurst (1835–50), Thomas Metternick (1838–63), and Susan Albertha (1843–74).
12. Wells and Malinda's children were Mary Ann (1832–45), Andrew Jackson (1833–1920), Volney (December 1837–August 1838), Thomas P. (1839–64), Benjamin D.

(1842–66), George (1844–46), Silvistin G. (1847–65), Francis Marion (1849–1929), and John Albert (1851–1913). Wells served as a bondsman for a performance bond on a small bridge John built in Russell County in 1853.

13. Cherry, "History of Opelika," 194.
14. The city council ordered the purchase of lumber from Lucas on July 9, 1833, and William Brooks operated Variety Works, a large river-powered saw- and planing mill in Columbus. As the executor of John Godwin's estate, Brooks later acted as a bondsman for Messina Godwin. John also sold property he owned south of Girard to Asa Bates in the 1840s and 1850; Bates and another builder, William Champion, in the 1860s defended Messina's sale of the Godwin Place. Had he not died before 1878, Brooks, who was apparently an active Unionist, would have been called to testify concerning King's Unionist inclination.
15. For specific references to the treatment of slaves in the Chattahoochee Valley see Williams, *Rich Man's War*, 18–23. Douglass quoted in Stampp, *The Peculiar Institution*, 89. *Digest of the Laws of the State of Alabama; . . . in January 1833*, 394.
16. *Columbus Enquirer*, July 7, 1832; Hatcher and McGehee Slave Book, 1858–1860, CSU Archives.
17. Minutes of Columbus City Council, March 18, 1837, September 9, 1834.
18. Godwin mortgaged the following slaves: Willis, Carpenter Henry White, Frank, Sandy, Pompy, Kate or Caty, John, Beccy, Hannah, Big Ben, Barbara, Charles, Mary, Waverly, Harriot, Peter, Will, Tome, and Simon. Russell County Deed Book B, 85 (1837), E, 105–106 (1842), G, 716 (1851); John Godwin's Estate Records, Russell County Probate Court.
19. Jemison to James Smith, July 8, 1845, Letterbook 1, Jemison Collection, UA.
20. Rachael Gould may have moved westward with the Creeks. For a female head of household to receive an allocation of land would have been unusual; Charles Tigner, "A Master Bridge Builder," Russell Remembrances column, *Phenix Citizen*, May 4, 1979, 4.
21. According to family tradition, Washington added the W. to his name; most of his contracts are signed in that manner.
22. The story of Ney's grave site was often told by John Lupold's mother, Dorothy McColl Lupold, a historian whose roots ran back to the Pee Dee area.
23. The infant Clarissa appeared in the John Godwin estate inventory in 1859 but not in the 1870 census. The local myth is recounted in W. O. Langley's account of Godwin and King (ca. 1941) that appears in Coulter, *"A People Courageous,"* 119–23.
24. "Martin vs. Reed," *Alabama Reports . . . Supreme Court of Alabama During . . . 1860 and . . . 1861* (Montgomery 1866), 37:198–200 (the point really was moot by the time it was published). Sellers, *Slavery in Alabama*, 231–32.
25. Russell County Deed Book P, 256 (April 4, 1872).

CHAPTER SEVEN. "Honest" John Godwin and Horace

1. Cherry, "History of Opelika," 194.
2. Local sources attribute the design of the 30-×-80-foot warehouses to A. J. Norris of New York City, which would be logical given the volume of the port's commerce with that metropolis. Personal communication, October 2003, with Willoughby Marshall, an Apalachicola preservation architect. Only Mueller tied Godwin to the

project, and he cited a typescript in the Apalachicola Chamber of Commerce, which is no longer extant, but it reflected the town's oral history. Mueller, *Perilous Journeys*, 28.

3. Local history Web sites credit construction of the Black Warrior bridge to Horace. See also Ben Windham, "Who Was Horace King?" in West Alabama section, *Tuscaloosa News*, January 19, 2003. The 650-foot structure, which included 200 feet of approach bridges, rested on 70-foot high piers. Jemison provided about 294,000 board feet for the Columbus span and 500,000 for the one for the Black Warrior River.
4. Jemison owned 135 and Seth King 105 of the 290 shares accounted for in 1861. Jemison to Freeman Dodd, April 11, 1861, Letterbook 9, and Jemison to Horace King, February 5, 1858, Letterbook 8, both in Jemison Collection, UA.
5. The contract and a plan specified the number of bricks in each wall, a Flemish bond for the exterior walls, the dimensions and placements of joists, the size and number of window lights, the location of interior cornice moldings and "fancy centre piece[s] of stucco work," and the design of the three-story spiral staircase.
6. The Godwins used a set of main front steps flanked by two sets of smaller curved ones attached to the sides of the portico. Delos Hughes, a retired Washington and Lee University professor of political science, suggested the possibility of a link between Mills's work in South Carolina and King's Lee County Courthouse. The Fairfield Courthouse in Winnsboro, S.C., has a set of "open arm" steps designed by Mills; Fairfield County buildings, Historic American Building Survey materials at the Library of Congress American Memory Web site. The Lee County version is discussed in chapter 11.
7. Minutes of Columbus City Council, June 9, 1838, January 25, 1840. Payment of $2,000 on May 18, 1839; $487.37 on May 18, 1839; $2,000 on August 24, 1839; $500 on October 19, 1839; and an advance of $1,500 on January 22, 1840.
8. The city paid $1,500 of the additional cost; the remainder fell to the county. Minutes of Columbus City Council, October 28, 1840.
9. Muscogee County Court of Ordinary Sitting for County Purposes, 1838–1857, July 13, 1842 (72), July 3, 1843 (74) October 23, 1845 (113), November 17, 1846 (140), and March 14, 1848 (173), CSU Archives. The deed book does not show this debt as ever being satisfied; Muscogee County Deed Book B, 134.
10. Muscogee County Court of Ordinary Sitting for County Purposes, 1838–1857, January 1 (52), May 10, 1842 (57), CSU Archives.
11. Coulter gives 1841 and Walker, 1842 as the completion date. This structure has not survived; the wooden two-story Tuckabatchee Masonic Lodge (ca. 1843) still stands and has been confused at times with the brick courthouse. Coulter, *"A People Courageous,"* 209, 113–15; Walker, *Backtracking in Russell County*, 113–15.
12. The eight men authorized to build the bridge were Abner McGehee, George Whitman, Edward Hancock, Sen. John Scott, Francis M. Gilmer, N. C. Benson, John C. Webb, and Charles R. Pearson. *Acts of the General Assembly of the State of Georgia*, 1835, 1:68; 1838, 1:124.
13. C. Smith, *History of Troup County*, 55–56; Fretwell, *West Point*, 21–22. Nick Tompkins supplied the lumber.
14. *Acts of the General Assembly of the State of Georgia*, 1849–50, 1:105.

15. Marshal Robert S. Crawford quoted in Green, *Politics of Indian Removal,* 175–77.
16. *Acts of the General Assembly of the State of Georgia,* 1837, 1:139.
17. The piers only extended thirty feet above the high-water mark; so, at high water, the river was fifty feet deep. Contract recorded Barbour County Deed Book B, 611–12.

    About a decade later, as a freedman, Horace worked in Alabama and Mississippi with a bridge builder identified only as Mr. Williams. On one occasion, Williams sought King to "attend to framing and raising a pier and sinking a crib at Columbus Bridge, Mississippi." Given the small size of the bridge building fraternity, Mr. Williams could very well have been Simon Williams. The initial letter in this sequence involves Horace trying to employ Mr. Williams. Robert Jemison Jr. to Horace King, March 10, July 10, and September 9, 1851, Letterbook 3, Jemison Collection, UA.

    Payments were to be made when the timber was on the ground, when the framing was done, when the bridge was "nailed,"and when the structure was completed.
18. Since Williams later asked Horace to build piers for one of his projects, a logical assumption would be that Horace erected these piers in Irwinton. "Bascom Dowling Writes Interestingly about the Old River Bridge Here," *Eufaula Daily Citizen,* May 6, 1925. The repair of the piers that revealed the log cribs is documented in J. L. Land, Chattahoochee Covered Bridge Photographs, 1923–25, and letter from Land to Milo Howard, December 31, 1968, SPP 32, Alabama Department of Archives and History, Montgomery.
19. *Irwinton Southern Shield* quoted in *Columbus Enquirer,* March 24, 1841.
20. Norton made a separate agreement with each Godwin for $1,150. Muscogee County Deed Book B, 399–400.
21. For a detailed account of banks within the valley see Willoughby, *Fair to Middlin',* 53–89; *Irwinton Shield* quoted in *Columbus Enquirer,* June 5, 1844.
22. Edward B. Young, plaintiff in error, vs. Kenneth McKenzie, James Harrison and Samuel Harrison, defendants in error. *Reports of Cases in Law and Equity. . . . Supreme Court of Georgia,* 1847, 31–46. Hereinafter referred to as *Georgia Reports,* which is how it is normally cited.
23. The original charter provided that if the owner and company could not agree the Randolph County Inferior Court would appoint arbitrators. That court sitting for ordinary purposes rather than the court sitting as a civil court appointed the arbitrators; that action provided the basis for the litigation. Such cases were rampant at that time with property owners who lost land to railroads through eminent domain claiming a share of the railroad profits. Most courts sided with the railroads, unlike the Georgia Supreme Court in this case. *Acts of the General Assembly of the State of Georgia,* 1847, 1:265.
24. Edward B. Young and John A. Calhoun, Intendant of the Town of Eufaula, Alabama, plaintiff in error, vs. James Harrison . . . defendants, *Georgia Reports,* 6:130–58. The final decision is buried in a volume in the basement or attic of one of the courthouses in the Superior Court circuit. Young . . . vs. Harrison, *Georgia Reports,* 1855, 17:30–46; 1857, 21:583–91.
25. Young argued that his company was engaged in this process when Harrison obtained an injunction to stop construction."Irwinton Bridge Company," typescript in

possession of the authors; Smartt, *History of Eufaula*, 54–55; Flewellen, *Along Broad Street*, 41–44.

26. Ledbetter and Braley, *Archeological and Historical Investigations*, 38–67.
27. *Acts of the General Assembly of the State of Georgia*, 1837, 1:3, 115, 264; *Georgia Mirror* of April 2, 1838, reproduced in Ledbetter and Braley, *Archeological and Historical Investigations*, 51. By comparison Columbus received about 25,000 bales of cotton in 1838 and 50,000 in 1839; Martin provided no figures for 1837. Martin, *Columbus, Georgia*, 1:108.
28. The physicians were not all practicing at the same time but their names appeared in newspaper ads from 1838 to 1841. Three of them even published a fee bill, a list of their charges, in 1839; such practices were common in larger cities at the time; Ledbetter and Braley, *Archeological and Historical Investigations*, 50–58; Terrill, *History of Stewart County*, 293–95.
29. *Columbus Enquirer*, July 19, 1838. This building was moved after twenty years and became the Methodist Church; in the early twentieth century it was relocated to Omaha, a few miles to the north, where it became a Masonic lodge and allegedly is still extant. Ledbetter and Braley, *Archeological and Historical Investigations*, 5–6; Terrill, *History of Stewart County*, 296.
30. *Georgia Mirror*, April 2, 1838. The small community of Jernigan still exists in southern Russell County at the site of his plantation.
31. Doster, "The Florence Bridge Company," 332–33.
32. *Columbus Enquirer*, August 5, 1840.
33. The pier failed on the Alabama side in the case of the Florence span. *Columbus Enquirer*, March 17, 1841.
34. Ledbetter and Braley, *Archeological and Historical Investigations*, 59.
35. Everett's name may be spelled Averitt in this deed, and the spelling might very well be in error in the Georgia law. Also, the amount of the sale might be $7,000; it is recorded in a nearly illegible cursive hand. Stewart County Deed Book L, 366–67.

CHAPTER EIGHT. Robert Jemison Jr. and Horace

1. *Columbus Enquirer-Sun*, May 30, 1885.
2. Meigs, "Life of Senator Robert Jemison, Junior," 1–9, 19–21; Daniels, "Entrepreneurship," 154–58.
3. Robert Jemison Jr. to James Smith, July 8, 1845, Letterbook 1, Jemison Collection, UA.
4. Jemison's tenure in the house spanned from 1837 until 1851 except for a three-year hiatus; he then served in the Alabama Senate until 1862, when he became a Confederate senator. Brewer, *Alabama*, 563–64.
5. Carter controlled the Coweta Falls Factory and built his own Carter Factory in Columbus, Georgia, but managerial problems and the rising price of raw cotton in the late 1840s led him to abandon his large-scale scheme to bring slave labor to Columbus.
6. He also had large holdings in Tuscaloosa. In 1864, when he valued his assets for state tax, he claimed $50,000 worth of city lots in addition to 102,820 rural acres worth $105,000. November 10, 1864, Letterbook 10, Jemison Collection, UA. Jemison to the

Honourable Court of Roads and Revenue of Pickens County, Alabama, January 3, 1846, Letterbook 1, Jemison Collection, UA.

7. Charles Lyon Wood, "Historic Lowndes, an Outline" MS, 1925, in Lowndes County Public Library, 18; *Southern Argus*, May 10, 1842, cited in Clement, "Bridges," 20, 62.
8. Jemison to H. B. Gevatheney, October 24, 1845, Letterbook 1, Jemison Collection, UA.
9. Jemison to Messrs. Jno. O. Cummins and Co., June 19, 1846, Letterbook 2, Jemison Collection, UA. The only feature remaining at the site of his Luxapalila mills is a portion of the dam (including some twentieth-century remnants) and the outline of the pond, now covered by woods. It is located south of the Luxapalila River and east of Gunshot Road. Note that Luxapalila is the modern spelling; Jemison consistently used the spelling Luxapelila.
10. The town of Columbus, Miss., was not chartered until 1821. The first settlers moving into the area believed they were in Alabama and that the Tombigbee River would be the boundary between Alabama and Mississippi. Instead the straight-line boundary between the states placed Columbus in Mississippi. Kaye, Ward, and Neault, "By the Flow of the Inland River," 75–98, typescript provided by authors; Map 15, Land Offering: Mississippi, in Young, *Redskins, Ruffle Shirts, and Rednecks*, 183.
11. Clement, "Bridges in the Upper Tombigbee River Valley," 109.
12. The other stockholders were W. L. Harris, Dr. J. H. Hand, Hardy Stevens, J. N. Mullen, John Estes, Eli Abbott, and E. F. Calhoun.
13. Jemison to John Godwin, April 17, 1845, Letterbook 1, Jemison Collection, UA.
14. "The firm of Godwin, Bates and King was dissolved about 1840, Godwin and King succeeding and continuing until about 1848, when they dissolved." Lane, Bridges, and Jones, "The King Family," 369.
15. The "hewed lumber" and "shear poles" appear to have cost from $18 to $20 per thousand, again a higher price that seems to indicate they were not readily available.
16. Jemison to John Godwin, July 4, 1845, Letterbook 1, Jemison Collection, UA.
17. Jemison to John Godwin, April 17, 1845, Letterbook 1, Jemison Collection, UA. Jemison had beseeched Calhoun to prepare the year's bridge account, because Jemison had to deal with Seth King. Jemison to E. F. Calhoun, November 27, 1844, Letterbook 1, Jemison Collection, UA.
18. Jemison to John Godwin, April 17, 1845, Letterbook 1, Jemison Collection, UA. "[E]very thing amongst the stockholders unsettled." Jemison to Ro. Kirkham, November 21, 1844, Letterbook 1, Jemison Collection, UA.
19. Quote is from Jemison to E. F. Calhoun, November 27, 1844, Letterbook 1, Jemison Collection, UA. The legislative charter called for traditional tolls: pedestrian, 6¼¢; man and horse, 12½¢; large livestock, 5¢ per head; small livestock, 3¢ per head; four-wheeled vehicle with two horses, 50¢; same with four horses, 75¢; two-wheeled vehicles, 25¢; all other vehicles, 50¢. Perhaps the practice at Blewitt's Bridge over the Luxapalila south of Columbus established a precedent of no or low fees. Major Blewitt, who built the span as a toll bridge, did not charge the county's residents. Consequently, residents might have expected similar treatment at this new bridge. Blewitt's became known as Green T. Hill Bridge, probably because of its proximity to Hill's house. Hill, Jemison's nephew, directed the construction of later bridges. Clement, "Bridges in the Upper Tombigbee River Valley," 105, 123.

   The company also faced collection on a $9,000 note it could not pay. Jemison to

E. F. Calhoun, November 27, 1844, Letterbook 1, Jemison Collection, UA. Jemison also had problems finding a satisfactory bridge keeper he could trust. Jemison to G. T. Hill, December 24, January 7, 1845, Letterbook 1, Jemison Collection, UA. Samuel H. Kaye, Carolyn B. Neault, and Rufus A. Ward Jr., "The Bridge at Bridge Street," typescript provided by authors.

20. Jemison to Green Hill, November 28, 1844; Jemison to Godwin, April 17, 1845; Jemison to Harris and Harrison, August 11, 1846, all in Letterbook 1, Jemison Collection, UA. Newspaper quote in Sherri Monteith, "Lowndes County Bridge History Colorful Since Early 1800s," *Commercial Dispatch*, October 12, 1980.
21. Lipscomb, *History of Columbus*, 71; Samuel H. Kaye, Carolyn B. Neault, and Rufus A. Ward Jr., "The Bridge at Bridge Street," typescript provided by authors, 4.
22. Samuel H. Kaye, Carolyn B. Neault, and Rufus A. Ward Jr., "The Bridge at Bridge Street," typescript provided by authors.
23. The first Jemison letter defined the dimensions of the bridge; the others document his and Seth King's involvement as stockholders. Jemison to James Smith, July 8, 1845, Letterbook 1; Jemison to J. C. Spencer, January 23, 1860; Jno. T. Taylor, May 24, 1860; Jno. Whiting, October 2, 1861; Jno. F. Matherson, Secr and Tr. Cheraw Br. Co., January 27, 1862, Letterbook 9, all in Jemison Collection, UA; Porter, *History of Wetumpka*, 123–26; Jackson, *Rivers of History*, 103–7.
24. The description of the dark bridge is from an unspecified article by Lynn Welden in the *Wetumpka Herald* cited by Porter in *History of Wetumpka*, 124.
25. Porter, *History of Wetumpka*, 123–26; Jackson, *Rivers of History*, 103–7.
26. Jemison to John Godwin, July 4, 1845, Letterbook 1, Jemison Collection, UA.
27. Quote is from Jemison to S. S. Franklin, July 26, 1845, Letterbook 1, Jemison Collection, UA. The final disposition of the insurance issue is unclear from the correspondence. See Jemison to John Godwin, Letterbook 1, July 4, August 5, 1845, Letterbook 1, Jemison Collection, UA.
28. Jemison to Godwin, July 4, 1845, Letterbook 1, Jemison Collection, UA.
29. Jemison to Godwin, July 25, 1845, Letterbook 1, Jemison Collection, UA.
30. Gillespie, *Free Labor in an Unfree World*, 132; Jemison to Godwin, July 4, 1845, Letterbook 1, Jemison Collection, UA. According to a memorandum for record, January 1, 1855, Jemison and H. B. Robinson "hired of E. F. Comages two Negroes. . . . Said Comages is to clothe said negroes and pay their taxes, Physicians and surgical Bills and if either of them shall at any one time from sickness or otherwise loose [*sic*] one week or more from work there shall be a pro rata deduction," Letterbook 5, Jemison Collection, UA.
31. Presumably, Godwin and Jemison were splitting the cost of transportation, with Godwin insuring that Horace reached Montgomery and Jemison paying his way to Lowndes County, Mississippi. Jemison to Dr. S. L. Franklin, July 12, 1845, Letterbook 1, Jemison Collection, UA.
32. Jemison to J. B. Greene, July 18, 1853, Letterbook 4, Jemison Collection, UA.
33. Jemison to his sister, July 28, 1853, Letterbook 4, Jemison Collection, UA.
34. In the rice-producing areas, Africans and their descendants outnumbered the whites in such large proportion that some slaves lived in a totally black society with only a white overseer and an occasional visit by the master's family. African American drivers controlled the slaves on a day-to-day basis.

35. Jemison to John Godwin, August 6, 1845, Letterbook 1, Jemison Collection, UA.
36. Jemison to John Godwin, July 25, 1845, Letterbook 1, Jemison Collection, UA.
37. Clement, "Bridges in the Upper Tombigbee River Valley," 20.
38. On August 9, 1845, Jemison informed Reynolds that he expected to "start over about this day week." Letterbook 1, Jemison Collection, UA.
39. A newspaper account called Green T. Hill "one of the local adept carpenters who did much for the bridge system in the county." Monteith, "Lowndes County Bridge History"; Clement, "Bridges in the Upper Tombigbee River Valley," 12.
40. In 1938 the county built an iron and steel Pratt truss at the Luxapalila site; it collapsed from the weight of a tractor-trailer rig in 1980. Clement, "Bridges in the Upper Tombigbee River Valley," 12; Sherri Monteith, "Bridge over Luxapelila Collapses," *Commercial Dispatch*, July 2, 1980.

CHAPTER NINE. Freedom

1. For an extended discussion of the prejudice against nonslave blacks, see Lawrence-McIntyre, "Free Blacks."
2. Mills also asserts that free blacks in Alabama enjoyed similar freedoms to those in Louisiana even though the laws were less restrictive in Louisiana. G. Mills, "Shades of Ambiguity," 161–86; quote, 165.
3. J. M. Glenn, letter to the editor, *Montgomery Advertiser*, September 16, 1948; Cherry, "History of Opelika," 194; Coulter, *"A People Courageous"* follows Cherry, 116–17.
4. The McKenzie-Wright bridge collapsed in 1863 and the trading community of Old Tallassee disappeared. The family connection between the Wrights and the McKenzies is noted in a letter from Edward Pattillo to Thomas L. French Jr., September 6, 1992; Golden, *History of Tallassee*, 14–17.
5. King Interrogatories, 8. The editor of the Union Spring journal would have known King from his work on the Mobile and Girard Railroad. The Union Spring's piece was copied by the *Dallas Gazette*, August 26, 1859.
6. Kilbourne, *Columbus, Georgia*, 243.
7. Postscript, Horace King to Robert Jemison Jr., April 6, 1871, Jemison Collection, UA.
8. This is one of many such examples that could be cited. Jemison to J. D. Watson, March 22, 1847, Letterbook 2, Jemison Collection, UA.
9. Jemison may have owed his sister money. Jemison to G. T. Hill [his nephew in Columbus, Miss.], December 17, 1854; Jemison to Helen, December 19, 1854, Letterbook 5; Jemison to John T. Taylor, January 28, 1860, Letterbook 9, all in Jemison Collection, UA.
10. This account, typical of the era, deals with loyal or humorous slaves. Works Progress Administration, "Lowndes County, Mississippi," pt. 1, 499–500, typescript, Lowndes County Public Library.
11. Jemison to Horace King, March 3, 1846, Letterbook 1, Jemison Collection, UA.
12. Bryan vs. Walton, 14, *Georgia*, 185, quoted in Sweat, "The Free Negro in Ante-Bellum Georgia," 95.
13. Trotter, *African American Experience*, 208; "An Act to Emancipate Horace King, a Slave," *Acts of the General Assembly of Alabama*, 1846, 207–8. This bond was not posted in the deed records of the probate judge, where John Godwin tended to

record his financial dealings. The other court records are either not extant or inaccessible.

14. Martin, *Columbus, Georgia*, 1:107, 124,130. Robinson's name presents some confusion. The name appears in print in the accounts about the Albany bridge as Dr. Alexander J. Robinson. In two other sources where his name appears in conjunction with King and Godwin, it is listed as Alexander Robinson and Alexander J. Robinson rather than Robison, but those are handwritten documents (Godwin's will and the Carroll County deed for King). John Martin's history shows him as Robison, and Martin must have known him. There is no evidence for two people with different names; the 1850 census shows him as A. I. Robison and the 1860 as Alexander J. Robinson; the two households include the same children. The city directory in 1860 lists Dr. A. Robison. His tombstone reads Alexander Irwin Robison, while his son is A. W. Robinson.

    The city directory of the same year listed Robison as a planter, who very typically also had an urban house. The census taker enumerated him as a farmer apparently living next to William H. Young, the city's most aggressive entrepreneur, presumably in Beallwood, an elite suburb north of Columbus. The 1850 census shows Alexander living with his wife, Amanda, while the 1860 record has them living in separate but relatively close houses. The following entry from the county records appears to place Robison's house or plantation on the south side of the city: The seventh bridge in a survey of the county's bridges "was Randalls across Upatoi on the Bald Hill Road just beyond Dr. A. J. Robinson." Inspection of county bridges by Jas. A. Bradford, Muscogee County Order Book, 1857–1871, August 29, 1859 (68).
15. Muscogee County Order Book, 1857–1871, September 16, 1862, (146).
16. Muscogee County Court of Ordinary Sitting for County Purposes, 1838–1857, March 5, 1849 (205), August 1, 1849, June 2, 1847 (151), November 2, 1847 (164), March 10, 1848 (171), February 2, 1849 (199), June 7, 1852 (265), November 6, 1854 (291), and December 4, 1854 (292); 1857–1871 volume, December 15, 1857 (37), April 20, 1864 (182), CSU Archives.
17. John received $2,085 for the jail. Perhaps King worked on these small jobs as well. The Reverend Cherry, in a notebook, indicated that King built the Russell County jail in 1853 or 1854. Book 5, 38, Folder 4B, Cherry Collection, Alabama Archives.

    The exact extent of local bridge building remains unknown, because only some records have survived. A Muscogee County Bridge Book, inventoried by the WPA Records Survey in the 1930s, has since disappeared. By this period Colonel Bates had moved to Russell County but kept working in Muscogee County. Between 1858 and 1864 he built at least ten spans in that Georgia county, and the county paid him an annual salary as a bridge builder in 1864. See Minutes of Russell County Commissioners, October 1850, April–December 1851; Minutes of Russell County Probate Court, June Term, 1853, and [Ann Godwin Estate], February 2, 1855; Russell County Bridge Book [1852–58], 4, 23, and 32.
18. If Horace controlled such a crew during the Civil War, then he certainly had access to such workers before 1861. See chapter 10 for his work at the Confederate Naval Iron Works.
19. Jemison to Horace King, September 9, 1851, Letterbook 3, Jemison Collection, UA.

20. Jemison to R. M. Patton, June 25, 1854, Box 1589, Jemison Collection, UA.
21. Jemison to Horace King, March 10, 1851, Letterbook 3, Jemison Collection, UA. King's letter did not survive; its content is derived from Jemison's return letter.
22. A recent account of the span cited its function as linking traffic with the Black Shoals on the Flint River, but that probably had little to do with its construction, since the river was not navigable at that point. In 2000 the Georgia Department of Transportation spent $176,253 to refurbish this bridge, which measures 391 total feet with a covered Town truss only spanning 127 feet, and over $1.5 million to stabilize ten covered bridges in the state. No documentary evidence links this bridge to King; it may have existed in the Meriwether County courthouse that burned in the 1980s. Pinkston, *Historical Account of Meriwether County*, 90–91; "State Touches Up Historic Bridge," *Augusta Chronicle*, March 30, 2000.
23. George A. Tierce received the charter in 1846, but the legislature granted George and G. H. Baker the right to construct a bridge across the North River in Tuscaloosa County in 1850. The 1846 job probably never materialized. *Acts of Alabama*, 1846, 135; and 1849–50, 425. Jemison to E. F. Calhoun, December 24, 1844, Letterbook 1; Jemison to John Godwin, April 17, 1845, Letterbook 1; Jemison to Harris and Harrison, August 11, 1846, Letterbook 2; Jemison to Horace King, February 5, 1858, Letterbook 8, all in Jemison Collection, UA. The February 5 letter to Horace discussed the specifics of the arbitration. "King vs. Jemison," *Reports of Cases Argued and Determined in the Supreme Court of Alabama during June Term, 1858, and January Term 1859*, 33:499–509.
24. The life of Dr. John R. Drish paralleled that of Jemison. A practicing physician, planter, and state legislator, Drish owned numerous slaves including "first-rate mechanics—masons, carpenters, plasterers, and blacksmiths." According to local tradition, Drish's craftsmen worked on Jemison's Tuscaloosa mansion. W. Smith, *Reminiscences of a Long Life*, 1:143–45.
25. Gagnon, "Transition to an Industrial South," 16; deTreville, "The Little New South," 41–80. The lack of correspondence between King and Jemison after March 3, 1846, might suggest that Horace was with Jemison but not necessarily. Also the skill level ascribed to Drish's slaves suggests they could have built the factory without Horace. They did not need to impart camber to a truss. As a source for this assertion, Daniels ("Entrepreneurship in the Old South") cited the reminiscences of James Robert Maxwell and William R. Smith. Neither mentioned Horace or the factory.
26. The Eagle Mill in Columbus remained very strong in the 1850s. Daniels, "Business, Industry, and Politics," 31–57; Daniels, "Entrepreneurship," 161–63; Jemison to John Whiting, June 1860; Jemison to William F. Plane, June 25, 1860, both in Letterbook 9, Jemison Collection, UA. In 1861 Jemison informed Pratt of the sale of the Tuskaloosa and North Port Cotton Mill machinery; perhaps this company was a reincarnation of the Warrior Mill that failed again. Jemison to John Whiting, June 1860, Jemison to Daniel Pratt, October 14, 1861, Letterbook 9, Jemison Collection, UA.
27. Jemison to R. M. Patton, June 25, 1854, Box 1589; Jemison to Horace King, September 9, 1851, Jemison Collection, UA. Smaller examples of double spiral stairs existed in many antebellum houses. King could have seen one in John Fontaine's house in Columbus. King built warehouses for Fontaine in 1855 and 1865. See interior

photographs of Fontaine's house [now destroyed] in John Fontaine House, Front Avenue, Muscogee County, Ga., HABS [Historic American Building Survey] Collection, Library of Congress. Photograph available online.

In the 1970s Thomas L. French Jr., while searching a group of miscellaneous records in the basement of the Alabama Department of Archives and History, found a voucher issued to Horace King that mentioned work on the capitol staircase. French gave that document to an archives staff member so it could be placed in a more appropriate collection. It was subsequently misplaced. Interpretative signs within the capitol (and on Web sites) suggest King did build the staircase, but the official architectural history of the capitol does not mention King as the builder. Gamble and Dolan, *Alabama State Capitol*, 9.

28. The statistic concerning nonslave heads of household is an approximation based on Boucher, "The Free Negro in Alabama"; he replicated all the census data for this group.
29. Minutes of Columbus City Council, March 18, 1837. Less than a month later the council lowered the rate to $6 for free blacks as compared to 46¢ for white males and slaves. Minutes of Columbus City Council, March 11, April 1, 1837.
30. King purchased the land, the east half of settler's lot number twenty-eight, from Walter B. Harris. See Russell County Deed Book H, 167.
31. The following letters were either sent to Tallassee or refer to it: Jemison to Horace King, December 9, 14, 1852; January 22, February 22, 1853, Letterbook 3, Jemison Collection, UA. The river falls forty feet in fifty yards at this site. These two local residents chartered the Tallassee Fall Manufacturing Company in 1841 and bought machinery from Daniel Pratt. Golden, *History of Tallassee*, 14–17; Wadsworth, "Tallassee," 17, 29, 43.
32. The role of Jemison is detailed in the following publications: Mellown, "Construction of the Alabama Insane Hospital," 83–104; Weaver, "Establishing and Organizing the Alabama Insane Hospital," 219–232; Mellown, "Bryce Hospital."
33. Jemison to Horace King, December 14, 1852, Letterbook 4, Jemison Collection, UA. Jemison did not inform Pratt of his real purpose, only that he was remodeling his steam mill. Jemison to Daniel Pratt, Letterbook 4, January 24, 1853; Jemison to James S. Bryant, February 1, 1853, Letterbook 4, both in Jemison Collection, UA.
34. Jemison to Horace King, January 22, January 30, February 22, 1853, Letterbook 4, Jemison Collection, UA.
35. Robinson laid brick sewers for the city of Columbus as early as 1837 and raised the ire of the council in 1845 by making bricks on the city's east common.
36. Jemison to Horace King, August 7, 1853, Letterbook 4, Jemison Collection, UA.
37. Sellers, *History of the University of Alabama*, 35; Wolfe, *The University of Alabama*, 60–61.
38. Jemison to R. M. Patton, June 25, 1854, Box 1589, Jemison Collection, UA.
39. The local evidence in Florence, Alabama, does not point to King's working there, but Horace's most accurate obituary (*LaGrange Reporter*, May 28, 1885) has him building bridges in Tennessee, which might have been one bridge on the Tennessee River in Alabama. Jackson, *Rivers of History*, 51–58, 59–60.
40. *Dallas Gazette*, April 28, 1854.
41. Ad and article, *Dallas Gazette*, August 31, 1855. The people of Cahaba knew Horace.

The *Dallas Gazette* copied the piece about him erecting a monument to Godwin from the *Union Springs Gazette* on August 26, 1859.

42. Ad and article, *Dallas Gazette*, October 5, 1855; obituary in *Dallas Gazette*, November 30, 1855.
43. The 1860 census listed James Mealer as a fifty-year-old carpenter worth $1,050; the 1870 one showed him as a sixty-three-year-old house carpenter; in the 1878 city directory he appears as a carpenter, not as a contractor, and he does not advertise in the directory. Fontaine entry in Georgia, vol. 23, p.112, R. G. Dun and Co. Collection, Baker Library, Harvard Business School. Worsley, *Columbus on the Chattahoochee*, 97, 118, 131–32, 195, 264, 377.
44. The geologic configuration of the river bottom created a natural waterpower site. No dam was ever built at this mill. A natural basin created a pond, and a rock ledge diverted the water to the head gates for the wheel and later for turbines. The natural fall was about forty to fifty feet. Julius R. Clapp owned forty-two slaves in 1860; since he does not appear in the agricultural census for that year, he must have been using his slaves in his factories.
45. Fontaine made dual signatures a requirement in 1865, when they rebuilt his cotton warehouse; presumably, he had the same requirement in 1855. Only four of these documents actually fulfilled that requirement. One (July 20, 1855) bears the signatures of both Meeler and King; the remaining ones have both Meeler's and King's names, apparently signed by Meeler. Two (July 23, 27) also have King's signature underneath Meeler and King, written by Meeler. The following description of the project comes from Correspondence, January–August 1855, Box 3, John Fontaine Papers (Ms 2014), Hargrett Rare Book and Manuscript Library, University of Georgia, Athens.

    King may not have been available to sign the later documents. After completing the heavy timber structure, he moved on to another job.
46. He bought the other mule from Robertson, or more probably Robison, in Columbus, Georgia.

CHAPTER TEN. Horace's Own Bridge

1. Information about this venture comes from two Carroll County deed transactions (1858 and 1869) noted below and from the 1878 Unionist interrogatories of King, Moore, and Jeddiah S. Miller, another neighbor. Moore dated his first meeting with King in 1855 or 1856, while Miller cited it as 1858. King Interrogatories. Information about Moore's property in Carroll County Deed Book D, 176; E, 104; and F, 5, and 349.
2. The 1850 census for Mabry's household lists a relative or boarder named Samuel Boggins as a teamster; perhaps he was involved in the lumber business. Stoneman ascertained this length as a result of the brief battle at the bridge. The total length must have included some land bridge. Maj. Gen. George Stoneman, Moore's Bridge, to Maj. Gen. William Sherman, July 13, 1864, *Official Records of the War of the Rebellion*, ser. 1, vol. 38, pt. 2, 912–13.
3. Carroll County Deed Book I, 104.
4. Carroll County Deed Book M, 130–31.

5. Moore's old house still stands at the top of the hill. It resembles a Carolina cottage similar to those built by Godwin around Cheraw, and some local oral sources claim Horace lived in Moore's house. The testimony of Moore and King defines the geographical and perhaps the social arrangement of the two houses.
6. Kolchin, *First Freedom*, 22–23. Horace's family members who remained on the Godwin place were his mother; his brother, Washington; his sister, Clarissa; and her husband, Henry Murray.
7. Samuel H. Kaye, Carolyn B. Neault, and Rufus A. Ward Jr., "The Bridge at Bridge Street," typescript provided by authors. This letter included a detailed description of the current litigation between Jemison and Seth King over the Tuscaloosa and Columbus bridges. They were arguing over a bridge that had already collapsed. Jemison to Horace King, February 5, 1858, Letterbook 8, Jemison Collection, UA.
8. *Acts of the Georgia Legislature*, 1807, 1:27; 1816, 1:75; 1817, 1:52; 1834, 1:51, 161; 1857, 1:156.
9. *Albany Patriot*, August 17, 1858; *Dictionary of Georgia Biography*, s.v., "Tift, Nelson"; *Acts of the Georgia Legislature*, 1851–52, 1:283.
10. Bogle neither cited nor remembered the source for those precise dates; the DAR history of Albany (1924) also mentions moving the timbers, but not the dates. Unfortunately none of the extant Milledgeville newspapers mention the bridge. Bogle, "Horace King," 33–35; Thronateeska Chapter, Daughters of the American Revolution, *History and Reminiscences of Dougherty County, Georgia* (Albany, Ga., 1924), 53–56.
11. Lee Formwalt, a historian who has done extensive research on the history of Albany, suggested less respectable uses for the bridge house including as a meeting place for the city's prostitutes and in the late nineteenth century for the Klan. When the bridge no longer existed, the building became a blacksmith shop of the Kennan family, which they evolved into an auto parts business. In 2001 the vacant building was considered for possible adaptive use as part of the refurbishing of the city's riverfront. Stephens, "Albany," 30–36; [Lee Formwalt] "Albany's Bridge Hall," a historical brochure for the Keenan Auto Parts Company. Formwalt is quoted in Ken Garner, "ATI Eyes Colorful, Old House," *Albany Herald*, February 5, 2001.
12. Formwalt, "Moving in That Strange Land of Shadows," 508.
13. Within a two-block area, between the present Fourteenth and Twelfth Streets, were the following multistory brick structures: Coweta Falls (textile) Mill; Carter's Factory, which housed a variety of small operations; Variety Work, a sawmill and planing operation located in the river; Howard (textile) Factory; and Eagle (textile) Mill. *Columbus Enquirer*, June 12, 1856; May 4, 1858.
14. *Columbus Enquirer*, January 6, 1852.
15. Olmsted, *Cotton Kingdom*, 213. In contrast Olmsted described Montgomery as "a prosperous town, with pleasant suburbs, and remarkably enterprising population."
16. *Columbus Enquirer*, June 12, June 17, 1856.
17. King family tradition claims Bates built the first bridge at this location, and the family calls it an open bridge. Untitled WPA typescript dealing with Horace and his family, Alva C. Smith Collection, CSU Archives. Minutes of Columbus City Council, November 22, 1858.
18. Mr. Wynn eventually became the lead contractor; William C. Gray, James Vernoy,

and others served in various capacities as either builders or suppliers. The *Daily Sun* on October 14, 1858, identified Mr. Wynn as the contractor. He was not listed in the city directory the next year and might have been an itinerant builder. Both Gray and Vernoy served on the council; one of their fellow aldermen challenged their right to participate in the negotiations because they had a vested interest. Minutes of Columbus City Council, November 29, 1859. The newspaper also commented on the filthy condition of the lower bridge, which needed a thorough cleaning. The misnaming of the location as Bryan Street was repeated in Martin's history and caused confusion about the location of the bridge. Other sources in the council minutes obviously place it at Franklin Street. *Columbus Enquirer,* April 24, 1858.

19. *Columbus Enquirer,* May 4, July 17, 1858.
20. *Sun,* November 10, 1858.
21. Minutes of Columbus City Council, November 22, 1858; *Sun,* November 8, 10, December 1, 3, 9, and 15; *Columbus Enquirer,* November 11, 12, 15, 16, 17, 23, 24, December 2, and 9.
22. Martin, *Columbus, Georgia,* 1:101–2.
23. *Columbus Enquirer,* February 20, 1862.
24. Russell County Deed Book F, 930–31; G, 202; and H, 223; Russell County Probate Court Minutes, November 12, 1856.
25. The Historic American Building Survey (HABS) documented the Yonges' house in 1934. At the same time, the Civil Works Administration was converting the site into a city park and Boy Scout camp. The house is similar in decoration and construction to the Edwards House in Opelika, which seems to weaken the claim that Godwin or Horace built it. For Spring Villa, see photographs, drawings, and text within HABS AL-508 available online through the Library of Congress's American Memory Page under Lee County, Alabama. Wonderful 1930s photographs of the Chewacla Lime Works can also be accessed through this site. For the Edwards House, see Jeane, ed., *Architectural Legacy,* 120–23.
26. "Hon. Arthur Yonge," *History and Biographical Record of North and West Texas,* 2:411–13. This brief sketch of the son also includes a detailed account of the Horace King story.
27. King testified that he was building bridges for this company during the Civil War; a logical assumption would be that he began working for them before 1861. King Interrogatories, 1.
28. One example of such overstatement follows: John Godwin "died penniless. . . . King paid Godwin's burial expenses and assumed a guardianship of the widow and children. He cared for the widow for the remainder of her life." Bailey, *Neither Carpetbaggers nor Scalawags,* 69.
29. Cherry, "History of Opelika," 195.
30. John Godwin Estate Papers, loose estate records, Probate Court Records, Russell County Courthouse. An estate sale of Theodora King's possessions in April 1989 included a bed, which Theodora identified as being Godwin's. It is now at the Columbus Museum.
31. Robinson probably saw to it that this transaction was posted. Russell County Deed Book K, 565–66.
32. This particular version appeared in the *Dallas Gazette,* August 26, 1859.

33. Columbus did not experience the same decline in its slave population as did the large southern cities from 1850 to 1860. Historians debate the causes of this decline: Wade cites the issue of control as the primary reason; Goldin points to rational economic reasons: owners gained more profits from them in the fields. Wade, *Slavery in the Cities*, 243–52; Goldin, *Urban Slavery*, 123–32.
34. Berlin, *Slaves without Masters*, 343–52.
35. *Semi-Weekly Mississippian (Jackson)*, May 21, 1858, quoted in Berlin, *Slaves without Masters*, 341; emphasis in original. Milledgeville *Tri-Weekly Recorder*, November 18, 1858.
36. By 1860 only 144 free blacks, most of them old, remained in the state. Berlin, *Slaves without Masters*, 373–74.
37. Ibid., 371–80.
38. "A Bill to Be Entitled an Act of the Relief of Horace King and His Family," in Bailey, *They Too Call Alabama Home*, app. L.
39. The Kings did not establish a tradition of erecting gravestones; Horace's family never marked his grave.

CHAPTER ELEVEN. The Reluctant Confederate

1. King Interrogatories, 3.
2. Turner, *Navy Gray*, 154; Web site for the Nineteenth Georgia Infantry. See http://www.fred.net/stevent/19GA/mabry.html; Moore, King Interrogatories, 9.
3. King Interrogatories, 3–6. For a neo-Confederate perspective, see Scott K. Williams, "Black Confederates Fact Page," at http://www.geocities.com/11thkentucky/blackconfed.htm.
4. The recorded testimony of King states he was working on the Mobile and Florida Railroad in Alabama. Apparently, the recorder made an error. No Mobile and Florida Railroad existed, and several local sources mention King working on the piers in Columbus in the early 1860s. See King Interrogatories, 1, and Miller, King Interrogatories, 13.
5. The aldermen wanted the company to retain the bridge and collect tolls until the company had enough money to rebuild the bridge. When the company could give the city the bridge and the necessary funds for its reconstruction, the aldermen would accept ownership of the span. No available gas pipes existed in the city by that date, and although King covered it before April 1865, it still had no illumination. Minutes of Columbus City Council, May 2, June 27, July 4, and October 3, 1864.
6. Turner, *Navy Gray*, 147–50; Standard, *Columbus, Georgia*, 27–45; S. Edwards, "'To do the manufacturing for the South,'" 538–54; DeCredico, *Patriotism for Profit*, 49–51, 61–62.
7. The actual construction had been delayed because, despite their fears, local planters would not divert their laborers until after the harvest. Turner, *Navy Gray*, 75–80.
8. King Interrogatories, 5–6.
9. Telegrams in John Gill Shorter Papers, Alabama Department of Archives and History, Montgomery.
10. Warrant 201 and 202, SG 16471, Alabama State Treasurer, Military Department,

Ledger 1861–1861, Alabama Department of Archives and History, Montgomery. Every planter who provided thirty slaves to the state was allowed to send an overseer, who was paid $2 a day. King's pay was probably double that of the white overseers. Governor Shorter had the legislature pass a law to force the impressments of slaves, with pay for the master, to work on the state's defensive works. Many of the owners complained about losing the services of their slaves and the slowness with which they were compensated. See Brannen, "John Gill Shorter," 43–49.

11. Warner's actual rank with the navy (both the U.S. and the Confederate) was chief engineer, but after the Iron Works workers were organized into military companies, he received the rank of major. Many local sources identify him as Major Warner.
12. Berlin, *Slaves without Masters*, 386.
13. Horace King to Jemison, March 23, 1864, Jemison Collection, UA.
14. For one he received $8 a day for twenty-six days and for another $9 a day for fifteen days. He also hired out one of his mules for the last five months of 1863 for $100.
15. Specifically he and his men provided 294 logs for piling @ $2.50 each, 181 pine logs @ $5.00 each, 48 trees @ $4.25 each, 57 oak knees @ $12 each, 8 white oak knees @ $15 each, 4 large oak knees @ $20 each, and 3,000 treenails @ $.05 each.
16. His loads included 2,781 feet of oak lumber, 234 pine logs ($5 each), and two long pine logs ($25 each) used for shears—the tall, two-legged device that supported a hoisting tackle. Archives, Civil War Naval Museum, Columbus, Ga.
17. Muscogee County Order Book, 1857–1871, September 16, 1862 (146).
18. Information about the Godwins' Civil War careers from Ken Smith's Web site. *Official Records of the War of the Rebellion*, ser. 1, vol. 38, pt. 1, 386. John Dill was a captain in Company G, 28th Battalion, Georgia Siege Artillery. He could have helped the Confederates build the structures used to support the artillery pieces at the Narrows. The structures were made from railroad trestles that were taken downriver and placed at this site. White, Knetsch, and Jones, "Archaeology, History, Fluvial Geomorphology," 142–43.
19. In his interrogatories King emphasized the need to carry a pass during the war. Stressing that requirement seems to indicate that before the war he did not routinely carry a pass or his freedom papers.
20. King probably used store receipts for these specific amounts. King Interrogatories, 17–20.
21. One of the best examples of a shrewd businessman is William H. Young, who built what became the Eagle and Phenix Mill village in Alabama during the war.
22. Stoneman suffered from several physical ailments, including hemorrhoids, which tended to slow down him—and his troops. The description of Stoneman and the battle at Moore's Bridge is based on David Evans's chapter "To Moore's Bridge and Back," on the *Official Records of the War of the Rebellion*, and on the interrogatories of King and other witnesses at King's 1878 hearing. Evans investigated these postwar claims against the federal government to document the local actions of the Union troops around Atlanta. Evans, *Sherman's Horsemen*, 47–66. The exiled women never returned to Georgia, at least not en masse.
23. John Thomas King's obituary (*LaGrange Reporter*, November 26, 1926) gave a convoluted description of his acting as a bridge keeper: "July 1864, John King kept toll bridge between Columbus and Girard, Ala., about the time of Stoneman's raid when

two thousand soldiers crossed over the bridge and destroyed it." John probably did serve as the toll-taker at Moore's Bridge where Stoneman struck, and he probably talked about it during his life. Family tradition and later the WPA accounts place his keeper's job at the Columbus bridge. The idea that a twenty-year-old black, even if he were light skinned, was allowed to collect tolls on the major bridge seems highly improbable. The income was seemingly too crucial for the city, and his job would have involved disciplining escaped or malingering slaves; such a responsibility would not have been entrusted even to a member of the respected King family.

24. The account of Frances's encounter with the Federals is based on King's recollection in 1878 of what his wife told him; she died in October 1864. King's Masonic membership is discussed in chapter 12.
25. Evans, *Sherman's Horsemen*, 61–62. On their return, the Federals even stole from the man who informed them of the existence of Moore's Bridge. On July 17 Jefferson Davis replaced Johnston with John Bell Hood, who then attacked Sherman in several decisive battles that led to a Confederate defeat and the Federal occupation of Atlanta.
26. Foote, *The Civil War*, 486–87.
27. King Interrogatories. Frances's date of death is recorded in the King family Bible.
28. Quoted in Turner, *Navy Gray*, 211.
29. *Official Records of the War of the Rebellion*, ser. 1, vol. 49, 429, quoted in Fretwell, *West Point*, 31–34.
30. Thoroughly absorbed in the tradition of the Lost Cause, locals originally called it simply the Last Battle, but they realized that a battle occurred later in Texas (Palmetto Ranch) and other conflicts occurred at sea, so they qualified their description. *Columbus Enquirer*, June 27, 1865. "Young Robison," according to the 1860 census, was Alexander W. Robison; he was eighteen years old in 1865, Martin, *Columbus Georgia*, 2:180.
31. King Interrogatories, 4.
32. Mary Kent Berry, compiler, "Records of Marriage, Baptism and Burial from the First Register of Trinity Parish, 1836–1903.

CHAPTER TWELVE. Economic Reconstruction

1. *Columbus Enquirer*, July 7, 1865.
2. *Columbus Sun*, October 28, 1865.
3. George Parker Swift, who operated mills in Upson County, Georgia, before 1865, moved to Columbus after the war and established Muscogee Mills. He was reported to have saved six hundred or seven hundred bales of cotton in 1865, which would have been worth at least $30,000. The foundation for one of the most important financial institutions in Columbus rested on successfully hiding cotton from the federal troops. In a similar fashion, another important local financier, J. P. Illges, "saved considerable cotton" and in January 1866 was "probably worth some 30 or 40" thousand dollars. Georgia, vol. 23, pp. 125, 1, R. G. Dun and Co. Collection, Baker Library, Harvard Business School.
4. City Foundry (Porter, McIlhenny and Co.), *Sun*, September 1, 1865; Phoenix Foundry and Machine Shop (L. Haiman and Co.), *Columbus Enquirer*, October 5,

1865; Georgia Iron and Nail Works, *Sun* April 5, 1866; John McIlhenny selling his foundry, *Columbus Enquirer,* April 27, 1866; Booker, Fee and Co. (tin, sheet iron, and copper ware), *Sun,* January 8, 1868; Georgia Iron Works (Porter and Fell), *Columbus Enquirer,* March 25, 1869; Thomas Gilbert Scrapbook, Gilbert Collection, Special Collections, University of Georgia; J. C. Porter (agricultural machinery), *Columbus Enquirer,* July 23, 1876.

5. In 1860 Columbus had the largest establishments among major southern industrial centers when measured in terms of per capita investment in capital. Lupold, "Industrial Reconstruction of Columbus, Georgia." DeCredico notes the reconstruction of Columbus industry after 1865, even though she has some mills being rebuilt, such as Palace, that were not burned. However, she does not mention the collapse of small producers between 1870 and 1880. DeCredico, *Patriotism for Profit,* 135–41.
6. *LaGrange Reporter,* June 4, 1885; Muscogee County Order Book, 1857–1871, April 1, 1871 (415); CSU Archives. All three of Horace's obituaries mention his financial problems.
7. Georgia, vol. 23, p.112, R. G. Dun and Co. Collection, Baker Library, Harvard Business School.
8. Contract between John Fontaine and W. H. Hughes, as owners, and James Meler and Horace King, as builders, undated; Fontaine Collection, Special Collections, Hargrett Library, University of Georgia. Meeler's name is spelled three different ways within the contract, but it is obviously Meeler's signature on the 1855 documents.
9. Based on the contract, the 1885 Sanborn Insurance Map, and the 1873 and 1886 Perspective Maps of Columbus.
10. The warehouse was definitely operating by February 1866. Georgia, vol. 23, p. 112, R. G. Dun and Co. Collection, Baker Library, Harvard Business School.
11. Worsely, *Columbus on the Chattahoochee,* biographical entry for Bradley and company entry for his firm. The economic components of W. C. Bradley's fortune included the Eagle and Phenix Mills, the Bibb Company, other local textile mills, Georgia Power, the Columbus Irons Works, as well as a major investment in Coca-Cola; their descendants include CB&T Bank, TSYS, Synovus, and CharBroil.
12. Credit agents considered Mott to be of "unexceptional char[acter] as a bus[iness] man," perhaps because he gained his initial fortune from his wife. Georgia, vol. 23, p. 63, R. G. Dun and Co. Collection, Baker Library, Harvard Business School. Letter dated May 15, 1865, in Minutes of Columbus City Council.
13. See Calhoon, "Building the South," 359–83.
14. Gray's letter appeared in Minutes of Columbus City Council, August 14, 1865. Calhoon, "Building the South," 80, 89, 101, 107, 375–76; *Sun,* September 22, 1865.
15. The ladder probably stood very close to the current site of the large Horace King plaque on the Chattahoochee Riverwalk under the present Dillingham Street Bridge.
16. *Sun,* September 23, November 26, 1865. The King family tradition asserts that John Thomas King collected tolls on this bridge in 1865, but the Civil War context of the reference appears to link this remembrance with Moore's Bridge rather than the Columbus one. If he did serve in Columbus, he probably worked for Hines.
17. *Sun,* January 28, 1866; the need for the bridge mentioned in the *Sun,* September 23, October 4, 1865.

18. Judge Iverson might have been Alfred Iverson Jr., but the age of his colleagues would seem to indicate that the older man was involved. The younger Iverson rose to the rank of brigadier general in the Confederate army, and his cavalry command captured Stoneman near Macon in 1864. *Acts of the Alabama Legislature*, 1865–66, 483. No single road connects the lower (Dillingham Street) bridge with the Crawford Road, if the route of Dillingham Street is extended to the west. This turnpike might have been the straight north-south road, now Broad Street between Dillingham and Thirteenth Street. But it seems unlikely that travelers would have paid a toll, equal to crossing the bridge, for that short passage.
19. Turner, *Navy Gray*, 250–55.
20. Telfair, *History of Columbus*, quoting R. M. Howard, a self-proclaimed, unreconstructed southerner, 151.
21. *Sun*, September 21, 1866.
22. In the early 1880s Bates had employed as many as 250 men while building railroads in Florida. He preferred not to mix the races on his work crews, and he voiced the white canard that blacks were better off under slavery, "the happiest race of people on earth." In 1882 the U.S. Senate's Committee on Education and Labor began an investigation into "the relations between labor and capital." They visited major cities and industrial centers in the North before they came south to visit Birmingham, Atlanta, Columbus, and Augusta. In Columbus they interviewed thirty people that included thirteen African Americans and only three women—operatives at the Eagle and Phenix Mills. The men were laborers, manufacturers, educators, ministers, an editor, and the mayor. They probably selected Bates because he used black laborers. U.S. Congress, Senate Committee on Education and Labor, *Report of the Committee*, 4:488–648; Bates's quote, 491; copy available CSU Archives.
23. Perspective maps of 1872 and 1886; Columbus *Sun*, September 21, 1866.
24. Mill no. 1 measured 204 by 56 feet, twice as long as Clapp's.
25. History of the factory written by George M. Clapp for Loretto Lamar Chappell, a longtime Columbus librarian, and a postcard, Chappell Collection, CSU Archives; Sanborn Map, 1885; *Sun*, April 24, 1867.
26. Jemison to Horace King, February 23, 1867, Letterbook 6, Jemison Collection, UA.
27. The extant county records show King receiving three payments between November 1867 and January 1869 of $1,528.00, $220.42, and $173.00. No contract exists for this construction. A volume entitled "The Register of Claims against Lee County" noted November 27 (1867—in another hand at the bottom of the page) to Horace King $1,528 "Building C.H. and Jail"; a separate page showed $220.42 to him allowed on February 1868, filed on March 7, 1868, and paid on February 1, 1869, and another entry $173 to him allowed in June 1868 (no date specified), filed on January 27, 1869, and paid on February 1, 1869. This record book ("gold ledger") is housed at the Museum of East Alabama, Opelika, Alabama. King's courthouse stood until 1896 when the county built the larger, ornate building, which is still extant.
28. The bridge entered Alabama at the extreme northeast corner of Russell County. In 1923 Girard and Phenix City—the successor to Browneville—consolidated into one town and the state legislature altered the county lines in order to place all of Phenix City within Russell County. Coulter, *"A People Courageous,"* 396–97.
29. The *Sun* reported that $8,000 had been subscribed for the bridge on October 18,

1866; *Sun*, September 18, 1867, October 8, 1867, March 8, 1868, May 5, 1868, *Enquirer*, September 8, 1869.

30. *Sun*, March 8, 1868. "The entire length is 919 feet, of which 650 feet is lattice work. It is nearly 100 feet longer than the lower bridge." *Sun*, May 5, 1868.
31. *Sun*, May 5, 1868. The editor also noted that King "has constructed or superintended every bridge which has yet been erected over the Chattahoochee at this point." King did build the post-1862 span at this site, but he does not appear to have played a major role in erecting the controversial 1858 version. See chapter 9 for a discussion of that particular bridge.
32. *Sun*, May 5, 1868; *Enquirer*, September 8, 1869.
33. A Mr. Gardner and a Mr. Clark were mentioned in regard to the masonry work. Mobile and Girard Railroad Minute Book of Board of Directors, July 11, 1860, May 18, 1868, Central of Georgia Railroad Collection, Georgia Historical Society, Savannah. The company "lost a good deal as a result of the raid." See company entry, *Muscogee*, vol. 23, p. 86, April 7, 1861, and Feb. 15, 1866, R. G. Dun and Company Collection, Baker Library, Harvard University.
34. R. G. Dun and Company Collection, Baker Library, Harvard University, February 12, 1867, and July 1, 1868. *Dictionary of Georgia Biography*, s.v., "Wadley, William Morrill."
35. Mobile and Girard Railroad Minute Book of Board of Directors, July 18, 1868.
36. *Georgia*, vol. 23, pp. 61, 42, 156, R. G. Dun and Company Collection, Baker Library, Harvard University; Lupold, Karfunkle, and Kimmelman, "Eagle and Phenix," 10.
37. Mobile and Girard Railroad Minute Book of Board of Directors, July 18, August 6, 1868; *Annual Report of the Officers of the Mobile and Girard Railroad Company for the Fiscal Year Ending May 31, 1869*, 12. Conversations with family members; Marjorie Bush Chitwood, "Family History of Asa Bates," *History of Russell County, Alabama*, by Russell County Historical Commission, F12, F13.
38. Jones leased the mill to D. A. Wynn, Jr. *Georgia*, vol. 23, p. 61, R. G. Dun and Company Collection, Baker Library, Harvard University.
39. *Columbus Enquirer*, July 7, 1869.
40. Kimmelman, Lupold, and Karfunkle, "City Mills," 1–26. When the Historic American Engineering Record (HAER) began studying City Mills in 1977, HAER's chief architect, Eric Deloney, focused the team's efforts on the 1890 flour mill, especially its power transmission equipment. Deloney in a brief walk-through of the building dismissed the Horace King corn mill as a twentieth-century addition. As the historians started their work, they were conscious of the 1869 newspaper account linking King with the site, but taking their lead from Deloney, they believed King's building had not survived. Investigations into company records showed a clear distinction between the corn mill and flour mill. The historians asked one of the longtime workers, probably "J. T.," which building was older. With a look that bordered on disdain for these young people asking stupid questions, he simply pointed to the northern structure, which is the corn mill. Donald Stevenson, the project architect, carefully examined the interior of the structure and declared it might have been built in the 1830s, a date consistent with New England buildings. The existing early mill is the same building that appears on the 1872 and 1886 perspective maps and the

1885 Sanborn Map. The entire City Mill complex was declared a National Historic Landmark as part of the Columbus Historic Riverfront Industrial District in 1978. In 1980 a firm leased the building and the waterpower rights and tried to refurbish the corn mill turbines in order to produce electricity at that site. This experiment failed.

41. Only one bridge appears in the County Order Book between 1865 and December 1869: J. H. Marshall was to build a Upatoi bridge but failed to act on the contract. Muscogee County Order Book, 1857–1871, August 4, 1868, September 1, 1868; CSU Archives. Several of the payments to Washington King were noted as being to George W. or G. W. King. While Washington did have a brother named George, George H. was too young to be building bridges at this point. Most likely some clerk simply presumed that Washington King was George Washington King. See Muscogee County Order Book, 1857–1871, March, 12, 1869 (332), April 2, 1869 (334), May 7, 1869 (339), June 2, 1869 (345), June 24, 1869, (345), July 22, 1869, (347), July 24, 1869, (348), July 27, 1869, (349), November 10, 1869, (358), April 13, 1870, (374), June 21, 1870, (380), August 9, 1870 (387), August 22, 1870 (389) September 20, 1870, (392), October 1, 1870, (393), October 5, 1870, (394), October 14, 1870, (395), November 1, 1870, (396), November 14, 1870, (397), November 5, 1870, (397), January 17, 1871, (405), January 20, 1871, (406), January 31, 1871, (407), February 18, 1871, (409), March 4, 1871, (413); Muscogee County Commissioner Court Minutes, May 19, 1871, (5), March 2, 1872, (37), March 9, 1872, (38), April 11, 1872, (41), April 17, 1872, (42), April 26, 1872, (43), June 22, 1872, (51), and June 28, 1872, (52).
42. Muscogee County Order Book, September 10, 1869, (356), March 21, 1871, (413), and April 1, 1871, (415); CSU Archives.
43. *LaGrange Reporter*, February 5, 1869.
44. Geologists such as William Frazier note that the Chattahoochee Valley experienced a late uplift that makes its topology different from most southern rivers, which tend to flow through flat land below the fall line. In terms of flora, the Fort Gaines area has species that usually occur at or above the fall line.
45. Coleman, *The Bluff at Fort Gaines*, 25–28; Walter E. Pierce III, "Four Generations of Bridges at Fort Gaines," *Columbus Ledger-Enquirer Sunday Magazine*, December 26, 1971; French and French, *Covered Bridges of Georgia*, 19.
46. Jemison's report from Luxapalila. See *Mississippi*, vol. 14, p. 18, R. G. Dun and Company Collection, Baker Library, Harvard University.
47. King apparently dictated his letters; they seem to be in different handwritings, and none of them match the calligraphy of his signature. His wife, Frances, may have written some of this correspondence. Horace King to Jemison, April 6, 1870, Jemison Collection, UA.
48. Jemison to Horace King, April 11, 1870, Letterbook 12; Horace King to Jemison, April 16, 1870, both in Jemison Collection, UA.
49. Jemison to Horace King, April 21, 1870, Letterbook 12, Jemison Collection, UA.
50. Ibid.
51. Jemison to Horace King, April 25, April 27, May 2, 1870, Letterbook 12, Jemison Collection, UA.
52. Horace King to Jemison, January 11, 1871, Jemison Collection, UA. This letter is

written in a different script from previous letters. The signature is not King's, but he obviously dictated the letter.

53. Horace King to Jemison, April 16, 1871, Jemison Collection, UA.

CHAPTER THIRTEEN. A Nominal Republican

1. The conflict between Congress and Johnson eventually culminated in the Senate's coming within one vote of removing him from office in May 1868. 1870 federal census; Bailey, *Neither Carpetbaggers nor Scalawags*, 32–33.

   Unless otherwise noted, Cherry's quotes in this chapter are from Cherry, "History of Opelika," 195–97.
2. Cash, "Alabama Republicans during Reconstruction," 81–85.
3. "J" [Thomas J. Jackson], "Old Recollections—No. 27, Reminiscences of Early Columbus Continued," *Columbus Enquirer-Sun*, August 11, 1895. Cartoons of black officeholders tended to show them in exaggerated spiked coattails that almost reached the floor and beaver hats that almost touched the ceiling. This imagery is part of the racist caricature used to stereotype the fancy dress of the black legislators during Reconstruction.
4. Wager Swayne to C. C. Sibley, August 24, 1867, Freedmen's Bureau Papers, Record Group 105, National Archives Microfilm 798, roll 19, frame 0675; Lee. W. Formwalt to Thomas L. French Jr., January 24, 1983; C. C. Sibley to Wager Swayne, August 29, 1867, Freedmen's Bureau Papers, Record Group 105, National Archives Microfilm 798, roll 6, frame 0157.
5. Quoted in Cimbala, *Under the Guardianship of the Nation*, 166, 176; Cimbala, "A Black Colony," 72–89.
6. Cimbala, *Under the Guardianship of the Nation*, 186–87.
7. Cash, "Alabama Republicans during Reconstruction," 85.
8. *Sun*, February 8, 18, 1868; Wiggins, *The Scalawag in Alabama Politics*, 36–39.
9. Columbus *Sun*, May 5, 1868. The editors of the *Columbus Enquirer* applied similar phrases to Horace when they discussed the rebuilding of City Mills the next year. See *Columbus Enquirer*, July 7, 1869.
10. Jemison to Horace King, June 23, 1868, Letterbook 12, Jemison Collection, UA; Rogers, Ward, Atkins, and Flynt, *Alabama*, 242.
11. A search of the House journal shows his name appearing in four roll call votes. These almost certainly represent errors by the clerk because on all four days when his name appeared, there were multiple votes, and his name only appeared on one of them.
12. *Journal of the House of Representatives during the Sessions Commencing in July, September, and November 1868*, 24, 266–67; T. L. Appleby to W. H. Smith, November 11, 1869, Smith Papers, Alabama Department of Archives and History, Montgomery, cited in Cash, "Alabama Republicans during Reconstruction," 248.
13. Information about the race and party of legislators is drawn from Wiggins, *The Scalawag in Alabama Politics*, 148–49; information about the number of bills introduced comes from the index of the *Journal of the House of Representatives, 1869–70*.

14. *Journal of the House of Representatives, 1869–1870,* 377, 396, 437, 543.
15. Information about the election results is drawn from returns cited in the *Enquirer,* November 10, 12, 1870, and the *Sun,* November 10, 12, 1870.
16. *Sun,* November 15, 1870.
17. Coulter, *"A People Courageous,"* 209.
18. The Albany bridge was restricted to the same tolls as the Columbus one, which provided the basis for Joiner's attack. Joiner was also a target of attacks by whites during the Camilla Riot in 1868. O'Donovan, "Philip Joiner," 56–71.
19. Bailey, *Neither Carpetbaggers nor Scalawags,* 67, 80; Loren Schweniger, "James T. Rapier."
20. Bailey, *They Too Call Alabama Home,* 230.
21. Grimshaw, *Official History of Freemasonry,* 189–213, 288–90; *Formation and Proceedings of the M. W. Grand Lodge of the Most Ancient and Honorable Fraternity of Free and Accepted Masons for the State of Alabama* (Mobile: Thompson and Powers, 1871), 19. Available at the Prince Hall Grand Lodge in Birmingham. The origins of the Prince Hall Masons are discussed in chapter 3.

CHAPTER FOURTEEN. The Kings in LaGrange

1. This chapter is based in large part on the work of F. Clark Johnson and Kaye Lanning Minchew of the Troup County Archives, especially on Kaye Minchew's draft of her biographical sketch of John Thomas King prepared for *African-American Architects.*
2. *Columbus Daily Enquirer-Sun,* May 30, 1885; *Atlanta Constitution,* May 30, 1885; *LaGrange Reporter,* June 4, 1885. The newspaper and the court cited Robert A. Forsyth as the defendant in the case and provide no more details about the litigation. No Robert A. Forsyth lived in Columbus or Girard according to the *City Directory,* 1873–74. A Robert E. Forsyth, according to the 1870 census, was a wagoner in Girard—an appropriate occupation for someone having an economic dispute with Horace. *Columbus Sun,* May 18, 1873; Muscogee County Superior Court Minute Book O, 138.
3. Lane, Bridges, and Jones, "The King Family," 3. The railroad at West Point would have been a logical destination for cotton bales. The rapids between West Point and Columbus would preclude transporting valuable cargo along that stretch of river. *LaGrange Reporter,* April 21, 1887. This account, written after Horace's death, used the story to argue for making the river navigable from Atlanta to LaGrange.
4. The description of the parcels' metes and bounds included sassafras and mulberry stakes and the river as a southern boundary. Carroll County Deed Book L, 32. The other two stockholders, James Moore and Charles Mabry, certified the transfer of Horace's stock. Carroll County Deed Book M, 130–31.
5. Whitesburg was formed by the crossroads of Five Notch Road, which followed the top of the ridge on the northwest side of the river, and Moore's Ferry Road coming from Moore's Bridge.
6. Georgia, vol. 5, p. 124–32, R. G. Dun and Co. Collection, Baker Library, Harvard Business School. Other businesses that started before 1872 (and some launched af-

ter that date) survived. The general stores that avoided embarrassment seemed to be branches of firms from larger towns.

7. Carroll County Deed Book N, 651. The county purchased the right to a bridge at this location or maybe just the site. Whether there was a bridge at that date is uncertain. *Acts and Resolutions of the General Assembly of the State of Georgia*, 1878–79, 338.
8. *LaGrange Reporter*, May 3, 1867; November 27, December 4, 1868; October 15, 1869; January 7, August 5, 1870; August 8, October 13, 1871; January 5, March 29, 1872; May 10, January 11, March 22, 29, 1877; October 31, 1878; June 12, 1879; January 13, October 13, April 7, 1881; February 1, August 9, September 13, and December 6, 1883; April 10, May 22, and August 21, 1884; September 10 and 17, 1885; December 12, 1890; and March 17, 1899; *Acts and Resolutions of the General Assembly of the State of Georgia*, 1870, 347 and 1872, 189.

   Mabry, who died in 1884, also organized a building and loan association, served as an incorporator of the North and South (or Columbus and Rome) Railroad in 1870 and the LaGrange and Barnesville Railroad in 1872, remained active within the Democratic Party, and encouraged agricultural diversity. His wife, the heir of several fortunes, operated a hosiery mill with women operatives and gained regional notice for her products. She also operated a creamery and urged farm wives to raise silk worms and grow strawberries. She died in 1899.
9. *LaGrange Reporter*, October 13, 1871.
10. Forrest Clark Johnson, "LaGrange History Speech," presented to the Georgia Trust for Historic Preservation, 1, available Troup County Archives; Johnson, *History of LaGrange*, 49, 74.
11. Johnson, *History of LaGrange*, 78–80. Although it would not have had an impact on King's decision, the impression that LaGrange was more dynamic stems from the textile history of the two communities, the yardstick of industrial progress in the South. Columbus was an older established mill town with significant economic development in the antebellum period because of its waterpower. The important textile mills in LaGrange more nearly resembled the stereotypical view of the South, with major mills funded by local capital starting in the 1880s. Thus, in a general sense, Columbus's infrastructure existed by the early 1870s; LaGrange's major facilities were just being built.
12. Major and Johnson, *Treasures of Troup County*, 76–77; *LaGrange Reporter*, April 18, 1873.
13. He must not have supplied the lumber for this span. *LaGrange Reporter*, May 23, June 20, 1873.
14. *LaGrange Reporter*, February 5, 1869. Reprinted in Chattahoochee Valley Historical Society, *Valley Historical Scrapbook*, CVHS Publication no. 9, September 1970, 3–4. Virginia Smith, "Chronology: From Covered Bridge of 1835 to John C. Barrow Bridge of 1977," *Columbus Ledger-Enquirer*, East Alabama Today, October 13, 1977. *LaGrange Reporter*, November 28, 1873; Janie Lovelace Heard, "Public Schools," 6–8, MS, Janie Lovelace Heard Collection (F 30), Cobb Memorial Archives, Valley, Ala.
15. Troup County Deed Book Q, 220.
16. George King to John T. King, March 28, 1888, Troup County Deed Book X, 167.
17. *Montgomery Daily Advertiser*, August 9, 1876. A careful search of newspapers out-

side of the Chattahoochee Valley may expand the geographical area in which he worked during the twilight of his career.

18. Troup County Deed Book R, 57; *LaGrange Reporter*, January 22, 1869; December 12, 1873; August 30, 1877; August 31, 1882; October 7, November 18, 1886. At that time, LaGrange lacked a town hall; the council rented meeting space from local businessmen.
19. Troup County Deed Book M, 101, R, 57.
20. Georgia, vol. 34, p. 18, R. G. Dun and Co. Collection, Baker Library, Harvard Business School.
21. *LaGrange Reporter*, December 18, 1877. A *LaGrange Reporter* envelope with $10 and $100 Confederate notes headed the list of items entombed, which also included ancient coins, local rocks, and contemporary information about the city and the college.
22. The college actually split into two, with Cox moving to College Park and establishing an institution bearing his own name. Major and Johnson, *Treasures of Troup County*, 33; Southern Female College–Cox College Archives, Special Collections, University of Georgia; Southern Female College Records, 1893–1937, Manuscripts, Georgia Department of Archives and History, Morrow.
23. Major and Johnson, *Treasures of Troup County*, 32.
24. Michael Cahalon vs. Horace King, Troup County Superior Court Minutes, December 12, 1873, Troup County Archives; Deposition of B. Atkinson from Carroll County and other records, Troup County Court Records, Box 97, November 1875; Troup County Superior Court Records, June 6, 1876, June 1, 1877; Major and Johnson, *Treasures of Troup County*, 32. In 1873 King had also provided no issuable defense in another minor case and was charged $372.19.
25. *LaGrange Reporter*, October 3, 1878.
26. *LaGrange Reporter*, April 3, 1879, August 7, November 21, 1877; *Hamilton Visitor*, February 26, 1880.
27. The family information in the following paragraphs comes from a handwritten account by Theodora Thomas and from the King family Bible, photocopies in Troup County Archives; several of the dates in the Bible appear to be in error. The 1880 census for Troup County listed Horace's wife as Sallie J.
28. *LaGrange Reporter*, September 25, 1879. His half-acre tract cost $60. Troup County Deed Book S, 613. Minutes of LaGrange City Council, March 7, 1888, Troup County Archives. Marshal's widow, Madaline Harrison, married George White in 1882; their son Walter White was a founder and longtime president of the NAACP.
29. Kenzer, *Enterprising Southerners*, 14, 34–66.
30. Sarah Jane's name only appears in the Troup County Deed Books as part of the joint transactions with her stepchildren. Exactly when Horace or his family sold the Girard property (and when Clarissa or Henry Murray sold the adjacent land) remains a mystery. Because of a conflict over where to place the county seat, which actually involved various factions moving records from one courthouse to another in the middle of the night, some early twentieth-century volumes of Russell County real estate records are missing. According to the oral tradition preserved by Theodora Thomas, when the family first arrived in LaGrange they resided in what is now a bed and breakfast on the south side of Greenville Street, the most imposing struc-

ture on the street. Reference to "where George and John now live" in Troup County Deed Book W, 228. George sold twenty-two acres to Dixie Mills in 1895 for $4,000; John's son, Horace H., sold a twenty-eight-by-forty foot lot to the mill in 1917 for $3,000. That property must have included a house, and it could have been the old Bean Place; Troup County Deed Books, Y, 384; and 18, 241. The last King family resident of LaGrange, Theodora Thomas, lived at 508 Greenville Street.

31. *LaGrange Reporter*, April 21, 1887, September 1, 1881.
32. *LaGrange Reporter*, April 21, 1887, August 4, 1881, July 20, 1882, May 8, 1884. The courtesy title "Mr." indicated the race of George Forbes; the absence of one showed George to be a black man.
33. The Mooty Bridge project consumed seventy thousand board feet of lumber.
34. *Columbus Enquirer-Sun*, May 30, 1885.
35. *Atlanta Constitution*, May 30, 1885; *LaGrange Reporter*, June 4, 1885. King's saw blade is discussed in chapter 11.
36. Account of Theodore Thomas told to Johnson. Johnson, "Horace King," 26. Theodora Thomas identified their graves in 1978 for Susie Fowler, whose efforts with the local historical society and the Troup County Archives resulted in the graves' being marked. Fowler documents her work in "Slave Earned His Freedom, Gained Fame as Builder," *LaGrange Daily News*, July 12, 1978, 7.

CHAPTER FIFTEEN. The King Legacy

1. They agreed to complete the crossing in four months according to their plan for $11 per foot, or according to the county's plan, which involved more lumber, for $12 per foot. Since the county made a $400 payment to W. W. King on October 31, 1883, he must have completed the project near the scheduled date. See Bartow County Commissioners Court Minutes, Book B, 174–75; Minutes of Circuit Court, Book B, 198.
2. French and French, *Covered Bridges of Georgia*, 23–24, 27–28, 43–46, 62, 64, 72.
3. Warren Temple remained affiliated with the northern Methodist Church and should not be associated with the CME churches, which were black congregations organized by the Methodist Episcopal Church South.
4. "History of the [Warren Temple] ME [Methodist Episcopal] Church," MS, [1923], Horace King Collection, Columbus Museum, 5, 7–8, 11, 16, 18, 22, 24, and 41. The church had lost all its records, and in 1923 John King chaired a committee that compiled this history based primarily on the quarterly records submitted to the district superintendent. This copy remained in the Kings' house and came with the items from Theodora Thomas's estate to the Columbus Museum.
5. Copy of a letter by unknown correspondent to Dr. P. J. Maveety, May 14, 1925, Freedmen's Aid Society Records, Atlanta University Center Archives; *LaGrange Reporter*, August 16 and October 25, 1877; Minutes of LaGrange City Council, September 12, 1876 and April 4, 1877; interviews with Theodora Thomas, May and June 1878; Johnson, *History of LaGrange*, 94. The church transferred the school property to the national Methodist Church Aid Society in 1878. It became a public school in 1903. "History of the [Warren Temple] ME Church," 91, 24, 91–93, and 101.
6. Johnson, *History of LaGrange*, 94–95.

7. *LaGrange Reporter*, September 18, 1884, January 31, 1884, June 30, 1887.
8. Clement, "Bridges," 2; Moneeck Jackson, "History's Lost and Found: The First Black Warrior Bridge Has Been Hidden by Vegetation," *Tuscaloosa News*, December 21, 2000; "King Iron Bridge Company," http://www.clevelandmemory.org/king/, the Cleveland Memory Project, Cleveland State University; *LaGrange Reporter*, June 3, July 22, 1886. The two major postwar wooden wagon bridges at Columbus lasted an average of forty years. Spans at West Point seemed to be more vulnerable. The 1919 flood coincided with the visit of Gen. John Pershing to Fort Benning and was thus labeled as his flood.
9. *LaGrange Reporter*, April 12, 1885.
10. Troup County Deed Book Q, 220; interviews with H. H. King by Thomas L. French Jr. in 1974, 1982, 1988.
11. *LaGrange Reporter*, March 18, 25, June 3, July 22, August 5, 1886.
12. *LaGrange Reporter*, July 7, September 29, 1887, January 23, 1890, January 19, February 9, March 12, July 12, August 9, December 6, 1888, May 2, 1889. The July 1888 article identified Mr. Borders as the contractor for most of the stonework.
13. *LaGrange Reporter*, March 26, 1888.
14. *LaGrange Reporter*, May 28, 1888; Troup County Deed Books, Y, 384.
15. *LaGrange Reporter*, March 28, 1890, April 8, 1886; Pinkston, *Historical Accounts of Meriwether County*, Book 5, 21.
16. *LaGrange Reporter*, January 23, March 28, November 21, 1890.
17. H. M. Park, "Troup County's 'Bridge of Folly,'" *LaGrange Reporter*, October 2, September 25, 1896; quote from *LaGrange Reporter*, October 2.
18. Mrs. Mary E. Glass was the source of this information; she also identified the bridge's architect as W. H. Armstrong. Tom Swint, "Glass Bridge Is Falling Down and with It, a Georgia Era," *Atlanta Journal*, March 14, 1955, reprinted in *Valley Daily Times-News*, March 18, 1955.
19. Theodora Thomas, "Horace Henry King," MS, Troup County Archives, and Horace King Collection, Columbus Museum; quote from *LaGrange Reporter*, August 20, 1897.
20. R. G. Dun and Co., *Georgia* (January 1909), Frank Schnell Collection, CSU Archives. King's credit listing as a contractor was K, 3½. The ratings were high, good, fair, and limited; the pecuniary strength had to reach a certain level before a ranking of high or good was possible. *LaGrange Graphic*, November 12, 1926, March 13, 1887; letter to "Sec'arys of the Freedm'aid Scty," November 16, 1921, Freedmen's Aid Society Records, Atlanta University Center Archives.
21. The church was scheduled to be completed by February or March 1903. See *LaGrange Reporter*, December 12, 1902.
22. In the early 1920s, as chair of the local board of trustees, King corresponded with the Freedman's Aid Society about the old school property and the city's assessment of paving fees for church property. Approximately twenty-five photographs relating to this school, apparently when John's daughter Olive Alice was a student there, can be found in the King Collection at the Columbus Museum. Peter A. Brannon, the longtime head of the Alabama Department of Archives and History and a Russell County native, wrote that one of King's children or grandchildren was married to the head of this school, which he called Cowalige; she must have been a grand-

child because she was not Horace's only daughter, Annie Elizabeth; Brannon to William J. Barr, February 9, 1938, Brannon Collection, ADAH. A postcard from the West Virginia Collegiate Institute in the Columbus Museum collection and the fact that a U.S. Representative from Institute, W. Va., introduced a bill to provide the King family with compensation for their losses during the Civil War seem to indicate that a family member taught at the school.

23. *Bulletin of Atlanta University*, 1895, 3; Cooper, *Cotton States*, 60.
24. *New York Observer*, June 27, 1895.
25. A search of all Washington's published letters and writings produced no mention of any of the Kings.
26. *LaGrange Daily News*, November 26, 1926, September 8, 1949.

CHAPTER SIXTEEN. The Legend of Horace King

1. In Columbus his lower railroad bridge was replaced with an iron and steel truss in the 1890s; the Fourteenth Street Bridge succumbed to a flood in 1902 and was superceded by an iron and steel through-truss; and a concrete span replaced the Dillingham Bridge in 1910–11.
2. The introduction in Alabama and other southern states of laws to enslave free blacks in 1859 is discussed in chapter 10.
3. Cherry, "History of Opelika," 193–97. Woodall claimed that Columbusites invented both the world's first breech-loading cannon and ice machine. Also, according to him, the first use of electricity to illuminate a mill occurred in Columbus. The following piece, published in Woodall's magazine, is an example of his Horace King articles, "Monument Built to Former Master by His Ex-Slave, Who Was Widely Known as a Builder of Bridges," *Columbus Magazine*, December 16, 1940, 17, 31. "Ripley's" is mentioned in W. O. Langley's discussion of King in Coulter, *"A People Courageous,"* 122–23.
4. Starobin, *Industrial Slavery*, 30, 107–8, 171–72, 173, 317.
5. For example, the White Oak and Red Oak Bridges in Meriwether County, Georgia, are doubtful, and he did not build the original bridge at Tuscaloosa, as is popularly believed.
6. Susie Fowler, "Slave Earned His Freedom, Gained Fame as Builder," *LaGrange Daily News*, July 12, 1978, 7.
7. The speech given by Bill Green on this occasion is recorded in Charles Tigner, "A Master Bridgebuilder," Russell Remembrances column, *Phenix Citizen*, May 4, 1979, 4.
8. The *Washington Star-News* article was accompanied by a picture, identified as Horace King, of a dark-complexioned man with negroid features; George Reasons and Sam Patrick, "Slave a Frontier Bridge Builder," *Washington Star-News*, September 23, 1972.
9. The following are examples of local articles about King: Sandra Blackman, "Ex-Slave Was Top Bridge-Builder," Special Sesquicentennial Supplement, *Columbus Ledger-Enquirer*, April 16, 1978, s-9; Charles Tigner, "A Master Bridgebuilder," Russell Remembrances column, *Phenix Citizen*, May 4, 1979, 4; Carolyn Danforth, "Noted Builder of Wooden Bridge Recalled," *Opelika-Auburn News*, June 27, 1982, c-12.

Eufaula and Montgomery, Alabama, featured Horace during Black History Month in 2002 and 2003. Newspapers from outside Horace's home area that have run articles about him include Patrick Johnston, "Black History Month Profile: Horace King; King Bridged Gap of Slavery, Freedom," *Eufaula Tribune*, February 17, 2002; Ben Windham, "Who Was Horace King?" West Alabama section, *Tuscaloosa News*, January 19, 2003.

10. Scott K. Williams, "Black Confederates Fact Page," at http://www.geocities.com/11thkentucky/blackconfed.htm. Hard copy in possession of the authors.
11. Ironically, Lenard's unproductive research in Ohio cast doubts on the family's account of Horace's freedom and his education. Lenard found the notation of Cherry's appointment with Horace in the Reverend's diary. Lenard's film is discussed in Bill Osinski, "Horace King," *Atlanta Journal-Constitution*, Sunday, December 7, 1997, F-3; this article is posted on the Troup County Archives Web site.
12. Priscilla Black Duncan, "King Deserves Remembrance" *Columbus Ledger-Enquirer*, February 19, 1988; Chris McCarter, "City Renames 14th Street Bridge for Former Slave," *Columbus Ledger Enquirer*, April 14, 1988, D-5.
13. Mayor Arnold D. Leak, "Horace King Memorial Bridge Project," City of Valley, Ala., Press Release, October 31, 2002.
14. Green's speech quoted in Tigner, "A Master Bridgebuilder," 4.

BIBLIOGRAPHY

This bibliography lists secondary sources only. Information on primary sources appears in the notes.

Allen, Richard Sanders. *Covered Bridges of the South.* New York: Bonanza, 1970.

Arfwedson, C. D. *The United States and Canada in 1832, 1833, and 1834.* London: R. Bentley, 1834.

Bailey, Richard. *Neither Carpetbaggers nor Scalawags: Black Officeholders during the Reconstruction of Alabama, 1867–1878.* 2nd ed. Montgomery, Ala.: R. Bailey, 1993.

———. *They Too Call Alabama Home: African American Profiles, 1800–1999.* Montgomery: Pyramid, 1999.

Berlin, Ira. *Slaves without Masters: The Free Negro in the Antebellum South.* New York: Oxford University Press, 1972.

Bogle, James G. "Horace King, 1807–1887." *Georgia Life* (spring 1980): 33–35.

Boucher, Morris Raymond. "The Free Negro in Alabama prior to 1860." Master's thesis, State University of Iowa, 1950.

Brannen, Ralph N. "John Gill Shorter: War Governor of Alabama, 1861–1863." Master's thesis, Alabama Polytechnic Institute [Auburn], 1956.

Brewer, Willis. *Alabama: Her History, Resources, War Record, and Public Men.* Spartanburg, S.C.: Reprint Company, 1975.

Brown, David J. *Bridges.* New York: Macmillan, 1993.

Calhoon, Margaret Obear. "Building the South: The Enterprise of William C. and John D. Gray." Ph.D. diss., Georgia State University, 2001.

Cash, William McKinley. "Alabama Republicans during Reconstruction: Personal Characteristics, Motivations, and Political Activity of Party Activists, 1867–1880." Ph.D. diss., University of Alabama, 1973.

Chattahoochee Valley Historical Society. *Valley Historical Scrapbook.* West Point, Ga.: Hester's Printing, 1970.

Cherry, F. L. ["Okossee"]. "The History of Opelika and Her Agricultural Tributary Territory, Embracing More Particularly Lee and Russell Counties, from the Earliest Settlement to the Present Date." *Alabama Historical Quarterly* 15, no. 2 (1953): 178–339, 15, nos. 3 and 4 (1953): 383–537. Reprinted from supplement to the *Opelika Times,* October 5, 1883–April 22, 1885.

Cimbala, Paul A. "A Black Colony in Dougherty County: The Freedmen's Bureau and the Failure of Reconstruction in Southwest Georgia." *Journal of Southwest Georgia History* 4 (fall 1986): 72–89.

———. *Under the Guardianship of the Nation: The Freedmen's Bureau and the Reconstruction of Georgia, 1865–1870.* Athens: University of Georgia Press, 1997.

Clement, Dan. "Bridges in the Upper Tombigbee River Valley." *Historic American Engineering Record* MS-11 (1983).

Coleman, James Edgar. *The Bluff at Fort Gaines, Georgia.* [Fort Gaines, Ga.:] J. E. Coleman, 1998.

Condit, Carl W. *American Building Art: The Nineteenth Century.* New York: Oxford, 1960.

Cooper, Walter G. *The Cotton States and International Exposition and South, Illustrated.* Atlanta: Illustrator, 1896.

Coss, Richard H. "On the Trail of Jim Henry." *Muscogiana* 3, nos. 3 and 4 (fall 1992).

Coulter, Harold S. *"A People Courageous": A History of Phenix City, Alabama.* Columbus, Ga.: Howard Printing, 1976.

Daniels, George H. "Business, Industry, and Politics in the Antebellum South: The View from Tuscaloosa, Alabama." In *Proceedings of the 150th Anniversary Symposium on Technology and Society: Southern Technology: Past, Present, and Future,* ed. Howard L. Hartman, 31–57. College of Engineering, University of Alabama, 1988.

———. "Entrepreneurship in the Old South: The Career of Robert Jemison, Jr., of Alabama." In *Looking South: Chapters in the Story of an American Region,* ed. Winfred B. Moore and Joseph F. Tripp, 153–69. New York: Greenwood, 1989.

Danko, George Michael. "The Evolution of the Simple Truss Bridge 1790 to 1850: From Empiricism to Scientific Construction." Ph.D. diss., University of Pennsylvania, 1979.

Davidson, William H. *Proudest Inheritance: A Bicentennial Tribute of the Chattahoochee Valley Historical Society.* Chattahoochee Valley Historical Society, 1975.

Davis, Edwin Adams, and William Ransom Hogan. *The Barber of Natchez.* Baton Rouge: Louisiana State University Press, 1954.

DeCredico, Mary A. *Patriotism for Profit: Georgia's Urban Entrepreneurs and the Confederate War Effort.* Chapel Hill: University of North Carolina, 1990.

Defebaugh, James Elliott. *History of the Lumber Industry of America.* Chicago: American Lumberman, 1906–7.

deTreville, John Richard. "The Little New South: Origins of Industry in Georgia's Fall-Line Cities, 1840–1865." Ph.D. diss., University of North Carolina, 1986.

A Digest of the Laws of the State of Alabama. Philadelphia: A. Towar, 1833.

A Digest of the Laws of the State of Georgia. Philadelphia, 1800.

Doster, Gary L. "The Florence Bridge Company of Georgia and Its Snygraphic Relics." *Paper Money* 17, no. 6 (1978): 332–33.

Dunaway, Wilma A. *The African-American Family in Slavery and Emancipation.* New York: Cambridge University Press, 2003.

Edgar, Walter B. *South Carolina: A History.* Columbia: University of South Carolina Press, 1998.

Edwards, Llewellyn N. "The Evolution of Early American Bridges." *Maine Technology Experiment Station, University of Maine,* paper no. 15, October 1934. Reprinted from *Transactions of the Newcomen Society,* vol. 13.

Edwards, Stewart C. "'To do the manufacturing for the South': Private Industry in Confederate Columbus." *Georgia Historical Quarterly* 85, no. 4 (winter 2001): 538–54.

Egerton, Douglas R. *He Shall Go Out Free: The Lives of Denmark Vesey.* American Profiles Series. Madison, Wisc.: Madison House, 1999.

Evans, David. *Sherman's Horsemen: Union Cavalry Operations in the Atlanta Campaign.* Bloomington: Indiana University Press, 1996.

Fitzgerald, Michael W. "The Union League Movement in Alabama and Mississippi: Politics and Agricultural Change in the Deep South during Reconstruction." Ph.D. diss., University of California, Los Angeles, 1986.

Fletcher, Robert Samuel. *A History of Oberlin College, from Its Foundation through the Civil War.* Vols. 1 and 2. Oberlin: Oberlin College, 1943; reprint, New York: Arno, 1971.

Flewellen, Robert H. *Along Broad Street: A History of Eufaula, Alabama, 1823–1984.* Eufaula: City of Eufaula, 1991.

Foote, Shelby. *The Civil War, A Narrative: Red River to Appomattox.* New York: Random House, 1974.

Formwalt, Lee W. "Moving in That Strange Land of Shadows, African American Mobility and Persistence in Post–Civil War Southwest Georgia." *Georgia Historical Quarterly* 82, no. 3 (fall 1998).

French, Thomas L., Jr., and Edward L. French. *Covered Bridges of Georgia.* Columbus, Ga., Frenco, 1984.

———. "Horace King, Bridge Builder." *Alabama Heritage* (winter 1989): 33–47.

Fretwell, Mark E. *West Point: The Story of a Georgia Town.* West Point, Ga.: Chattahoochee Valley Historical Society, 1987.

Gagnon, Michael John. "Transition to an Industrial South: Athens, Georgia, 1830–1870." Ph.D. diss., Emory University, 1999.

Galer, Mary Jane. *Columbus, Georgia: Lists of People in the Town, 1828–1852, and Sexton's Reports to 1866.* Columbus: Iberian, 2000.

Gamble, Robert S., and Thomas W. Dolan. *The Alabama State Capitol: Architectural History of the Capitol Interiors.* [Montgomery, Ala.:] Alabama Historical Commission, 1984.

Gerber, David A. *Black Ohio and the Color Line, 1860–1915.* Chicago: University of Illinois Press, 1976.

Gibbons, Faye. *Horace King: Bridges to Freedom.* Birmingham: Crane Hill, 2002.

Gies, Joseph. *Bridges and Men.* Garden City, N.Y.: Doubleday, 1963.

Gillespie, Michele. *Free Labor in an Unfree World, White Artisans in Slaveholding Georgia, 1789–1860.* Athens: University of Georgia, 2000.

Golden, Virginia Noble. *A History of Tallassee for Tallasseeans.* [Tallassee:] Tallassee Mills, 1949.

Goldin, Claudia Dale. *Urban Slavery in the American South, 1820–1860, A Quantitative History.* Chicago: University of Chicago Press, 1976.

Green, Michael D. *The Politics of Indian Removal: Creek Government and Society in Crisis.* Lincoln: University of Nebraska Press, 1982.

Gregg, Alexander. *History of the Old Cheraws: Containing an Account of the Aborigines of the Pedee. . . .* Greenville, S.C.: Southern Historical Press, 1991.

Grimshaw, William H. *Official History of Freemasonry among the Colored People in North America.* New York: Broadway, 1903.

Hall, Basil. *Travels in North America in the Years 1827 and 1828.* London: Simpkins and Marshall, 1830.

Hartman, Howard L., ed. *Proceedings of the 150th Anniversary Symposium on Technology and Society: Southern Technology: Past, Present, and Future.* College of Engineering, University of Alabama, 1988.

Heath, Milton Sydney. *Constructive Liberalism, The Role of the State in Economic Development in Georgia to 1860.* Cambridge: Harvard University Press, 1954.

Hindle, Brooke. "Introduction: The Span of the Wooden Age." In *America's Wooden Age: Aspects of Its Early Technology,* ed. Hindle, 3–12. Tarrytown, N.Y.: Sleepy Hollow Restorations, 1975.

The History of Clay County, Georgia. Ed. Mrs. Donald (Priscilla Neves) Todd. [Fort Gaines, Ga.: Clay County Library Board, 1976].

"Hon. Arthur Yonge." *A Twentieth-Century History and Biographical Record of North and West Texas.* Vol. 2. New York: Lewis, 1906.

Hopkins, H. J. *A Span of Bridges: An Illustrated History.* New York: Praeger, 1970.

Horace: The Bridge Builder King. Prod. and dir. Tom C. Lenard. Auburn, Ala.: AU Telecom/ETV, 1996.

Hyde, Samuel C., Jr., ed. *Plain Folk of the South Revisited.* Baton Rouge: Louisiana State University Press, 1997.

Jackson, Harvey H., III. *Rivers of History: Life on the Coosa, Tallapoosa, Cahaba, and Alabama.* Tuscaloosa: University of Alabama Press, 1995.

Jeane, D. Gregory, ed. *The Architectural Legacy of the Lower Chattahoochee Valley in Alabama and Georgia.* University: Published for the Historic Chattahoochee Commission by the University of Alabama Press, 1978.

"The John Godwin Bible." *Tap Roots—The Genealogical Society of East Alabama Quarterly* 2, no. 3 (January 1965).

Johnson, Forrest Clark, III. *A History of LaGrange, Georgia, 1828–1900.* Vol. 1, *Histories of LaGrange and Troup County, Georgia.* LaGrange, Ga.: Family Tree, 1987.

———. "Horace King." *Troup County Georgia and Her People* 1, no. 1 (October 1981): 24–26.

Johnson, Michael P., and James L. Roark. *Black Masters: A Free Family of Color in the Old South.* New York: W. W. Norton, 1984.

Jones, Jacqueline. *Soldiers of Light and Love: Northern Teachers and Georgia Blacks, 1865–1873.* Chapel Hill: University of North Carolina Press, 1980.

Jones, James Pickett. *Yankee Blitzkrieg: Wilson's Raid through Alabama and Georgia.* Athens: University of Georgia Press, 1976.

Kenzer, Robert C. *Enterprising Southerners: Black Economic Success in North Carolina, 1865–1915.* Charlottesville: University Press of Virginia, 1997.

Kilbourne, Elizabeth Evans. *Columbus, Georgia, Newspaper Clippings* (Columbus Enquirer). Savannah, Ga.: E. E. Kilbourne, 2000.

Kimmelman, Barbara, John S. Lupold, and J. B. Karfunkle, "City Mills." *Historical American Engineering Record* GA-25 (summer 1977).

Kletzing, H. F., and W. H. Crogman. *Progress of a Race; or, The Remarkable Advancement of the Afro-American Negro. . . .* With an introduction by Booker T. Washington. Atlanta: J. L. Nichols, 1898.

Kolchin, Peter. *First Freedom: The Responses of Alabama's Blacks to Emancipation and Reconstruction* Westport, Conn.: Greenwood, 1972.

Kranakis, Eda. *Constructing a Bridge: An Exploration of Engineering Culture, Design, and Research in Nineteenth-Century France and America.* Cambridge, Mass.: MIT Press, 1997.

Lane, Mills. *The Rambler in Georgia.* Savannah: Beehive, 1973.

Lane, Sarah, Lucile Bridges, and J. M. Jones. "The King Family." In *The American Slave:*

A Composite Autobiography, ed. George P. Rawick, 368–71. Supp., ser. 1, vol. 4, *Georgia Narratives*, pt. 2. Westport, Conn.: Greenwood, 1977.

Lawrence-McIntyre, Charshee Charlotte. "Free Blacks: A Troublesome and Dangerous Population in Antebellum America." Ph.D. diss., State University of New York at Stony Brook, 1985.

Ledbetter, R. Jerald, and Chad O. Braley. *Archeological and Historical Investigations at Florence Marina State Park, Walter F. George Reservoir, Stewart County, Georgia*. Athens: Southeastern Archeological Services, 1989.

Linder, Suzanne C. *Medicine in Marlboro County, 1736–1980*. Baltimore: Gateway, 1980.

Lindsay, Bobby L. *The Reasons for the Tears: A History of Chambers County, Alabama, 1832–1900*. West Point, Ga.: Hester Printing, 1971.

Lipscomb, W. L. *A History of Columbus, Mississippi, during the Nineteenth Century*. Birmingham, Ala.: Dispatch Printing, 1909.

Love, Edgar F. "Registration of Free Blacks in Ohio: The Slaves of George C. Mendenhall." *Journal of Negro History* 69 (winter 1984): 38–74.

Lupold, John S. "The Industrial Reconstruction of Columbus, Georgia, 1865–1881." Paper read at the Georgia Historical Society Meeting, Columbus, Georgia, October 1975.

Lupold, John S., and Thomas L. French Jr. "Horace King." In *African-American Architects: A Biographical Dictionary, 1865–1945*, ed. Dreck Spurlock Wilson. New York: Routledge, 2003.

Lupold, John S., J. B. Karfunkle, and Barbara Kimmelman. "Eagle and Phenix." *Historical American Engineering Record* GA-20 (summer 1977).

Macmillan Encyclopedia of Architects. London; New York: Macmillan, 1982.

Major, Glenda, and Forrest Clark Johnson III. *Treasures of Troup County: A Pictorial History*. LaGrange, Ga.: Troup County Historical Society, 1993.

Martin, John H. *Columbus, Georgia, from Its Selection as a 'Trading Town' in 1827, to Its Partial Destruction by Wilson's Raid in 1865*. Columbus: Thomas Gilbert, 1874.

McFeely, William S. *Frederick Douglass*. New York: W. W. Norton, 1991.

Medley, Mary L. *History of Anson County, North Carolina, 1750–1976*. Wadesboro, N.C.: Anson County Historical Society, 1976.

Meigs, Paul Avery. "The Life of Senator Robert Jemison, Junior." Master's thesis, University of Alabama, 1928.

Mellown, Robert O. "Bryce Hospital Historic Structures Report." [Tuscaloosa, Ala.: Heritage Commission of Tuscaloosa County], 1990.

———. "The Construction of the Alabama Insane Hospital, 1852–1861." *Alabama Review* 38, no. 2 (April 1985): 83–104.

Merrell, James H. *The Indians' New World: Catawbas and Their Neighbors from European Contact through the Era of Removal*. Chapel Hill: University of North Carolina Press and the Institute of Early American History and Culture, 1989.

Miller, Randall M., et al. *The World of Daniel Pratt: Essays on Industry, Politics, Art, Architecture, Reform, and Town-Building in Alabama*. Montgomery, Ala.: Black Belt, 1999.

Mills, Gary B. "Shades of Ambiguity: Comparing Free People of Color in 'Anglo' Alabama and 'Latin' Louisiana." In *Plain Folk of the South Revisited*, ed. Samuel C. Hyde Jr., 161–86. Baton Rouge: Louisiana State University Press, 1997.

Mills, Robert. *Atlas of the State of South Carolina*. Columbia, S.C.: N.p., 1825.

Minchew, Kaye Lanning. "John Thomas King," In *African-American Architects: A Biographical Dictionary, 1865–1945*, ed. Dreck Spurlock Wilson. New York: Routledge, 2003.

Moore, Winifred B., and Joseph F. Tripp, eds. *Looking South: Chapters in the Story of an American Region.* New York: Greenwood, 1989.

Mueller, Edward A. *Perilous Journeys: A History of Steamboating on the Chattahoochee, Apalachicola, and Flint Rivers, 1828–1928.* Eufaula, Ala.: Historic Chattahoochee Commission, 1990.

Myers, John B. "Human Capital: The Freedmen and the Reconstruction of Labor in Alabama, 1860–1880." Ph.D. diss., Florida State University, 1974.

Newton, Roger Hale. *Town and Davis Architects, Pioneers in American Revivalist Architecture, 1812–1870.* New York: Columbia University Press, 1942.

O'Donovan, Susan E. "Philip Joiner, Southwest Georgia Black Republican." *Journal of Southwest Georgia History* 4 (fall 1986): 56–71.

———. "Transforming Work: Slavery, Free Labor, and the Household in Southwest Georgia, 1850–1880." Ph.D. diss., University of California, San Diego, 1997.

Olmsted, Frederick Law. *The Cotton Kingdom.* New York: Modern Library, 1984.

Pease, George B. "Timothy Palmer, Bridge-Builder of the Eighteenth Century." *Essex Institute Historical Collections* 83, no. 2 (April 1947): 97–111.

Pinkston, Regina P., comp. *Historical Accounts of Meriwether County, 1827–1974.* Greenville, Ga.: Meriwether Historical Society, 1974.

Porter, Elizabeth. *A History of Wetumpka.* Wetumpka, Ala.: Wetumpka Chamber of Commerce, 1957.

Potter, Dorothy William. *Passports of Southeastern Pioneers, 1770–1823.* Baltimore, Md.: Gateway, 1982.

Power, Tyrone. *Impressions of America, during the Years 1833, 1834, and 1835.* London: R. Bentley, 1836.

Pursell, Carroll W., Jr. *Early Stationary Steam Engines in America: A Study in the Migration of a Technology.* Washington: Smithsonian Institution Press, 1969.

Rawick, George P., ed. *The American Slave: A Composite Autobiography.* Westport, Conn.: Greenwood, 1977.

Regosin, Elizabeth Ann. *Freedom's Promise: Ex-Slave Families and Citizenship in the Age of Emancipation.* Charlottesville: University Press of Virginia, 2002.

Report of the Committee of the Senate upon the Relations between Labor and Capital and Testimony Taken by the Committee. 4 vols. Washington, D.C., 1885; reprint, New York: Arno, 1976.

Rogers, George C., Jr. *The History of Georgetown County, South Carolina.* Columbia: University of South Carolina Press, 1970.

Rogers, William Warren, Robert David Ward, Leah Rawls Atkins, and Wayne Flynt. *Alabama: The History of a Deep South State.* Tuscaloosa: University of Alabama Press, 1994.

Rudisill, Horace Fraser. *Doctors of Darlington County, South Carolina, 1760–1912.* Darlington, S.C.: Darlington County Historical Society, 1962.

Russell County Historical Commission, *The History of Russell County, Alabama.* Dallas, Texas: National ShareGraphics, 1982.

Sangster, Tom, and Dess L. Sangster. *Alabama's Covered Bridges.* Montgomery: Coffeetable Publications, 1980.

Schwartz, Marie Jenkins. *Born in Bondage: Growing Up Enslaved in the Antebellum South.* Cambridge, Mass.: Harvard University Press, 2000.
Schweniger, Loren. *James T. Rapier and Reconstruction.* Chicago: University of Chicago Press, 1978.
———. "James T. Rapier of Alabama and the Noble Cause of Reconstruction." In *Southern Black Leaders of the Reconstruction Era,* ed. Howard N. Rabinowitz, 79–99. Urbana: University of Illinois Press, 1982.
Sellers, James Benson. *History of the University of Alabama.* University: University of Alabama Press, 1953.
———. *Slavery in Alabama.* Tuscaloosa: University of Alabama Press, 1994.
Smartt, Eugenia Persons. *History of Eufaula, Alabama, 1930.* Birmingham: Roberts and Son, 1933; reprint, 1995.
Smith, Clifford. *History of Troup County.* Atlanta: Foote and Davies, 1933.
Smith, William R. *Reminiscences of a Long Life: Historical, Political, Personal, and Literary.* Washington, D.C.: William R. Smith, 1889.
Southerland, Henry deLeon, Jr., and Jerry Elijah Brown. *The Federal Road through Georgia, the Creek Nation, and Alabama, 1806–1836.* Maps by Charles Jefferson Hiers. Tuscaloosa: University of Alabama Press, 1989.
Stampp, Kenneth M. *The Peculiar Institution: Slavery in the Ante-Bellum South.* New York: Vintage, 1956.
Standard, Diffee William. *Columbus, Georgia, in the Confederacy: The Social and Industrial Life of the Chattahoochee River Port.* New York: William-Frederick, 1954.
Starobin, Robert S. *Industrial Slavery in the Old South.* New York: Oxford, 1970.
Stephens, Pauline Tyson. "Albany." *Georgia Review* 4, no. 1 (spring 1950): 30–36.
Sweat, Edward F. *Economic Status of Free Blacks in Antebellum Georgia.* Atlanta: Southern Center for Studies in Public Policy, Clark College, [1974?].
———. "The Free Negro in Ante-Bellum Georgia." Ph.D. diss., Indiana University, 1957.
Swift, C. J. *The Last Battle of the Civil War.* Columbus, Ga.: Gilbert Printing, [1915].
Telfair, Nancy [Louise Jones DuBose]. *History of Columbus, Georgia, 1828–1928.* Columbus, Ga.: Historical Publishing, 1929.
Terrill, Helen Eliza. *History of Stewart County, Georgia.* Columbus, Ga.: Columbus Office Supply, 1958.
Ticknor, Frank O. *The Poems of Frank O. Ticknor, M.D.* New York: J. B. Lippincott, 1879.
Trotter, Joe William, Jr. *The African American Experience.* Boston: Houghton Mifflin, 2001.
Turner, Maxine. *Navy Gray: A Story of the Confederate Navy on the Chattahoochee and Apalachicola Rivers.* Tuscaloosa: University of Alabama Press, 1988.
U.S. Congress. Senate Committee on Education and Labor. *Report of the Committee of the Senate upon Relations between Labor and Capital and Testimony Taken by the Committee.* Vol. 4. Washington, D.C.: Government Printing Office, 1885.
The Very Worst Road: Travellers' Accounts of Crossing Alabama's Old Creek Indian Territory, 1820–1847. Comp. Jeffrey C. Benton. Eufaula, Ala.: Historic Chattahoochee Commission of Alabama and Georgia, 1998.
Wade, Richard C. *Slavery in the Cities: The South, 1820–1860.* New York: Oxford University Press, 1964.

Wadsworth, Erwing W. "Tallassee: Some Aspects of Southern Textile and Hydro-Electric Development." Master's thesis, Alabama Polytechnic Institute [Auburn], 1941.

Walker, Anne Kendrick. *Backtracking in Russell County.* Richmond, Va.: Dietz, 1950.

Weaver, Bill L. "Establishing and Organizing the Alabama Insane Hospital, 1846–1861." *Alabama Review* 48, no. 3 (July 1995): 219–232.

White, Nancy Marie, Joe Knetsch, and B. Calvin Jones. "Archaeology, History, Fluvial Geomorphology, and the Mystery Mounds of Northwest Florida." *Southeastern Archaeology* 18, no. 2 (winter 1999).

Whitney, Charles S. *Bridges: Their Art, Science, and Evolution.* New York: Greenwich House, 1983.

Wiggins, Sarah Woolfolk. *The Scalawag in Alabama Politics, 1865–1881.* University: University of Alabama Press, 1977.

Wik, Reynold M. *Steam Power on the American Farm.* Philadelphia: University of Pennsylvania Press, 1953.

Williams, David. *Rich Man's War: Class, Caste, and Confederate Defeat in the Chattahoochee Valley.* Athens: University of Georgia Press, 1998.

Willoughby, Lynn. *Fair to Middlin': The Antebellum Cotton Trade of the Apalachicola/Chattahoochee River Valley.* Tuscaloosa: University of Alabama Press, 1993.

Wolfe, Suzanne Rau. *The University of Alabama: A Pictorial History.* University: University of Alabama Press, 1983.

Worsley, Etta Blanchard. *Columbus on the Chattahoochee.* Columbus, Ga.: Columbus Office Supply, 1951.

Young, Mary Elizabeth. *Redskins, Ruffleshirts, and Rednecks: Indian Allotments in Alabama and Mississippi, 1830–1860.* Norman: University of Oklahoma Press, 1961.

Index

GAYLORD RG